Taxing Soda for Public Health

Yann Le Bodo · Marie-Claude Paquette
Philippe De Wals

Taxing Soda for Public Health

A Canadian Perspective

Springer

Yann Le Bodo
Quebec Heart and Lung Institute Research Centre
Laval University (Université Laval)
Quebec City, QC
Canada

Philippe De Wals
Department of Social and Preventive Medicine
Laval University (Université Laval)
Quebec City, QC
Canada

Marie-Claude Paquette
Department of Nutrition, Faculty of Medicine
University of Montreal (Université de Montréal)
Montreal, QC
Canada

ISBN 978-3-319-81565-7 ISBN 978-3-319-33648-0 (eBook)
DOI 10.1007/978-3-319-33648-0

Printed on acid-free paper

This Springer imprint is published by Springer Nature
The registered company is Springer International Publishing AG Switzerland

Foreword

In 2000, American academic Kelly Brownell, together with activist Michael Jacobson published a paper calling for soda taxes. Since then, the proposal has been subject to continuous debate. In recent years, this has translated into action: several countries now have soda taxes, or tax a range of products that include soda. In 2015 alone, three countries introduced new soda taxes.

This book asks the following: Is now the time for Canada to implement a soda tax? It assesses the why—the rationale for taxation—the what—the evidence of impact of the taxes—and the how—the feasibility of soda taxes in Canada.

The book is a highly useful and comprehensive overview of these issues in what is a dynamic policy area. The policy terrain—and the evidence base—will continue to move. Fortunately, Canadian policy makers now have this book that reviews the current state of knowledge as it stands today to inform their own decisions.

The book's conclusion is noteworthy: taxing soda is feasible and likely to be effective in Canada. The question now is, whether the Canadian government will take on this recommendation?

Corinna Hawkes
Professor of Food Policy
Centre for Food Policy
City University London

Preface

Although life expectancy in Canada is among the highest worldwide, the country is concerned by the progression of non-communicable diseases (NCDs): cancers, cardiovascular diseases, chronic respiratory diseases and diabetes are predominant causes of mortality. Obesity, which contributes to some of these conditions, has reached an alarming prevalence in both Canadian youth and adults. This situation places an unprecedented sanitary and economic burden on our society. To tackle the issue, there is strong evidence that, alongside regular physical activity, adopting and maintaining a healthful diet is an important part of the solution. According to the most recent nutrition surveys, Canadian eating habits leave room for improvements. Adopting a healthy diet is a challenge: dietary behaviours are subject to a wide array of conditions at individual and environmental levels including strong physical, political, economic and sociocultural influences. Consequently, the implementation of multiple healthy eating promotion strategies involving a diversity of stakeholders is generally recommended. In a given context, however, the best "mix" of cost-effective and applicable interventions to be implemented is debatable.

Of these interventions, this book addresses the rationale, the effectiveness and the applicability of one of the most controversial: taxing soda. It emerged from a comprehensive literature review performed by the first author as the first step of a Ph.D. thesis in community health at Laval University (*Université Laval*), Quebec City, Canada. The original aim of the thesis was to analyse the decision process that led to the adoption of a soda tax in France in December 2011 and to draw lessons that could inform stakeholders and decision-makers in the province of Quebec.

It appeared very quickly that the issue was relevant for a much larger audience and that a comprehensive review adopting a public health perspective would be of interest for the scientific community and public health authorities in Canada and many other countries. Indeed, the topic has been extensively covered in peer-reviewed scientific journals and in the media across the world, but there have been few attempts: (1) to summarize available evidence in a book format, (2) to comprehensively assemble evidence from a diversity of disciplines, (3) to contextualize evidence at a country level (e.g. in Canada) and (4) to provide an evidence-based conceptual framework to support governmental decision-making. Along the writing process, a multidisciplinary team was built up to elaborate this book, which attempts to bridge these gaps for several reasons:

- First, obesity and chronic diseases have been described as urgent but complex public health problems. Therefore, summarizing evidence in this area is required on a continuous basis, especially on issues deemed controversial and where the volume of evidence is rapidly growing: this is precisely the case of soda taxes.
- Secondly, the literature available on soda taxation is dispersed across a wide variety of journals namely in public health (soda and health), ethics (tax justification), economics (impact evaluations and behavioural simulations), law (tax adoption and implementation) and communication (tax political and social acceptability). Each domain brings a critical piece of evidence. However, we often miss the "big picture", i.e. the gathering of all these pieces to support decision-making.
- Thirdly, taxation proposals often unleash passions as it threatens economic actors' interests, touches upon moral and ethical concerns (e.g. autonomy of choice), questions governmental intrusion into people's lives, while attempting to promote healthier societies. This situation contributes to a somewhat cacophonic debate on soda taxation in the political sphere, the civil society and even in the scientific community. In this context, this book proposes a balanced point of view of the current evidence without tipping into sensationalism or an ideological standpoint.
- Finally, the debate on soda taxation may change significantly from one country to another because of varying overweight and obesity prevalence, soda consumption patterns, existing prevention efforts, existing taxes, jurisdiction prerogatives, etc. All these aspects that alter the relevancy to tax soda from a public health perspective need to be accounted for. This book addresses the issue specifically from a canadian perspective, although our framework is transferable to other countries and many of the ideas put forward are universal in nature.

Overall, the book adopts a knowledge-breaking perspective, deemed critical in the field of obesity prevention. Inspired by the US Institute of Medicine's LEAD framework (for Locate the evidence, Evaluate the evidence, Assemble the evidence, and inform Decisions), it attempts to summarize conveniently scientific and contextual evidence in a way that will inform decision-making. It covers not only tax effectiveness on behaviours and health but also tax justification and applicability concerns. A multidimensional evidence-based perspective is proposed at the end of the document, which is a distinctive feature.

The material in this book can be meaningful for stakeholders interested in healthy eating promotion, particularly for those interested in policies. The writing tries to avoid scientific jargon so that researchers, students and stakeholders interested in the issue could easily go through its contents, without necessarily requiring further reading to understand the key concepts used and messages conveyed.

After an introduction including a brief presentation of the methods, the book is subdivided into three main sections focused on: soda taxation rationale (Part I); soda taxation impacts (Part II); and soda taxation applicability (Part III). The conclusion proposes a graphic representation of a multidimensional evidence-based perspective of soda taxation to improve the public's health.

Acknowledgements

Yann Le Bodo received a Ph.D. fellowship from the Quebec Research Funds—Society and Culture (*Fonds de recherche du Québec—Société et culture*) in the framework of his research on sugar-sweetened beverage taxation led as a Ph.D. student in community health at Laval University (*Université Laval*)'s Faculty of Nursing. This work was undertaken at the Evaluation Platform on Obesity Prevention (EPOP), supported by the Quebec Heart and Lung Institute, Laval University (*Université Laval*) and the Lucie and André Chagnon Foundation.

Yann Le Bodo is grateful to his Ph.D. research director Philippe De Wals for his continued support, as well as to his EPOP colleagues, in particular Nathalie Dumas, Luc Ricard, José Massougbodji, Michael Gandonou, Maude Dionne and Ramona Fratu for their advice and contributions to various aspects of his Ph.D. research. Yann would also like to thank sincerely Philippe De Wals, Marie-Claude Paquette, Emmanuel Guindon and Barbara von Tigerstrom for their most valuable contributions to this book. He also thanks Natalie Alméras, Maggie Vallières, Frédéric Bergeron, Fabrice Étilé, France Gagnon, Chantal Blouin and Johanne Laguë for their inspiring collaboration to connected parts of his Ph.D. research. Thanks also go to the Quebec working group on sugar-sweetened beverages led by the Ministry of Health and Social Services (*Ministère de la Santé et des Services sociaux*), for having occasioned fruitful discussions and information-sharing on the issue in 2015.

This book consists in a review of a vast amount of studies and publications performed by many scholars and organizations to whom the authors are grateful for having provided much valuable information, concepts and ideas as well as inspiration to organize and deliver this synthesis. Some of them occasionally shared key publications and punctually commented specific sections of the manuscript, which was very appreciated. From January 2012 to December 2015, parts of the work in progress were presented to various local, national and international audiences during classes, meetings and conferences. Informal discussions held with students, professionals and experts on these occasions were very beneficial. The authors also sincerely thank Corinna Hawkes for having welcomed their invitation to write the foreword of the book. Thanks also go to Martine Six for having helped revise the final version of the manuscript and to Springer, in particular to Janet Kim, for their confidence, responsiveness and professionalism along the editing and publishing process.

Contents

Part III How: Applicability of Sugar-Sweetened Beverage Taxation

Authors and Reviewers

About the Authors

Yann Le Bodo, M.Sc., Ph.D. student
Ph.D. student at the Faculty of Nursing, Laval University (Université Laval), and Project manager at the Evaluation Platform on Obesity Prevention (EPOP), Quebec Heart and Lung Institute Research Center, Quebec City (QC), Canada.

Since 2012, Yann Le Bodo is a project manager at the Evaluation Platform on Obesity Prevention (EPOP) set-up by Laval University (*Université Laval*) and the Quebec Heart and Lung Institute (Canada). He is in charge of research projects in the area of healthy eating and physical activity policies. He is also a Ph.D. student in community health at Laval University (*Université Laval*)'s Faculty of Nursing. An important part of his research has focused on sugar-sweetened beverage taxation in Canada. Before joining the EPOP, Yann worked as a project manager in health education at community level in France (2007), and as European and international coordinator of the EPODE childhood obesity prevention network (2008–2011). He holds a M.Sc. in food innovation from Angers School of Agriculture (France, 2005) and a M.Sc. in diet-related social sciences from Toulouse II University (France, 2006). He is co-author of several peer-reviewed publications, as well as co-editor and co-author of the books "Preventing childhood obesity: EPODE European Network recommendations" (Lavoisier 2011) and "L'Expérience québécoise en promotion des saines habitudes de vie et en prévention de l'obésité: comment faire mieux?" (Presses de l'Université Laval 2016). Email: yann.lebodo@criucpq.ulaval.ca.

Marie-Claude Paquette, Dt.P., Ph.D.
Associate Professor, Faculty of Medicine, Department of nutrition, University of Montreal (Université de Montréal), Montreal (QC), Canada.

Marie-Claude Paquette is an expert and researcher at the Quebec National Public Health Institute (*Institut national de santé publique du Québec*). She is a registered dietician from University of Montreal (*Université de Montréal*) in Quebec, where she also obtained a Master degree in nutrition. She completed her Ph.D. in community nutrition at Cornell University, which was followed by a postdoctoral position at the University of Alberta in nutrition and health promotion. She has worked for the past 10 years at the Quebec National Public Health Institute (*Institut national de santé publique du Québec*) in a multidisciplinary team focusing on environmental and policy approaches for the promotion of healthy lifestyle habits and the prevention of weight-related problems. Her expertise lies mainly in the food environment and the sociocultural environment. She has been involved in characterizing the school food environment and more recently in developing indicators using supermarket food purchases. Over the last few years, she has developed an expertise on sugar-sweetened beverages following a governmental mandate on this topic. She is the author of many government publications on topics associated with nutrition and the sociocultural dimensions of eating. Finally, she is also an associate professor at the Department of nutrition of University of Montreal (*Département de nutrition de l'Université de Montréal*). Email: marie-claude.paquette.1@umontreal.ca.

Philippe De Wals, MD, Ph.D.
Professor at the Department of Social and Preventive Medicine, Faculty of Medicine, Laval University (Université Laval), Scientific Director of the Evaluation Platform on Obesity Prevention (EPOP), Quebec City (QC), Canada.

Philippe De Wals is professor at the Department of Social and Preventive Medicine, Laval University (*Université Laval*), Quebec City, Canada. He gained his medical degree and a doctorate in public health at the Louvain Catholic University, in his home country of Belgium. His research interests include the epidemiology of infectious diseases, birth defects and chronic conditions including obesity, as well as the evaluation of public health programs. He is the author of over 150 published articles in scientific journals and has contributed several chapters to textbooks. In 2011, he was appointed as scientific director of the Evaluation Platform on Obesity Prevention at the Quebec Heart and Lung Institute. Currently, Professor De Wals is a medical advisor to the Quebec National Public Health Institute (*Institut national de santé publique du Québec*) and to several national and international organizations. In 1990, he was awarded the Jean Van Beneden Prize in recognition of his excellent work in the public health field, and in 2005, he was elected to the Royal Academy of Medicine of Belgium. Email: philippe.dewals@criucpq.ulaval.ca.

Reviewers

The authors are most grateful to Emmanuel Guindon and Barbara von Tigerstrom for having reviewed and provided insightful comments and suggestions on several draft versions of this book. It definitely contributed to improving the structure and contents of the final product, whose authors remain responsible.

G. Emmanuel Guindon, M.A., Ph.D.
Assistant Professor, Department of Clinical Epidemiology and Biostatistics, McMaster University, Hamilton (ON), Canada.

G. Emmanuel Guindon is the inaugural Centre for Health Economics and Policy Analysis (CHEPA)/Ontario Ministry of Health and Long-Term Care Chair in Health Equity, an assistant professor in the Department of Clinical Epidemiology and Biostatistics and an associate member of the Department of Economics at McMaster University. Emmanuel's research interests include the economics of health behaviours, equity in health and health care, health services research and empirical health economics and policy. Through his research, education and service activities, Emmanuel works closely with decision-makers in provincial governments as well as in national and international non-governmental organizations. Email: emmanuel.guindon@mcmaster.ca.

Barbara von Tigerstrom, M.A., Ph.D.
Professor in the College of Law at the University of Saskatchewan, Saskatoon (SK), Canada.

Barbara von Tigerstrom is Professor in the College of Law at the University of Saskatchewan, where she teaches in the areas of health law and ethics, administrative law, tort law and international law. She holds a law degree from the University of Toronto and a Ph.D. in law from the University of Cambridge. Her current research focuses on legal issues in chronic disease prevention and the regulation of drugs and medical devices. She has acted as principal investigator on projects funded by the Canadian Institutes of Health Research (CIHR) and the Stem Cell Network, and co-investigator on projects funded by Genome Canada. She is a member of the CIHR Stem Cell Oversight Committee and an external research fellow of the University of Alberta Health Law Institute. Email: barbara.vontigerstrom@usask.ca.

Abbreviations

AICR	American Institute for Cancer Research
AIDS	Acquired immunodeficiency syndrome
BC	British Columbia
BMI	Body mass index
CAD	Canadian Dollar(s)
CAI	Canadian Children's Food and Beverage Advertising Initiative
CBA	Canadian Beverage Association
CCHS	Canadian Community Health Survey
CDC	Centers for Disease Control and Prevention (in the United States of America)
CEFQ	Commission d'examen sur la fiscalité québécoise (Quebec Tax Review Board)
CIHI	Canadian Institute for Health Information
CNW	Canada Newswire
CQPP	Coalition québécoise sur la problématique du poids (Quebec Weight Coalition)
CRIUCPQ	Centre de recherche de l'Institut universitaire de cardiologie et de pneumologie de Québec (Quebec Heart and Lung Institute Research Centre)
DKK	Danish Krones (currency)
EPOP	Evaluation Platform on Obesity Prevention (in the province of Quebec)
EQSJS	Enquête québécoise sur la santé des jeunes du secondaire 2010–2011 (Quebec Secondary School Students' Health Survey 2010–2011)
FNB	Fédération nationale des boissons (French National Beverages Federation)
FQSE	Fédération québécoise du sport étudiant (Quebec University Sports Federation)
FTC	Federal Trade Commission (in the United States of America)
GST	Goods and services tax (in Canada)
HSF	Heart and Stroke Foundation of Canada

HST	Harmonized Sales Tax (in Canada)
IARC	International Agency for Research on Cancer
IEPS	Impuesto Especial sobre Producción y Servicios (Mexican special tax on production and services)
INSP	Instituto Nacional de Salud Pública (Mexican National Public Health Institute)
INSPQ	Institut national de santé publique du Québec (Quebec National Public Health Institute)
IOM	Institute of Medicine (in the United States of America)
LEAD	Locate the evidence, Evaluate the evidence, Assemble the evidence, and inform Decisions (framework of the US Institute of Medicine)
NCD(s)	Non-communicable disease(s)
NCSB(s)	Non-calorically sweetened beverage(s)
NHANES	National Health and Nutrition Examination Survey (in the United States of America)
NIHD	National Institute for Health Development (in Hungary)
OECD	Organization for Economic Cooperation and Development
PEI	Prince Edward Island
PHAC	Public Health Agency of Canada
PHE	Public Health England
PHPT	Public Health Product Tax (in Hungary)
PST	Provincial Sales Tax (in Canada)
QST	Quebec Sales Tax
RESQ	Réseau du sport étudiant du Québec (Quebec university sports network)
SSB(s)	Sugar-sweetened beverage(s)
WCRF	World Cancer Research Fund
WHO	World Health Organization

Chapter 1
Introduction

Abstract Taxes altering the prices of less healthful foods have been increasingly recommended as part of non-communicable disease and obesity prevention strategies. Progressively, sodas or sugar-sweetened beverages (SSBs) have become a target of particular interest and, over the last ten years, taxes have been reinforced or newly imposed on these beverages in a growing number of countries. Twenty-one cases of soda taxes are presented in this introduction. However, such taxes remain controversial. Based on a realist review of literature and knowledge synthesis techniques, this book attempts to provide answers to the following question: Is it relevant to tax soda for public health in Canada? The findings are subdivided into three parts focusing on: SSB taxation rationale (Part I), SSB taxation impacts (Part II) and SSB taxation applicability (Part III).

1.1 Background

In 2011, for the second time in United Nations' history,[1] the General Assembly met to address an alarming public health issue: the prevention and control of non-communicable diseases (NCDs). The political declaration of this high-level meeting called for an urgent international response (United Nations 2011). NCDs were deemed to cause 68 % of the world's 56 million deaths in 2012. Cardiovascular diseases, cancer, diabetes and chronic respiratory diseases account for 82 % of this burden (World Health Organization (WHO) 2014a). Although life expectancy in Canada is among the highest worldwide, NCDs remain a threat as they are estimated to represent 88 % of total deaths. Cancers (30 % of deaths), cardiovascular diseases (27 %), chronic respiratory diseases (7 %) and diabetes (3 %) are predominant causes (WHO 2014b). About one in five Canadian adults live with at least one of these chronic diseases, including one in 10 suffering from diabetes (Public health agency of Canada (PHAC) 2015). Obesity is known to increase the risk of a number of these conditions (American Institute for Cancer Research

[1]After the special session held in 2001 on the human immunodeficiency virus (HIV)/acquired immunodeficiency syndrome (AIDS).

Y. Le Bodo et al., *Taxing Soda for Public Health*,
DOI 10.1007/978-3-319-33648-0_1

(AICR)/World Cancer Research Fund (WCRF) 2007; WHO 2011). In Canada, according to statistical data collected in 2012–2013, obesity prevalence reached 26 % in adults and 13 % in youth aged 5–17[2] (PHAC 2015). These figures have doubled over the last 30 years. The Canadian economic burden related to obesity and its comorbidities has been estimated to be between $4.6 billion and $7.1 billion in 2008[3] (Canadian Institute for Health Information (CIHI)/PHAC 2011). To tackle the issue, there is strong evidence that a healthful diet,[4] regular physical activity, moderate alcohol use and no smoking are beneficial (AICR/WCRF 2007; Mozaffarian 2014). According to the Canadian Community Health Surveys, eating habits of Canadians leave room for improvement (Garriguet 2007; PHAC 2015). Yet, adopting and maintaining a healthy diet is a challenge. Health-related behaviours are subject to a wide array of conditions at individual and environmental levels including physical, political, economic and sociocultural influences (Kremers et al. 2006; Butland et al. 2007; Ministère de la Santé et des Services sociaux 2012). According to Swinburn et al. (2011), the fast increase in obesity prevalence in North America over the last decades has been predominantly related to significant increases in the food supply and marketing environments promoting high intake of energy-dense foods. In the meantime, environmental moderators (e.g. sociocultural norms, socioeconomic conditions, transportation modes, etc.) as well as individual factors (e.g. genetics, cognition) probably "amplified" or "attenuated" the phenomena within and between populations (Swinburn et al. 2011). Curbing this evolution requires tremendous efforts: a systemic approach combining multiple healthy eating policies and programs, and involving a diversity of stakeholders at all levels has been widely recommended (AICR/WCRF 2009; United Nations 2011).

In a given context, however, designing the best portfolio of cost-effective and applicable interventions to be implemented remains a challenge. So far, nutritional policies aimed at informing consumers about the benefits of a healthful diet have led to modest impacts on consumers' behaviours and health. Therefore, it has been suggested (including in Canada) to reinforce the implementation of food policies and dissuasive regulations aimed at changing the market environment (Réquillart and Soler 2014; Tourigny et al. 2014; Duhaney et al. 2015). Among other options,

---

[2]These prevalences were established from direct measurements conducted for the Canadian Health Measures Survey (CHMS). Here, obesity is defined according to the body mass index (BMI), which is an indicator calculated by dividing an individual's weight (kilograms) by its height (metres) squared. In adults (age $\geq$ 18), obesity is defined by a BMI over 30 kg/m^2. In youth, the same BMI calculation applies, but the obesity threshold varies by age and sex to account for growth and maturation. In this case, the 2007 WHO BMI age-/sex-specific cut-offs have been used to define obesity threshold and calculate obesity prevalence in youth (Public health agency of Canada 2015).

[3]Depending on the diseases included in the model.

[4]Mozafarian (2014, p. 1307), commenting recent scientific evidence on cardiovascular health, considers that such diets tend to be "neither extreme nor exceptional, but reasonable and consistent with dietary guidelines (e.g. $\approx$5 daily servings of fruits and vegetables, 4 daily servings of whole grains, and 2 weekly servings of fish)".

taxes altering the prices of less healthful foods have been increasingly recommended as part of comprehensive NCD and obesity prevention[5] strategies (Gortmaker et al. 2011; Mozaffarian et al. 2012; WHO 2015b). The concept of health-related food taxation emerged in the USA in the 1990s, in a context of rapid and important increase of NCD prevalence. It referred to taxes imposed on the manufacture or consumption of foods and beverages particularly dense in energy, fat or sugar, with the explicit intention to earmark the proceeds for health promotion purposes (Jacobson and Brownell 2000).

Progressively, sodas appeared as ideal candidates for taxation because of their high sugar content, their poor nutritive value, the health risks related to their overconsumption,[6] the high intakes often observed in youth, and the aggressive marketing practices of manufacturers (Brownell and Frieden 2009; Brownell et al. 2009; Lustig et al. 2012; Block and Willett 2013; Mytton 2015). Generally, tax proposals particularly focus on sugar-sweetened beverages (SSBs). The precise definition of these beverages varies in the literature, but generally converges to designate non-alcoholic beverages containing added sugars.[7]

Over the last ten years, soda taxes[8] have been reinforced or newly introduced in a growing number of countries. Table 1.1 includes soda taxes that have been attached to health considerations, be this in governmental position statements, budgetary speeches or legislative texts.[9] This means that other excise duties and

[5]In this book, obesity prevention is used as a generic term encompassing all kinds of interventions on lifestyle habits associated with weight problems, including physical activity and nutrition, as well as environmental, social and other factors influencing physical activity and nutrition (EPOP 2015). Thus, as in other key references, the prevention of overweight ($25 \leq$ BMI < 30, see footnote 1) is implicitly included in the prevention of obesity (BMI ≥ 30) (Institute of Medicine (IOM) 2012, p. 289).

[6]At individual level, we refer to "SSB overconsumption" as the daily consumption of SSB resulting in a daily intake of added sugars that exceeds 10 % of daily energy intake (see Box 2.1). This threshold is congruent with WHO recent recommendation to maintain the intake of free sugars to fewer than 10 % of total energy intake (WHO 2015a).

[7]In this book, please note that "sugar-sweetened beverages" (SSBs) include carbonated drinks, fruit drinks, flavoured waters, energy drinks and sports drinks, as long as these beverages contain added sugars. Our definition excludes bottled water, 100 % fruit juice, milk, sugar-sweetened milks, non-calorically sweetened beverages (NCSBs), hot chocolate, and hot coffee-/tea-based beverages (see Sect. 2.1).

[8]In the context of nutrition policies, the term "soda," although originally designating carbonated beverage, frequently and commonly refers to sugar-sweetened beverage (SSB). In this book, both terms are used in an interchangeable manner.

[9]This list is not necessarily comprehensive. It was established along our realist review of evidence (see Sect. 1.2). Cross-checks were made with other lists identified (European Commission 2012; OECD 2012; Ecorys 2014a; WHO 2015b; WCRF 2015) to favour exhaustiveness, accuracy and the most up-to-date information. Documents referenced in English, French or Spanish have been considered.

Table 1.1 Cases of health-related taxes targeting non-alcoholic beverages across the world (dated December 2015)

Jurisdiction (world region)	Adoption[a]	Health-related considerations	Targeted beverages[b]	Tax type[c] and rate[d]	Allocation of tax revenues	References
French Polynesia (Australasia)	2002	Introduction of various taxes (on sweetened drinks, beer, confectionery, ice cream) to establish a "preventive health fund" addressing nutrition-related issues and NCDs	Sweetened drinks	Excise taxation of 40 Pacific Francs (CFP) (eq. 0.51 CAD)/litre for sweetened drinks manufactured locally; import tax of 60 CFP (eq. 0.76 CAD)/ litre for imported sweetened drinks	2002–2005: funding of preventive health and "citizenship projects" 2008: 80 % of tax revenues earmarked for the Ministry of Health's general budget	Thow et al. (2011), Snowdon and Thow (2013)
Fiji (Australasia)	2006, removed in 2007	Excise taxes enacted as part of a budgetary initiative. However, prior to the implementation of the tax, Thow et al. (2011, p. 57) highlight that "informal discussions [occurred] between the Ministry of Health and Finance policy makers regarding the possibility of a soft drink tax in the context of the multisectoral national Non-Communicable Disease committee"	Soft drinks	5 % import excise tax and domestic excise tax of 0.05 Fiji dollars (eq. 0.03 CAD)/litre both introduced in 2006, but the domestic tax was replaced in 2007 with a 3 % fiscal import duty on raw materials "for the purpose of fairer administration and collection of the duty across industry" (Thow et al. 2011, p. 57)	Revenues collected by the Fiji Islands' Revenue and Customs Authority, with no specific information identified in terms of earmarking	Thow et al. (2011)

(continued)

Table 1.1 (continued)

Jurisdiction (world region)	Adoption[a]	Health-related considerations	Targeted beverages[b]	Tax type[c] and rate[d]	Allocation of tax revenues	References
Nauru (Australasia)	2007	Introduction of a "sugar levy" for both financial and health reasons, including an attempt to "discourage excessive consumption of sugar" (Thow et al. 2011, p. 59)	New tax on all high-sugar drinks (including carbonated soft drinks, cordials, flavoured milk and drink mixes) and removal of a tax on bottled water	30 % customs and excise duty based on existing fiscal schemes applied to tobacco, alcohol and petrol	Contribution to the general budget of the state	Thow et al. (2011), Snowdon and Thow (2013)
Samoa (Australasia)	1984, revised in 1998 and 2008	Official objective to raise revenues for the general budget of the state, although an objective to "further support the Government's drive to improve health outcomes for the community" has also been mentioned by the finance minister in 2008 (Thow et al. 2011, p. 58)	Soft drinks	Excise taxation of 0.40 Tala (eq. 0.21 CAD)/litre on domestic production and imported beverages (≈5–8 % of purchase price)	No specific information identified	Thow et al. (2011)

(continued)

Table 1.1 (continued)

Jurisdiction (world region)	Adoption[a]	Health-related considerations	Targeted beverages[b]	Tax type[c] and rate[d]	Allocation of tax revenues	References
Denmark (Europe)	2009, abolished in 2014	Main objective to raise fiscal revenues and secondary objective to decrease sugar consumption from beverages in order to improve public health	Soft drinks and juices (100 % or not) with sugar content (threshold ≥ 0.5 g/100 ml)	Increase from 0.91 Danish Krones (DKK) (eq. 0.19 CAD)/litre (since 2007) to 1.08 DKK (eq. 0.22 CAD)/litre[a] of the excise tax on sugary soft drinks, and decrease from 0.91 DKK (since 2007) to 0.57 DKK (eq. 0.12 CAD)/litre of the excise tax on sugar-free soft drinks [a]*Further increased by 0.5 and 0.06 DKK/litre in 2012 and 2013, respectively, before being halved in July 2013 and abolished in 2014*	No specific information identified	Danish Ministry of Taxation (2009), European Commission (2012), Landon and Graff (2012), Smed and Robertson (2012), Bomsdorf (2012), Scott-Thomas (2013), Ecorys (2014b)
Finland (Europe)	2011	Increase and/or restoration of excise taxes motivated by health considerations and included within	Soft drinks (including sugar- and non-sugar-sweetened carbonated drinks), non-alcoholic fruit- and	Increase in the excise taxation of targeted beverages from 0.045 € (eq. 0.06 CAD)/ litre (since 1999) to	Tax mainly aimed to provide revenues to the Finnish government. No evidence of	Finnish Ministry of Finance (2012, 2013), OECD (2012), Ecorys (2014a)

(continued)

Table 1.1 (continued)

Jurisdiction (world region)	Adoption[a]	Health-related considerations	Targeted beverages[b]	Tax type[c] and rate[d]	Allocation of tax revenues	References
		the "Act on excise duty on sweets, ice cream and soft drinks". This act targets confectionery (e.g. sweets and similar products, chocolate, other products that contain cocoa), ice cream and non-alcoholic beverages (e.g. waters, juices and soft drinks)	vegetable-based drinks whether or not containing added sugar or other sweetening matter (e.g. fruit juice, syrups, lemonade and nectar), waters (including natural or artificial mineral waters and aerated waters, not containing added sugar or other sweetening matter nor flavoured) and some slightly alcoholic beverages (0.5 % vol. < alcoholic strength ≤ 1.2 % vol.)	0.075 € (eq. 0.11 CAD)/litre in 2011 and to 0.11 € (eq. 0.17 CAD)/litre in 2012. In 2014, increase in the excise taxation to 0.22 € (eq. 0.33 CAD)/litre for sugar-sweetened beverages and juices (containing more than 0.5 g of sugar/100 g or 100 ml) (the rate remained unchanged for sweetener-based soft drinks and waters)	earmarking for public health-oriented interventions	
Hungary (Europe)	2011	"Public Health Product Tax" (PHPT) applied to pre-packaged foods containing high salt and sugar contents, including crisps, salted nuts,	Sugar-sweetened beverages (including syrups or soft drink concentrates) containing more than 8 g of added sugar/100 mL (From January 2012, exemption for drinks containing more than 25 %	Tax at 5 Hungarian forint (HUF) (eq. 0.02 CAD)/litre in 2011, increased to 7 HUF (eq. 0.03 CAD)/litre in 2012, payable by manufacturers for	Data suggesting a redistribution of tax revenues as bonuses granted to healthcare workers exposed to " […] *increased physical and psychological*	Hungarian Ministry of Human Resources (2011), Holt (2011), Landon and Graff (2012), Ecorys (2014b), Bíró (2015), WHO (2015b)

(continued)

Table 1.1 (continued)

Jurisdiction (world region)	Adoption[a]	Health-related considerations	Targeted beverages[b]	Tax type[c] and rate[d]	Allocation of tax revenues	References
		chocolates, sweets, biscuits, ice creams, various alcoholic and non-alcoholic beverages in order to promote healthy nutrition, to improve the financing of health services and to promote product reformulation by manufacturers	of fruit or vegetable content and for beverages containing at least 50 % of milk-based raw material)	beverages produced in Hungary for the domestic market, and by the first domestic seller for imported products	*burdens as a result of working directly in patient care, or in contact with other potential health risks*" (Hungarian Ministry of Human Resources 2011). According to Ecorys (2014b, p. 214) the allocation of tax revenues to fund health-related services (e.g. salary increases for health professionals, ambulance stations) has been confirmed.	
France (Europe)	2011	The initial objective of the tax, as elaborated by the government, was to influence behaviours and prevent obesity. It was finally better recognized by the	Beverages containing added sugars and non-caloric beverages containing artificial sweeteners (including carbonated drinks, fruit juices, fruit beverages, flavoured milk)	Increase in the excise taxation from 0.0054 € to 0.0716 € (eq. 0.11 CAD)/litre in 2012 (≈ +6 % of average purchase price according to Berardi et al. 2012). The rate is yearly	Part of the revenue earmarked for the National Health Insurance System and another part to the general budget of the State (*NB*: Since 2013, all the revenue goes to the National	French Government (2011), French Constitutional Council (2011), French Ministry of Budget (2012), Berardi et al. (2012), Chauliac (2014), Ecorys (2014a),

(continued)

Table 1.1 (continued)

Jurisdiction (world region)	Adoption[a]	Health-related considerations	Targeted beverages[b]	Tax type[c] and rate[d]	Allocation of tax revenues	References
		legislator as a way to raise fiscal revenues.		adjusted in line with the growth rate of the consumer price index. In 2016, the soda tax rate reached 0.0753 €/litre	Health Insurance System)	Ministère des finances et des comptes publics (2015)
Cook Islands (Australasia)	2012	New import duty included in the 2012/2013 budget and presented "[…] primarily as a public health measure aimed at increasing the cost of sweetened drinks to encourage behavioural change" (Brown 2012, p. 53)	Sugar-sweetened drinks	Import duty of 15 % with a subsequent 2 % rise per year. Not applicable to the small local producers	No specific information identified	Brown (2012), Snowdon and Thow (2013)
Algeria (North Africa)	2012	Establishment of a national "Fund against cancer", partially funded from taxes on alcohol, tobacco-based products, cell phone operators and carbonated drinks	Carbonated drinks	Tax at a rate of 0.5 % on the revenue of manufacturers and importers of carbonated drinks	Fiscal revenues earmarked for cancer prevention, early screening and treatment	Official Journal of the Algerian Republic (2012), Zikara (2012), Ferdi (2012)

(continued)

Table 1.1 (continued)

Jurisdiction (world region)	Adoption[a]	Health-related considerations	Targeted beverages[b]	Tax type[c] and rate[d]	Allocation of tax revenues	References
Mauritius (Indian Ocean)	2012	To " […] discourage bad [lifestyle] habits that increase the risk of non-communicable diseases" (Duval 2012, p. 32)	Sugary drinks including any carbonated beverage, any "syrup", any fruit squash, cordial or drink. Exemption of soft drinks containing only artificial sweeteners, bottled water, fruit and/or vegetable juice	New excise tax of 2 cents per gram of sugar content in the targeted drinks (whether imported or locally produced), increased to 3 cents per gram in 2014 (1 Mauritius rupee (MUR) eq. 0.038 CAD)	No specific information identified	Mauritius Revenue Authority (2012), Duval (2012), WCRF (2015)
Mexico (Latin America)	2013	Specific tax on SSBs enacted jointly with the 8 % "special tax on production and services" (IEPS) on energy-dense foods (except basic groceries) with the aim to raise fiscal revenues but also to decrease consumption	Flavoured beverages containing added sugars (including carbonated drinks, fruit beverages, syrups and excluding milk and yogurt)	Specific tax of 1 peso (eq. 0.08 CAD)/litre of beverage, collected at the manufacturer or importer level and estimated to represent 9 % of average pre-tax soda price (Grogger 2015)	According to WCRF (2015), the tax revenues were supposed to be earmarked for public health nutrition-related efforts but are currently allocated to the general budget	Prodecon (2013), KPMG (2013), Astudillo (2014), WCRF (2015), Rosenberg (2015), Grogger (2015)
Tonga (Australasia)	2013	Health-related food tax	Soft drinks containing sugar or sweeteners	Excise tax of 1 Tonga Pa'anga (eq. 0.61 CAD) per litre	No specific information identified	WCRF (2015)

(continued)

Table 1.1 (continued)

Jurisdiction (world region)	Adoption[a]	Health-related considerations	Targeted beverages[b]	Tax type[c] and rate[d]	Allocation of tax revenues	References
St Helena (South Atlantic Ocean)	2014	Excise duty aimed to make sugar-sweetened beverages more expensive than diet beverages, to encourage healthier beverage consumption, to contribute preventing obesity, diabetes and promoting dental health	High-sugar carbonated drinks, defined as drinks containing $\geq$ 15 g of sugar per litre	Excise duty £0.75 (eq. 1.56 CAD) per litre	Excise tax part of an annual budget planning new investments in healthcare services (but specific earmarking of SSB tax revenues was not specified)	Owen (2014), Pipe (2014), WCRF (2015)
India (Asia)	2014, replaced in 2015	From a "human and fiscal health" perspective, in the 2014 Budget, the Ministry of Finance announced an extra excise duty on "aerated waters containing added sugar", as it is the case for tobacco products (Business-standard 2015)	"Waters", including "mineral water and aerated water containing added sugar or other sweetening matter or flavoured".	Creation of an extra 5 % excise duty in the 2014 Budget, abolished and replaced by an increase of the central excise duty on aerated drinks and flavoured water from 17.5 to 18 % in the 2015 budget	No specific information identified	Bhaumik (2014), The Telegraph (2014), Business-standard (2015)

(continued)

Table 1.1 (continued)

Jurisdiction (world region)	Adoption[a]	Health-related considerations	Targeted beverages[b]	Tax type[c] and rate[d]	Allocation of tax revenues	References
City of Berkeley (California, US, North America)	2014	To " […] diminish the human and economic costs of diseases associated with the consumption of sugary drinks by discouraging their distribution and consumption in the city of Berkeley" (City of Berkeley 2014, p. 2)	Sugar-sweetened beverages, e.g. sodas, energy drinks, pre-sweetened teas (exemption of diet drinks, milk products, 100 % juice, baby formula, alcohol, or drinks taken for medical reasons)	In addition to any other taxes imposed by the city, specific excise taxation of 0.01 $US (eq. 0.014 CAD) per fluid ounce of sugar-sweetened beverage products in the city: according to Cawley and Frisvold (2015), it was deemed to represent an average price increase of 11, 31 and 25 % for a 20-oz.bottle, a 2-l bottle, and a 12-pack of 12-oz. cans, respectively	Revenue allocated to "[…] the general governmental needs" of the Berkeley population (City of Berkeley 2014, p. 2). In January 2016, the City Council voted the funding (from the general budget) of large public health and nutrition programs, in particular to tackle SSB overconsumption	City of Berkeley (2014), Falbe et al. (2015), Cawley and Frisvold (2015), Raguso (2016)
Navajo Nation (US, North America)	2014	With a goal of NCD prevention, the "Healthy Diné' Nation Act" introduced a tax targeting sweetened beverages as well as other energy-dense	Sweetened beverages are defined as carbonated or non-carbonated beverages including soda; flavoured water; sports drinks; energy drinks; fountain drinks; iced coffee; iced tea; flavoured drinks; drinks containing	2 % special sales tax levied on the retailer, i.e. the person receiving the gross receipts from the sale of the targeted items	All revenues earmarked for the Community Wellness Development Projects Fund and used for projects such as farming and vegetable gardens,	Navajo Nation Council (2014), Barclay (2015), Morales (2015), WCRF (2015)

(continued)

Table 1.1 (continued)

Jurisdiction (world region)	Adoption[a]	Health-related considerations	Targeted beverages[b]	Tax type[c] and rate[d]	Allocation of tax revenues	References
		and micronutrient-poor food items. The Act is presented as a way to raise awareness of unhealthy foods, to create healthier environments (e.g. making healthy food available and easily accessible) and to fund health promotion programs	sugar with natural fruit or vegetable juice, fruit or vegetable juice; powder or base product as a liquid for sale		healthy eating promotion initiatives, new sports and physical activity infrastructure, healthy cooking classes, etc.	
Chile (Latin America)	2015	Contribution to public health objectives, as in the case of tobacco and alcohol taxes	All non-alcoholic drinks with added sweeteners (including energy drinks and waters), i.e. with sugar content greater than 6.25 g of sugar per 100 mL	Before 2015, a 13 % special tax was applied on all non-alcoholic beverages. In 2015, this tax rate was lowered at 10 % except for sugar-sweetened beverages whose tax rate was raised at 18 %	The additional tax on SSBs is part of a large fiscal reform aimed at ensuring the sustainability of structural reforms in various domains including education, health and social protection	Ferrando (2014), Lezaeta (2014), WCRF (2015), Ministerio de Hacienda (2015), Servicio de Impuestos Internos (2016)

(continued)

Table 1.1 (continued)

Jurisdiction (world region)	Adoption[a]	Health-related considerations	Targeted beverages[b]	Tax type[c] and rate[d]	Allocation of tax revenues	References
Barbados (Caribbean)	2015	Encouragement to healthier beverage consumption patterns in order to contribute preventing NCDs (particularly type 2 diabetes)	Sugary drinks, including carbonated soft drinks, juice drinks, sports drinks and others. Drinks exempt from the tax include 100 % natural fruit juice, coconut water, plain milk and evaporated milk	10 % excise tax on locally produced and imported items (levied on the value of the product before application of the value-added tax)	Tax expected to generate $10 million (for the first fiscal year) earmarked for the health sector	Sinckler (2015), WCRF (2015)
Dominica (Caribbean)	2015	New tax targeting drinks and foods with high sugar content (including sweets, candy, chocolate bars) in order to discourage consumption, contribute to preventing NCDs, and to promote health	Drinks with high sugar content (including soft drinks and energy drinks)	10 % excise tax	Contribute to financing the "production" and "maintenance" of the national "Get Healthy" campaign	Skerrit (2015), WCRF (2015)
Belgium (Europe)	2015	In the framework of the 2016 budget, the government announced a large fiscal reform ("*tax shift*") including an	Sugar-sweetened beverages and non-calorically sweetened beverages	New excise tax of 0.03 € (eq. 0.045 CAD)/litre	No specific information identified	Belgian Federal Government (2015), Le Soir (2015), Radio Télévision Belge Francophone (2015)

(continued)

Table 1.1 (continued)

Jurisdiction (world region)	Adoption[a]	Health-related considerations	Targeted beverages[b]	Tax type[c] and rate[d]	Allocation of tax revenues	References
		additional excise tax on sugary drinks connected to the national nutrition plan				

[a]Depending on the case, the actual implementation of the tax may have started the following year

[b]The level of detail and the terminology used in this column reflect the information collected in the references. When the scope of the tax explicitly includes non-caloric beverages and calorically sweetened beverages, both categories have been reported in the list of targeted beverages

[c]Excise taxes are applied to specific types of goods and are generally levied on the manufacturer, as a fixed amount per unit of product. See Table 12.1 to consult full definitions of terms related to SSB taxation mechanisms

[d]Conversions to Canadian dollars (using nominal rate) were calculated in December 2015 via the Daily Converter of the Bank of Canada available at: http://www.bankofcanada.ca/rates/exchange/daily-converter/

general sales taxes applied for many years on SSBs were not included in this table, unless health-related considerations were explicitly put forward.[10]

The first cases of health-related soda taxes have been described in the early 2000s in Pacific Islands (Fiji, Samoa, Nauru and French Polynesia), a region where obesity and related chronic diseases have been highly prevalent for a long time. The fact that their governments have early been prone to soda taxation may partly be explained by their need to find new fiscal revenues in a context of trade tariffs decline over the last years (Thow et al. 2011). From 2009 to 2011, a wave of four European countries followed (Denmark, Finland, Hungary, France), in the context of a global economic downturn. It should be noticed that the Danish tax on soft drinks and juices was actually abolished in 2014, probably dragged down with the rollback of the world-first tax on saturated fat, deemed ineffective and administratively burdensome (Scott-Thomas 2013; Bødker et al. 2015). In Finland, the soda tax increased in 2011, and was substantially boosted in 2012 and 2014. It is actually part of a large excise duty targeting sweets, ice cream and soft drinks. This is not the case in France, the first large jurisdiction to single out soda as a specific non-alcoholic food category for taxation. A distinctive and controversial feature of this tax is the inclusion of non-calorically sweetened beverages (NCSBs) in the tax scope, essentially for budgetary reasons (see Sect. 13.6).

In 2012, following the United Nations' summit for the prevention and the control of NCDs (United Nations 2011), international recommendations for health-related food taxes became stronger (WHO 2011; De Schutter 2011; Institute of Medicine (IOM) 2012). New proposals for soda taxes were adopted in 2012 and 2013, with explicit public health motives (i.e. in the Cook Islands, Algeria, Mauritius, Mexico, Tonga). The Mexican case was extensively covered in the international media, being one of the countries where obesity and NCDs are the most prevalent, where soft drink consumption levels are among the highest, but also since Mexico was the largest and first Latin American country ever to adopt a soda tax (Rosenberg 2015). In 2014, the case of the soda tax in Saint Helena (part of the British Overseas Territories) raised particular attention as it was enacted at the particular moment when such a policy was again debated in the United Kingdom, without sufficient political support to be passed (Pipe 2014). Since then, institutional support has grown in the UK. In July 2015, the British medical association (BMA) strongly supported a 20 % tax on SSBs. Three months later, Public Health

[10]For instance, in the USA, soft drinks and snack food taxes applied for years in many states and localities are not included. Revenues generated from these taxes are not officially allocated to obesity prevention efforts, although a few of them have generated funds earmarked for health- or social-related causes (Jacobson and Brownell 2000; Powell and Chriqui 2011). In Canada, basic groceries are often exempt from general sales tax applied on food and beverages at federal and provincial levels. Depending on the jurisdiction, soda is sometimes included in this category, sometimes not (and thus taxed at a higher rate). In all cases, nutrition and health criteria have not been explicitly taken into considerations (see Chap. 12). Therefore, taxes currently applied on soda in Canada are not part of the inventory presented in Table 1.1.

England[11] (PHE) released a report on sugar reduction, recommending the "introduction of a price increase of a minimum of 10–20 % on high-sugar products through the use of a tax or levy such as on full-sugar soft drinks" (BMA 2015; PHE 2015, p. 8). Soda taxes are also controversial in Asia. In India, the enactment of an extra 5 % excise duty on aerated drinks in 2014 was followed by important counter-advocacy efforts. The Indian Beverage Association asserted its significant contribution to agriculture and employment as well as the value of carbonated soft drinks in a country where "options of safe, convenient and hygienic beverages are rather limited" (Indian Beverage Association 2014; The Economic Times 2014). These movements likely contributed to replacing this duty by a less soda-stigmatizing surtax in the 2015 budget. Maybe not for a long time: in December 2015, India's Chief Economic Advisor proposed a 40 % tax on aerated beverages (India Resource Center 2015; Ambwani 2015).

In 2014, another turning point has been the approval of an excise tax on the distribution of SSBs by public referendum in the city of Berkeley, California. About 15 years after the birth of the "soda tax" concept in the USA and dozens of bills across the country (Jacobson and Brownell 2000; Wang et al. 2012), this vote has created a precedent in North America. It was followed by the enactment of the "Healthy Diné' Nation Act" by the Navajo Nation, a US community comprising over 250,000 inhabitants on a territory extending into Utah, Arizona and New Mexico (Navajo Nation Government 2011). Beyond sweetened beverages, this act targets a broad scope of "minimal-to-no-nutritional-value food items" (Barclay 2015). In 2015, Chile, the country where the growth of SSB sales has recently been shown to be the fastest (Popkin and Hawkes 2016),[12] also adopted a tax on non-alcoholic beverages containing added sweeteners. In the same year, soda taxes spread to the Caribbean (Barbados, Dominica). Finally, like its French neighbour a few years before, Belgium recently enacted a soda tax whose scope includes both SSBs and NCSBs as part of a broad fiscal reform (see Table 1.1).

In the meantime, the institutional support for soda taxes has grown up. In January 2016, the WHO Commission on ending childhood obesity issued a set of 36 recommendations aimed at promoting healthy eating and physical activity. Among those appears the "implementation of an effective tax on sugar-sweetened beverages" (WHO 2016, p. 18). These growing extension and recognition should not make us underestimate the controversies surrounding the relevancy to tax SSBs. This issue remains debated in the scientific literature (Chaloupka et al. 2011; Fletcher et al. 2011; Sarlio-Lähteenkorva and Winkler 2015) and, until now, many governments have remained hesitant or reluctant to tax SSBs in the Americas (e.g. in Columbia), in Europe (e.g. in the UK, Ireland, Italy, Romania), in South-east Asia/Australasia (e.g. in Thailand, Malaysia, Australia, New Zealand) and Africa

---

[11]Public Health England is "an operationally autonomous executive agency of the Department of Health" (PHE 2015, p. 2).

[12]Popkin and Hawkes' calculations are based on the Euromonitor Passport International database, for the period 2009–2014.

(e.g. in South Africa) (Dinero.com 2015; Campbell 2015; Briggs et al. 2013; European Public Health Alliance 2012; National News Bureau of Thailand 2013; Vandevijvere et al. 2015; Swire 2012; Back 2015; Myers et al. 2015).

Soda taxes are also debated in Canada where, over the last years, significant pro-taxation advocacy efforts have faced a fierce opposition from the beverage industry and pro-business organizations. So far, political leaders have generally been reluctant to adopt a soda tax either at federal or provincial level (see Chap. 13). This may change as the country and some provinces recently observed significant political turnovers. In such circumstances, bringing evidence to the table is essential. The scientific literature is rich in simulations analysing potential effects of an SSB tax on consumption and health, but integrated evidence-based multidimensional frameworks aimed to support decision-making are scarcer. Based on a realist review of literature and knowledge synthesis techniques (Pawson et al. 2005; Morestin et al. 2011), this book attempts to advance knowledge on the following question: Is it relevant to tax soda for public health in Canada[13]?

1.2 Methods

This section describes the methods we have used to build a Canadian perspective on the relevance to tax soda for public health. We expose the theoretical basis that we have adopted, the type of review that we have conducted and, finally, the data collection and analysis that we have performed.

1.2.1 Theoretical Basis

As emphasized by Marshall and Rossman (2011, p. 207), "the process of bringing order, structure, and interpretation to a mass of collected data is messy, ambiguous, time-consuming, creative and fascinating. It does not proceed in a linear fashion." Indeed, an obvious challenge of our research, aimed to inform a policy decision, is to cover a wide spectrum of evidence. Another challenge lies in our ability to make sense of a high volume of heterogeneous information, anchored across different disciplines and theoretical standpoints. We relied on conceptual frameworks in the area of knowledge synthesis and knowledge transfer to tackle this endeavour.

These conceptual frameworks generally require combining rigorous scientific evidence (e.g. systematic reviews of experimental research) with contextual

---

[13]It is noteworthy that Canada is a federal state with powers belonging to the federal government, or to provincial and territorial governments, or shared by those different entities. Canadian regions and municipalities (in some cases) also have a role in the protection of public health. Therefore, a Canadian perspective could be of interest for a diversity of governmental agencies and public health authorities worldwide.

scientific evidence (e.g. judgments about modifying factors, needs, values, costs and resources) to address 3 key questions: (1) Why should we do something about the health issue considered and why does the policy at stake deserve attention? (2) What could be the impact of this policy? (3) Would this policy be applicable in Canada, and how? (Lomas et al. 2005; Pawson et al. 2005; Oxman et al. 2009). Basically, the first question refers to the normative justification and the objectives of the policy (in comparison to alternatives). The second question refers to the intended and unintended effects of the policy (i.e. its effectiveness, equity and potential undesirable effects). Finally, the third question refers to feasibility concerns (administratively and technically speaking) including the costs associated with the adoption and implementation of the policy, potential gains (e.g. revenues from a tax and associated investments), as well as the political and social acceptability of the policy (Swinburn et al. 2005; Morestin et al. 2011). Other dimensions have sometimes been added to these conceptual frameworks such as the evaluability of the targeted policy, i.e. the possibility for the policy to be evaluated in a reliable and credible way (Erickson et al. 2005; Leclerc 2012).

In the particular area of obesity prevention, the LEAD framework (for Locate the evidence, Evaluate the evidence, Assemble the evidence, and inform Decisions), developed by the Institute of Medicine (Kumanyika et al. 2010), provides a comprehensive picture of the types of questions to be addressed to inform decisions in a systems-oriented perspective. Chatterji et al. (2014) have applied this framework to the particular case of SSB taxation and distinguish three series of questions to be answered:

- The "**WHY**" **questions**, referring to the public health situation (e.g. health burden and disparities in the population, epidemiological links between SSB and health, SSB consumption trends in the population, social and environmental determinants of SSB overconsumption) and to the social and monetary costs related to the health burden under consideration.
- The "**WHAT**" **questions**, referring to potential intervention logics (e.g. discouraging SSB consumption by increasing prices; discouraging production and consumption by stigmatizing SSBs; raising revenue to fund health promotion initiatives); referring also to the potential effects and circumstances of effectiveness of the policy (e.g. on prices, consumption, energy balance and health) based on the empirical studies and simulations; and, finally, referring to potential undesirable effects of the policy (e.g. compensatory consumption of unhealthy foods and beverages; strategies used by marketers to circumvent the tax or its effects).
- The "**HOW**" **questions**, referring to the legal and administrative conditions needed to make different types of taxes work and to the possibility to build on existing tax frames; referring to equity concerns and opposition from various stakeholders; referring to the political circumstances and advocacy efforts that may be needed for an SSB tax to be adopted; referring to the revenues that could be generated from the tax and the purposes for which these revenues could be earmarked; and referring to the costs related to the implementation of the tax.

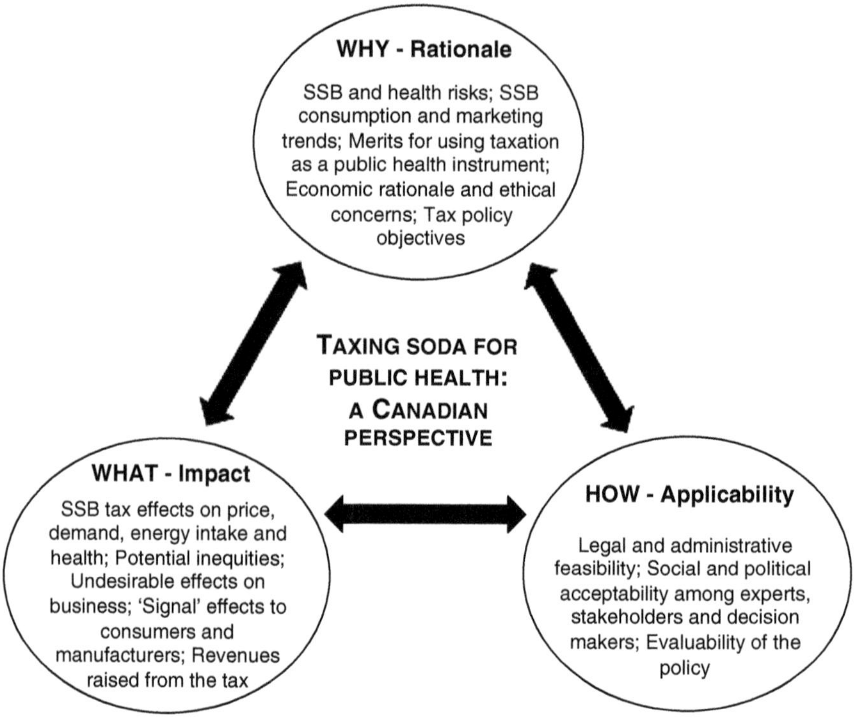

Fig. 1.1 Conceptual framework adapted from Chatterji et al. (2014) to guide the collection, analysis and gathering of evidence on soda taxes for public health, from a Canadian perspective

We judged this framework relevant and inspiring to guide our data collection and analysis. Consequently, we used it as a starting point and adapted it iteratively as we discovered new pieces of evidence. This heuristic process allowed for exploring new areas and themes when they emerged from the literature, and to adjust this framework to our particular research question. Figure 1.1 outlines the organization and diversity of themes covered in this book. Using such a multidimensional approach, we attempt to provide "a broad view of the whole picture rather than a narrow or isolated focus on the component parts" (Chatterji et al. 2014, p. 4).

1.2.2 A "Realist" Review of Evidence

In the context of the LEAD framework and obesity prevention, the term "evidence" refers to "a body of systematically gathered and screened evidence relevant to addressing a user question for making a decision on a population-relevant obesity issue" (Chatterji et al. 2014, p. 6). From this perspective, we conducted a "realist" review of evidence, specifically aimed to address complex public health issues, to

Table 1.2 Domains and types of evidence considered to address our research question

Questions	Types of evidence considered
WHY-Rationale	Large-scale Canadian population health surveys; Epidemiological reviews, meta-analyses and recent experimental studies on SSB and health; surveillance data on SSB consumption and marketing trends; comparative evaluation of SSB-related interventions; parallel evidence on tobacco taxes; analyses on economic and ethical justifications for SSB taxation
WHAT-Impact	SSB tax effectiveness research on prices, consumption and health from experimental studies and simulations; evaluation of large-scale SSB taxation policies in "real-world" conditions; economic impact assessments; data on revenues raised from SSB taxes in other countries; any data suggesting "signal" effects of health-related taxation policies
HOW-Applicability	Legal and technical analyses of SSB taxation feasibility in Canada; experts and stakeholders consensus meetings on the acceptability of SSB taxation; indications (including press clips) referring to the political and social acceptability of SSB taxation; opinion polls; evaluability assessments; methodological analyses referring to evaluation requirements

Sources Chatterji et al. (2014), Morestin et al. (2011)

analyse a diversity of evidence type, and to document "what works, for whom, in what circumstances, in what respects and how" (Pawson et al. 2005, p. 21). In our case, addressing the wide array of themes linked to SSB taxation called for an interdisciplinary approach including various types of data from nutrition, public health, health economics, health policy, law and communication. Across these disciplines, the "realist" approach aims to identify barriers and facilitators to the implementation of an intervention and to inform decision-making (Pawson et al. 2005). In our review, the types of evidence that we considered are summarized in Table 1.2.

1.2.3 Data Collection

Due to the very broad scope of our review, we used a cumulative and dynamic research approach over four years (December 2011–December 2015), until reaching a satisfying degree of saturation, i.e. a moment when we considered that the discovery of new evidence relevant to our research question was marginal and did not affect significantly the content of our analysis and our conclusions (Lapaige 2012). This section describes our methodological choices.

Due to the vast amount of information on food taxes in general, the choice was initially made to prioritize the exploration of data specific to the taxation of non-alcoholic beverages (excluding, a priori, documents specific to other taxes, e.g. fat taxes, alcohol taxes, tobacco taxes). Since health-related SSB taxation is a relatively recent policy option, no restriction was put on the search period. However, considering that SSB taxation has met a growing interest in both research and practice over the last years, a particular attention was given to the methodological developments and the quality of evidence published in certain domains (e.g. epidemiology of SSB and health, simulations of SSB taxation effectiveness). Due to the lack of specific evidence on SSB taxation in Canada, it was considered essential to extend the search to evidence from other countries, more specifically from high-income countries presenting similarities with Canada from a cultural, economic and political standpoint. When using databases, keywords in English were used. However, evidence in French and Spanish was also punctually considered. The main components of our data collection are summarized here below:

- We conducted an initial PUBMED search (December 2011) using various combinations of the following keywords: (sweetened beverage(s), soda(s), sugar-sweetened beverage(s), soft drink(s), sugared beverage(s), sugared drink (s), soft drink(s)) AND (subsidies, subsidy, tax, taxes, taxation, taxing, price(s)). This initial search led us to identify 194 references of interest. From here, via a snowballing approach,[14] we iteratively explored reference lists, consulted other databases and sources, and considered expert advice. Some of the main steps and components of this process are described below.
- From March 2012 to December 2015, we systematically screened Google Scholar alerts using a combination of the following keywords: (beverage, soda, sugar, food) AND (tax).
- We regularly consulted a diversity of blogs and newsletters in the area of public health nutrition (e.g. Food Politics,[15] Weighty Matters,[16] World Obesity,[17] etc.).
- Our work was concomitantly enriched by related projects of the Evaluation Platform on Obesity Prevention (EPOP), including: (1) a review of international recommendations on healthy eating and physical activity promotion (Le Bodo et al. 2016); (2) a study examining the correlates of reviews' conclusions on SSB and body weight (Massougbodji et al. 2014); (3) a comparative analysis of clinical, epidemiological, marketing and public health evidence on SSB and tobacco (Le Bodo et al. 2015b); (4) a case study analysis of the French soda tax (Le Bodo et al. 2014); (5) a Canadian press review (2008–2013) of stakeholders' positions on sugar-sweetened beverage taxation to prevent obesity (Le Bodo et al. 2015a).
- Presentations and discussions took place at a multidisciplinary one-day scientific conference organized by the EPOP in Quebec City in April 2012 on the following topic: "Health-related tax? The SSB case study."[18]
- A second PUBMED search has also been performed in March 2015 once the synthesis was almost complete to make sure additional relevant studies had not

---

[14]The snowballing approach refers to "pursuing references of references," in an iterative manner, to enrich the documentary search (Pawson et al. 2005, p. 29).

[15]http://www.foodpolitics.com/.

[16]http://www.weightymatters.ca/.

[17]http://www.worldobesity.org/news/.

[18]https://www.evaluation-prevention-obesite.ulaval.ca/cms/site/pepo/lang/fr/page83598.html.

been left out. This search was based on various combinations of the following keywords: (soda, soft drink, sugar-sweetened beverage, sugary drink) AND (tax*).

- The work in progress was presented to and challenged by various local, national and international audiences from January 2012 to December 2015.[19]
- During the same period, the research team attended scientific meetings and conferences and benefited from informal discussion with public health and economic experts in taxation policies.

The title and abstracts of identified sources were screened and, as appropriate, the full text was considered. To organize and rank evidence properly, greater importance was generally given to systematic and comprehensive reviews, to most recent studies, as well as to contextual evidence highly relevant to our research question (e.g. Canadian surveillance data on SSB consumption and health, legal analysis of SSB taxation feasibility in Canada, SSB tax acceptability among Canadian stakeholders). In some specialized domains (e.g. epidemiology, behavioural economics), priority was given to the peer-reviewed literature. In other cases, where the evidence was scarce, descriptive or observational, we had more propensities to consider "grey literature" (e.g. data on "real-life" SSB taxation experiences across the world, Canadian law documents, Canadian press clipping).

1.2.4 Data Analysis

In the preliminary stages of the data collection, a data extraction matrix was developed to map the diversity of themes that emerged from the literature and to report the main findings and conclusions from the documents. Between January and April 2013, a narrative knowledge synthesis was produced on the basis of the evidence identified (unpublished). At this stage, the matrix was used to subdivide this intermediary report into 12 research subquestions referring to the "WHY-Rationale," "WHAT-Impact," and "HOW-Applicability" of SSB taxation (see Fig. 1.1). From April 2013 to October 2015, the literature search was pursued (see Sect. 1.2.3) and each document deemed relevant for the analysis was systematically considered (in alphabetical order). At this stage, main findings and conclusions were not reported in the data extraction matrix, but directly included, as

[19]Canadian Obesity Student Meeting (COSM 2012), Edmonton (Canada); 23rd and 24th scientific days of the Quebec Heart and Lung Institute Research Centre (CRIUCPQ 2013, 2014), Quebec City (Canada); EPOP symposium on SSB taxation (2013), Vancouver (Canada); Obesity Week (2013), Atlanta (US); Research seminar on obesity-metabolism, Quebec Heart and Lung Institute Research Centre (CRIUCPQ 2013), Quebec City (Canada); Scientific day of the faculty of nursing, Laval University (*Université Laval*) (FSI, 2014), Quebec City (Canada); *Table de concertation nationale Québécoise en prévention/promotion* (2014), Quebec City (Canada); Quebec Scientific Committee on Obesity Prevention (2015), Montreal (Canada); Montreal Public Health Directorate (*Direction de santé publique de Montréal*) (2015) (Canada).

appropriate, in successive new draft versions of the report. This incremental process, inspired from realist and narrative synthesis techniques (Lapaige 2012), led to the production of the present book.

This cumulative and dynamic approach presented several advantages: (1) it ensured that all the evidence collected would be considered to some extent; (2) it allowed flexibility in the exploration of the data and the navigation (back and forth) across various contents and reference lists; (3) it facilitated logical connections across the diversity and heterogeneity of evidence considered; (4) it allowed fast triangulation across various sources as well as easy adjunction of newly identified documents in our analysis, as appropriate; (5) overall, it helped to make sense from a diversity of observations and to fine-tune the conclusions of our review.

On the other hand, our approach presents several limitations: (1) this approach heavily relies on the ability and skills of the research team to make relevant connections across readings and consultation of various pieces of evidence. Recurrent presentation of the work in progress to experts and stakeholders and discussions within the research team were used as mechanisms to control for the consistency and robustness of the process; (2) the mass and diversity of information consulted as well as the long-term developmental process of writing made it difficult to provide comprehensive data extraction tables. However, the book provides detailed and specific bibliographic references for presented evidence. Ultimately, it reflects the "archaeology" of decisions that we have made and the "interpretative trail" that we have followed (Pawson et al. 2005, p. 31).

Finally, as will be reminded in our conclusion, we pretend in no way that this review is definitive per se. Researchers' role is not to decide on the adoption and implementation of policies, but to inform decision-makers of the pros and cons of various options under consideration. From this perspective, we consider this research as a starting point for further discussions, consultations, contextualization and deliberation. As Pawson and colleagues highlight, "room for further rounds of negotiation must be left open about whether an unforeseen chink in the implementation chain deserves closer inspection". In a realist review, "the progress made (…) is not one from ignorance to answer, but from some knowledge to some more knowledge" (2005, p. 32).

Key Messages

- Promoting healthy eating habits is essential for the prevention of NCDs.
- Soda taxes are part of the strategies increasingly recommended and adopted over the last 10 years across the world.
- However, the relevancy to tax soda remains debated on a scientific and political ground, including in Canada.
- This book presents findings from a realist review of evidence addressing soda taxation rationale (part I), impacts (part II) and applicability (part III) from a Canadian perspective.

References

Ambwani, M. V. (2015, December 11). Coca Cola India says may shut some factories if "sin tax" is implemented. *The Hindu Business Line*. New Delhi. http://www.thehindubusinessline.com/companies/sin-tax-on-aerated-beverages-will-make-industry-unviable-cocacola-india/article7976813.ece. Accessed January 5, 2016

American Institute for Cancer Research (AICR) & World Cancer Research Fund (WCRF) (Ed.). (2007). *Food, nutrition, physical activity and the prevention of cancer: A global perspective: A project of World Cancer Research Fund International*. Washington, D.C: American Institute for Cancer Research.

American Institute for Cancer Research (AICR), & World Cancer Research Fund (WCRF) (Ed.). (2009). *Policy and action for cancer prevention: food, nutrition, and physical activity: A global perspective*. Washington, D.C: American Institute for Cancer Research.

Astudillo, O. (2014). Country in focus: Mexico's growing obesity problem. *The Lancet Diabetes and Endocrinology*, *2*(1), 15–16. doi:10.1016/S2213-8587(13)70160-8

Back, A. (2015, October 13). Sugar tax inevitable in Australia's fight against obesity, says public health researcher Boyd Swinburn. *The Canberra Times*. http://www.canberratimes.com.au/act-news/sugar-tax-inevitable-in-australias-fight-against-obesity-says-public-health-researcher-boyd-swinburn-20151013-gk7ryw.html. Accessed January 22, 2016

Barclay, E. (2015, April 1). Navajos fight their food desert with Junk food and soda taxes. *npr.org. The salt. What's on your plate*. http://www.npr.org/sections/thesalt/2015/04/01/396607690/navajos-fight-their-food-desert-with-junk-food-and-soda-taxes. Accessed December 17, 2015

Belgian Federal Government. (2015). Accord Budget 2016: ce qui change pour le portefeuille des Belges. http://www.rtbf.be/info/belgique/detail_accord-tax-shift-budget-2016-voici-quelques-elements?id=9104399. Accessed December 20, 2015

Berardi, N., Sevestre, P., Tepaut, M., & Vigneron, A. (2012). *The Impact of a "Soda Tax" on Prices: Evidence from French Micro Data* (Banque de France Working Paper No. 415) (p. 29). Banque de France Eurosystème. Direction Générale des Études et des Relations internationales. http://ssrn.com/abstract=2192470. Accessed February 24, 2016

Bhaumik, S. (2014). The public health threat from sugary drinks in India. *BMJ (Clinical research ed.)*, *349*, g6216.

Bíró, A. (2015). Did the junk food tax make the Hungarians eat healthier? *Food Policy*, *54*, 107–115. doi:10.1016/j.foodpol.2015.05.003

Block, J. P., & Willett, W. C. (2013). Taxing sugar-sweetened beverages: Not a "Holy Grail" but a cup at least half comment on "Food Taxes: A New Holy Grail?" *International Journal of Health Policy and Management*, *1*(2), 183–185. doi:10.15171/ijhpm.2013.33

Bødker, M., Pisinger, C., Toft, U., & Jørgensen, T. (2015). The rise and fall of the world's first fat tax. *Health Policy (Amsterdam, Netherlands)*, *119*(6), 737–742. doi:10.1016/j.healthpol.2015.03.003

Bomsdorf, C. (2012, November 11). Denmark scraps much-maligned "fat tax" after a year. *The Wall Street Journal*. http://online.wsj.com/article/SB100014241278873238947045781131206227 63136.html. Accessed March 1, 2013

Briggs, A. D., Mytton, O. T., Madden, D., O'Shea, D., Rayner, M., & Scarborough, P. (2013). The potential impact on obesity of a 10 % tax on sugar-sweetened beverages in Ireland, an effect assessment modelling study. *BMC Public Health*, *13*(1), 860. doi:10.1186/1471-2458-13-860

British Medical Association (BMA). (2015). *Food for thought: Promoting healthy diets among children and young people* (executive summary) (p. 12). London. http://www.bma.org.uk/working-for-change/improving-and-protecting-health/food-for-thought. Accessed January 22, 2016

Brownell, K. D., Farley, T., Willett, W. C., Popkin, B. M., Chaloupka, F. J., Thompson, J. W., et al. (2009). The public health and economic benefits of taxing sugar-sweetened beverages. *New England Journal of Medicine*, *361*(16), 1599–1605. doi:10.1056/NEJMhpr0905723

Brownell, K. D., & Frieden, T. R. (2009). Ounces of prevention—The public policy case for taxes on sugared beverages. *New England Journal of Medicine, 360*(18), 1805–1808. doi:10.1056/NEJMp0902392

Brown, M., & Cook Islands Government Budget Estimates. (2012/2013). Part 1—Appropriation bill. Appropriations and commentary. Hon. Mark Brown. Ministry of Finance. June 2012. (2012). http://www.mfem.gov.ck/mfemdocs/treasury/budget-downloads/2012-13-1/130-budget-book-i-budget-estimates-scn/file. Accessed December 18, 2015

Business-standard.com. (2015, March 2). Despite tax hike, soft drinks to retain fizz. *business-standard.com.* Mumbai. http://www.business-standard.com/article/companies/despite-tax-hike-soft-drinks-to-retain-fizz-115030200020_1.html. Accessed December 18, 2015

Butland, B., Jebb, S., Kopelman, P., McPherson, K., Thomas, S., Mardell, J., et al. (2007). *Foresight. Tackling obesities: Future choices—project report.* (2nd Edn.). Government Office for Science. https://www.gov.uk/government/uploads/system/uploads/attachment_data/file/287937/07-1184x-tackling-obesities-future-choices-report.pdf. Accessed February 24, 2016

Campbell, D. (2015, October 22). Ministers urged to introduce sugar tax in childhood obesity report. *TheGuardian.com.* http://www.theguardian.com/society/2015/oct/22/sugar-report-delayed-jeremy-hunt-tax-radical-action-obesity. Accessed January 22, 2015

Canadian Institute for Health Information (CIHI)., & Public Health Agency of Canada (PHAC). (2011). *Obesity in Canada a joint report from the Public Health Agency of Canada and the Canadian Institute for Health Information.* Ottawa: Public Health Agency of Canada : Canadian Institute for Health Information.

Cawley, J., & Frisvold, D. (2015). *The incidence of taxes on sugar-sweetened Beverages: The case of Berkeley, California* (No. w21465). Cambridge, MA: National Bureau of Economic Research. http://www.nber.org/papers/w21465.pdf. Accessed February 25, 2016

Chaloupka, F. J., Powell, L. M., & Chriqui, J. F. (2011). Sugar-sweetened beverages and obesity prevention: Policy recommendations. *Journal of Policy Analysis and Management: (the Journal of the Association for Public Policy Analysis and Management), 30*(3), 662–664.

Chatterji, M., Green, L. W., & Kumanyika, S. (2014). L.E.A.D.: A framework for evidence gathering and use for the prevention of obesity and other complex public health problems. *Health Education and Behavior, 41*(1), 85–99. doi:10.1177/1090198113490726

Chauliac, M. (2014, December 17). *Faut-il taxer les boissons sucrées pour contribuer à prévenir le surpoids et l'obésité: l'expérience française.* Quebec Heart and Lung Institute, Quebec city (Quebec), Canada.

City of Berkeley. (2014). Imposing a general tax on the distribution of sugar-sweetened beverage products. Pub. L. No. Measure D § Ballot question.

Danish Ministry of Taxation. (2009). *Danish tax reform 2010.* Paper to the OECD WP 2 meeting November 2009. http://www.skm.dk/foreign/9276/taxationindenmark/8013.html. Accessed March 1, 2013

De Schutter, O. (2011). *Report submitted by the Special Rapporteur on the right to food, Olivier De Schutter. United Nations General Assembly A/HRC/19/59* (Human Rights Council. Nineteenth session. Agenda item 3). http://www.unscn.org/en/announcements/conferences/?id=650. Accessed April 25, 2012

Dinero.com. (2015, November 19). ¿Colombia necesita un impuesto a las bebidas azucaradas? *Dinero.* http://www.dinero.com/economia/articulo/impuesto-bebidas-azucaradas-colombia/216187. Accessed January 22, 2016

Duhaney, T., Campbell, N., Niebylski, M. L., Kaczorowski, J., Tsuyuki, R. T., Willis, K., et al. (2015). Death by diet: The role of food pricing interventions as a public policy response and health advocacy opportunity. *The Canadian Journal of Cardiology, 31*(2), 112–116. doi:10.1016/j.cjca.2014.09.005

Duval, X. L. (2012, November 9). Rising to the challenges of a world in transition. Mauritius National Budget Speech 2013. http://www.investmauritius.com/media/37909/Budget Speech2013.pdf. Accessed February 25, 2014

Ecorys. (2014a). *Food taxes and their impact on competitiveness in the agri-food sector. Final report* (No. Ref. Ares(2014)2365745—16/07/2014). Rotterdam: Client: DG Enterprise and Industry.

Ecorys. (2014b). *Food taxes and their impact on competitiveness in the agri-food sector. Annexes to the Main report* (Ref. Ares (2014)2604304—06/08/2014). Rotterdam: Client: DG Enterprise and Industry.

Erickson, L. J., De Wals, P., & Farand, L. (2005). An analytical framework for immunization programs in Canada. *Vaccine, 23*(19), 2470–2476. doi:10.1016/j.vaccine.2004.10.029

European Commision. (2012). *Flash report. High level group on nutrition and physical activity of 2 February 2012. Conclusions of the chair.* http://ec.europa.eu/health/nutrition_physical_activity/events/index_en.htm. Accessed March 1 2012

European Public Health Alliance (EPHA). (2012). Food taxation in Europe: Evolution of the legislation—Update October 2012. http://www.epha.org/4814. Accessed January 22, 2016

Evaluation Platform on Obesity Prevention (EPOP). (2015). Mission. https://www.evaluation-prevention-obesite.ulaval.ca/cms/site/pepo/lang/en/page83590.html;jsessionid=3D0BC3F7DFEAE6DE62FA5ADFAA75D4E9. Accessed September 25, 2015

Falbe, J., Rojas, N., Grummon, A. H., & Madsen, K. A. (2015). Higher retail prices of sugar-sweetened beverages 3 months after implementation of an excise tax in Berkeley, California. *American Journal of Public Health, 105*(11), 2194–2201. doi:10.2105/AJPH.2015.302881

Ferdi, S. (2012, January 3). Des taxes pour la lutte contre le cancer L'alcool, le tabac et les boissons gazeuses plus chers. *Algérie 360°*. http://www.algerie360.com/algerie/des-taxes-pour-la-lutte-contre-le-cancer-lalcool-le-tabac-et-les-boissons-gazeuses-plus-chers/. Accessed February 25, 2014

Ferrando, K. (2014, April 6). El alza del tributo a las bebidas azucaradas sin alcohol, ¿para qué sirve? *La Tercera.* http://www.latercera.com/noticia/negocios/2014/04/655-572803-9-el-alza-del-tributo-a-las-bebidas-azucaradas-sin-alcohol-para-que-sirve.shtml. Accessed December 17, 2015

Finnish Ministry of Finances. (2012). *Budget review 2012.* http://www.vm.fi/vm/en/04_publications_and_documents/01_publications/01_budgets/20120123Budget/BKtammikuu_enkku_netti.pdf. Accessed March 1, 2013

Finnish Ministry of Finances. (2013). *Excise duties on sweets, ice cream and soft drinks.* http://www.vm.fi/vm/en/10_taxation/05_excise_duty/04_excise_duty_on_sweets_ice_cream_and_soft_drinks/index.jsp. Accessed March 1, 2013

Fletcher, J. M., Frisvold, D. E., & Tefft, N. (2011). Are soft drink taxes an effective mechanism for reducing obesity? *Journal of Policy Analysis and Management: (the Journal of the Association for Public Policy Analysis and Management), 30*(3), 655–662.

French Constitutional Council. (2011). *Décision No2011-644DC. Article, consolidation, travaux parlementaires.* http://www.conseil-constitutionnel.fr/conseil-constitutionnel/francais/les-decisions/acces-par-date/decisions-depuis-1959/2011/2011-644-dc/decision-n-2011-644-dc-du-28-decembre-2011.104235.html. Accessed March 1, 2012

French Government. (2011). Mesures pour la réduction du déficit public. Dossier de Presse. Création d'une taxe sur les boissons sucrées. Web Portal. http://www.gouvernement.fr/presse/mesures-pour-la-reduction-du-deficit-public. Accessed 1 March 2012

French Ministry of Budget. (2012). Contributions sur les boissons et préparations liquides pour boissons sucrées et édulcorées. Contributions indirectes. Circulaire du 24 janvier 2012. NOR : BCRD 1202351C.

Garriguet, D. (2007). Canadians' eating habits. *Health Reports, 18*(2), 17–32.

Gortmaker, S. L., Swinburn, B. A., Levy, D., Carter, R., Mabry, P. L., Finegood, D. T., et al. (2011). Changing the future of obesity: Science, policy, and action. *Lancet (London, England), 378*(9793), 838–847. doi:10.1016/S0140-6736(11)60815-5

Grogger, J. (2015). *Soda taxes and the prices of sodas and other drinks: Evidence from Mexico* (No. w21197). Cambridge, MA: National Bureau of Economic Research. http://www.nber.org/papers/w21197.pdf. Accessed February 28, 2016

Holt, E. (2011). Hungary to introduce broad range of fat taxes. *Lancet (London, England), 378* (9793), 755.

Hungarian Ministry of Human Resources. (2011). Hungary's public health tax on food products a hit: 42,000 health care workers to receive bonuses this year. http://www.kormany.hu/en/ministry-of-human-resources/news/hungary-s-public-health-tax-on-food-products-a-hit-42-000-healthcare-workers-to-receive-bonuses-this-year. Accessed March 1, 2013

Indian Beverage Association (IBA). (2014, July 10). Indian beverage association reacts to the excise duty hike on aerated drinks in the union budget 2014–2015. *Coca-Cola India*. https://www.coca-colaindia.com/indian-beverage-association-reacts-excise-duty-hike-aerated-drinks-union-budget-2014-2015/. Accessed January 5, 2015

India Resource Center. (2015, December 8). Move to tax sugar beverages in India welcomed by public health advocates. http://www.indiaresource.org/news/2015/1032.html. Accessed January 5, 2016

Institute of Medicine (U.S.), Glickman, D., Committee on Accelerating Progress in Obesity Prevention, & Food and Nutrition Board (Ed.). (2012). *Accelerating progress in obesity prevention: Solving the weight of the nation*. Washington, DC: National Academies Press.

Jacobson, M. F., & Brownell, K. D. (2000). Small taxes on soft drinks and snack foods to promote health. *American Journal of Public Health, 90*(6), 854–857.

KPMG. (2013). *Mexico—Tax reform 2014, overview of general provisions* (TaxNewsFlash Americas).

Kremers, S. P. J., de Bruijn, G.-J., Visscher, T. L. S., van Mechelen, W., de Vries, N. K., & Brug, J. (2006). Environmental influences on energy balance-related behaviors: A dual-process view. *The International Journal of Behavioral Nutrition and Physical Activity, 3*, 9. doi:10.1186/1479-5868-3-9

Kumanyika, S. K., Parker, L., Sim, L. J., & Committee on an Evidence Framework for Obesity Prevention Decision Making, Food and Nutrition Board, & Institute of Medicine (IOM) (Eds.). (2010). *Bridging the evidence gap in obesity prevention: A framework to inform decision making*. Washington, D.C: National Academies Press.

Landon, J., & Graff, H. (2012). What is the role of health-related food duties? In *A report of a National Heart Forum meeting*. London: National Heart Forum. http://www.worldobesity.org/site_media/uploads/UKHF_duties.pdf. Accessed February 28, 2016

Lapaige, V. (2012, March 23). *L'enjeu méthodologique des synthèses en science de transfert et de l'application des connaissances*. Presented at the Conférence à la faculté des Sciences Infirmières de l'Université Laval, Québec (Québec).

Le Bodo, Y., Blouin, C., Dumas, N., Laguë, J., & De Wals, P. (2016). *L'Expérience québécoise en promotion des saines habitudes de vie et prévention de l'obésité : COMMENT FAIRE MIEUX?* (Plateforme d'évaluation en prévention de l'obésité (PEPO) and Institut national de santé publique du Québec (INSPQ)). Québec (Québec): Presses de l'Université Laval. http://iucpq.qc.ca/fr/comment-faire-mieux. Accessed February 28, 2016

Le Bodo, Y., Dionne, M., & De Wals, P. (2015a, November 2). *Is it time for a sugar-sweetened beverages tax in Canada and may Québec come first?* Poster presented at the Obesity Week 2015, Los Angeles (California).

Le Bodo, Y., Étilé, F., & De Wals, P. (2014, April 8). *Conditions influençant l'adoption d'une politique de taxation des boissons sucrées en prévention de l'obésité : réflexions et perspectives à partir du cas français*. Presented at the 2ème Journée scientifique de la Faculté des Sciences Infirmières de l'Université Laval, Québec (Québec).

Le Bodo, Y., Paquette, M.-C., Vallières, M., & Alméras, N. (2015b). Is sugar the new tobacco? Insights from laboratory studies, consumer surveys and public health. *Current Obesity Reports, 4*(1), 111–121. doi:10.1007/s13679-015-0141-3

Leclerc, B.-S. (2012). Utilité de l'évaluation de l'évaluabilité des politiques gouvernementales de lutte contre le tabagisme: l'expérience québécoise des centres d'abandon du tabagisme. *The Canadian Journal of Program Evaluation, 24*(3), 61–73.

Le Soir. (2015, October 13). L'industrie du soda dénonce la nouvelle taxe sur les boissons sans sucre. *Le Soir (Belgium)*. http://www.lesoir.be/1015603/article/economie/2015-10-13/lindustrie-du-soda-denonce-nouvelle-taxe-sur-boissons-sans-sucre. Accessed December 20, 2015

Lezaeta, M. (2014, September 26). Hacienda: Nuevos impuestos a bebidas alcohólicas y tabaco regirán desde el 1 de octubre. *emol.com (Chili)*. Santiago. http://www.emol.com/noticias/economia/2014/09/26/682026/hacienda-nuevos-impuestos-a-bebidas-alcoholicas-y-tabaco-regiran-desde-el-1-de-octubre.html. Accessed February 28, 2016

Lomas, J., Culyer, T., McCutcheon, C., McAuley, L., & Law, S. (2005). *Conceptualizing and Combining Evidence for Health System Guidance: final report*. Canadian Health Services Research Foundation. http://www.cfhi-fcass.ca/migrated/pdf/insightAction/evidence_e.pdf. Accessed February 28, 2016

Lustig, R. H., Schmidt, L. A., & Brindis, C. D. (2012). Public health: The toxic truth about sugar. *Nature, 482*(7383), 27–29. doi:10.1038/482027a

Marshall, C., & Rossman, G. B. (2011). *Designing qualitative research* (5th ed.). Los Angeles: Sage.

Massougbodji, J., Le Bodo, Y., Fratu, R., & De Wals, P. (2014). Reviews examining sugar-sweetened beverages and body weight: Correlates of their quality and conclusions. *The American Journal of Clinical Nutrition, 99*(5), 1096–1104. doi:10.3945/ajcn.113.063776

Mauritius Revenue Authority. (2012, December 11). Communiqué. Excise tax on sugar content of soft drink. http://mra.gov.mu/download/CommuniqueSoftDrinks2012.pdf. Accessed February 25, 2014

Ministère de la Santé et des Services sociaux. (2012). *Pour une vision commune des environnements favorables à la saine alimentation, à un mode de vie physiquement actif et à la prévention des problèmes reliés au poids* (Document rédigé en collaboration avec Québec en Forme et l'Institut national de santé publique du Québec) (p. 24). Québec, Canada: Gouvernement du Québec. http://publications.msss.gouv.qc.ca/acrobat/f/documentation/2012/12-289-03.pdf. Accessed November 18, 2015

Ministère des Finances et des comptes publics. (2015). Tarifs des droits sur les alcools. Les boissons alcooliques et les boissons non alcooliques applicables au 1er Janvier 2016. Circulaire du 18 Décembre 2015. http://circulaires.legifrance.gouv.fr/pdf/2015/12/cir_40344.pdf. Accessed January 22, 2016

Ministerio de Hacienda (Chili). (2015). Reforma Tributaria. Principales modificaciones. http://reformatributaria.gob.cl/principales-modificaciones.html. Accessed December 17, 2015

Morales, L. (2015, April 9). The Navajo nation's tax on junk food splits reservation. *npr.org*. http://www.npr.org/sections/codeswitch/2015/04/08/398310036/the-navajo-nations-tax-on-junk-food-splits-reservation. Accessed December 17, 2015

Morestin, F., Gauvin, F.-P., Hogue, M.-C., & Benoit, F. (2011). *Method for synthesizing knowledge about public policies*. [Montréal]: Centre de collaboration nationale sur les politiques publiques et la santé, Institut national de santé publique Québec. http://collections.banq.qc.ca/ark:/52327/2066075. Accessed February 28, 2016

Mozaffarian, D. (2014). The promise of lifestyle for cardiovascular health. *Journal of the American College of Cardiology, 64*(13), 1307–1309. doi:10.1016/j.jacc.2014.06.1191

Mozaffarian, D., Afshin, A., Benowitz, N. L., Bittner, V., Daniels, S. R., Franch, H. A., et al. (2012). Population approaches to improve diet, physical activity, and smoking habits: A scientific statement from the American heart Association. *Circulation, 126*(12), 1514–1563. doi:10.1161/CIR.0b013e318260a20b

Myers, A., Fig, D., Tugendhaft, A., Mandle, J., Myers, J., & Hofman, K. (2015). Sugar and health in South Africa: Potential challenges to leveraging policy change. *Global Public Health*, 1–18. doi:10.1080/17441692.2015.1071419

Mytton, O. (2015). Time for a sugary drinks tax in the UK? *Journal of Public Health (Oxford, England), 37*(1), 24–25. doi:10.1093/pubmed/fdu033

National News Bureau of Thailand. (2013, February 12). Thailand's biggest health watchdog proposes higher tax on sweetened soda drinks. *National News Bureau of Thailand*. http://www.thaivisa.com/forum/topic/618706-thailands-biggest-health-watchdog-proposes-higher-tax-on-sweetened-soda-drinks/. Accessed January 22, 2016

Navajo Nation Council. (2014). President Shelly signs Healthy Dine' Nation Act of 2014 into law. Resolution of the Navajo Nation Council. 22nd Navajo Nation Council, Fourth Year, 2014. http://www.navajo-nsn.gov/News%20Releases/OPVP/2014/nov/Healthy%20Dine%20Nation%20Act%20of%202014.pdf. Accessed December 17, 2015

Navajo Nation Government. (2011). Navajo Nation Government's official website. Section "History." http://www.navajo-nsn.gov/history.htm. Accessed December 17, 2015

Official Journal of the Algerian Republic. (2012). Conventions et accords internationaux – loi et décrets, arrêtés, décisions, avis, communications et annonces. Parution Numéro 52, 51ème année, 19 Septembre 2012. http://www.joradp.dz/FTP/jo-francais/2012/F2012052.pdf. Accessed March 1, 2013

Organisation for Economic Co-operation and Development (OECD). (2012, February 21). Obesity update 2012. Policy brief. http://www.oecd.org/document/10/0,3746,en_2649_37407_38334282_1_1_1_37407,00.html. Accessed March 1, 2012

Owen, C. (2014, March 21). St Helena Government budget speech 2014–15. http://www.sainthelena.gov.sh/wp-content/uploads/2013/01/Budget-Speech-2014-15-21-March-2014.pdf. Accessed December 17, 2015

Oxman, A. D., Lavis, J. N., Lewin, S., & Fretheim, A. (2009). SUPPORT tools for evidence-informed health policymaking (STP) 1: What is evidence-informed policymaking? *Health Research Policy and Systems / BioMed Central*, *7*(Suppl 1), S1. doi:10.1186/1478-4505-7-S1-S1

Pawson, R., Greenhalgh, T., Harvey, G., & Walshe, K. (2005). Realist review—A new method of systematic review designed for complex policy interventions. *Journal of Health Services Research and Policy*, *10*(Suppl 1), 21–34. doi:10.1258/1355819054308530

Pipe, S. (2014, February 28). St Helena shows UK the way with fizzy drinks tax. *St Helena online*. http://sthelenaonline.org/2014/03/28/st-helena-shows-uk-the-way-with-fizzy-drinks-tax/. Accessed December 17, 2015

Popkin, B. M., & Hawkes, C. (2016). Sweetening of the global diet, particularly beverages: Patterns, trends, and policy responses. *The Lancet. Diabetes and Endocrinology*, *4*(2), 174–186. doi:10.1016/S2213-8587(15)00419-2

Powell, L. M., & Chriqui, J. F. (2011). Food taxes and subsidies: Evidence and policies for obesity prevention. Chapter 38. In J. Cawley (Ed.), *The Oxford handbook of the social science of obesity*. New York: Oxford University Press.

Prodecon. (2013). *Impuesto a bebidas saborizadas con azúcares añadidas y alimentos con alta densidad calórica — Ley del impuesto especial sobre producción y servicios — documento técnico – investigación subprocuraduría de análisis sistémico y estudios normativos. Dirección general de estudios jurídicos e interpretación normativa. Diciembre 2013*. http://prodecon.gob.mx/estudios_tecnicos/estudio_ieps_bebidas_azucardas_y_comida_chatarra_publicar.pdf. Accessed February 25, 2014

Public Health Agency of Canada (PHAC). (2015). Chronic disease and injury indicator framework. Quick stats overview. http://infobase.phac-aspc.gc.ca/cdiif/. Accessed February 16, 2016

Public Health England (PHE). (2015). *Sugar reduction. The evidence for action*. London. https://www.gov.uk/government/uploads/system/uploads/attachment_data/file/470179/Sugar_reduction_The_evidence_for_action.pdf. Accessed January 22, 2016

Radio Télévision Belge Francophone (RTBF). (2015, October 12). Budget: la taxe "soda" de Maggie de Block fâche les producteurs. http://www.rtbf.be/info/belgique/dossier/gouvernement-michel/detail_budget-la-taxe-soda-de-maggie-de-block-injuste-selon-les-producteurs?id=9106143. Accessed December 20, 2015

Raguso, É. (2016, January 20). Council approves $1.5 M to fight soda consumption. *Berkeleyside*. http://www.berkeleyside.com/2016/01/20/berkeley-council-approves-1-5m-to-fight-soda-consumption/. Accessed January 22, 2016

Réquillart, V., & Soler, L.-G. (2014). Is the reduction of chronic diseases related to food consumption in the hands of the food industry? *European Review of Agricultural Economics*, *41*(3), 375–403. doi:10.1093/erae/jbu010

Rosenberg, T. (2015, November 3). How one of the most obese countries on earth took on the soda giants. *TheGuardian.com*. http://www.theguardian.com/news/2015/nov/03/obese-soda-sugar-tax-mexico. Accessed December 16, 2015

Sarlio-Lähteenkorva, S., & Winkler, J. T. (2015). Could a sugar tax help combat obesity? *BMJ*, h4047. doi:10.1136/bmj.h4047

Scott-Thomas, C. (2013, April 25). Denmark to scrap decades-old soft drink tax. *FoodNavigator.com*. http://www.foodnavigator.com/Legislation/Denmark-to-scrap-decades-old-soft-drink-tax. Accessed February 27, 2014

Servicio de Impuestos Internos (Chili). (2016). Impuestos indirectos. Información vigente, actualizada el 4-01-2016. http://www.sii.cl/aprenda_sobre_impuestos/impuestos/impuestos_indirectos.htm. Accessed January 22, 2016

Sinckler, C. P. (2015). Presentation of the financial statement and budgetary proposals 2015 delivered by the hon. Christopher P. Sinckler to the House of Assembly, Monday June 15, 2015, 4 pm. http://www.barbadosparliament.com/uploads/document/d1efb84aac6a7abe4c6c0efcf8ceedd2.pdf. Accessed December 17, 2015

Skerrit, R. (2015). *Budget address 2015–2016—"Keeping it Real" by Honourable Roosevelt Skerrit*. http://finance.gov.dm/index.php/budget/budget-addresses/file/22-budget-address-2015-2016-keeping-it-real-by-honourable-roosevelt-skerrit-July-24-2015. Accessed December 17, 2015

Smed, S., & Robertson, A. (2012). Are taxes on fatty foods having their desired effects on health? *BMJ*, *345*(16), e6885–e6885. doi:10.1136/bmj.e6885

Snowdon, W., & Thow, A. M. (2013). Trade policy and obesity prevention: Challenges and innovation in the Pacific Islands. *Obesity Reviews: An Official Journal of the International Association for the Study of Obesity*, *14 Suppl 2*, 150–158. doi:10.1111/obr.12090

Swinburn, B. A., Sacks, G., Hall, K. D., McPherson, K., Finegood, D. T., Moodie, M. L., et al. (2011). The global obesity pandemic: Shaped by global drivers and local environments. *Lancet (London, England)*, *378*(9793), 804–814. doi:10.1016/S0140-6736(11)60813-1

Swinburn, B., Gill, T., & Kumanyika, S. (2005). Obesity prevention: A proposed framework for translating evidence into action. *Obesity Reviews: An Official Journal of the International Association for the Study of Obesity*, *6*(1), 23–33. doi:10.1111/j.1467-789X.2005.00184.x

Swire, M. (2012, December 4). Malaysia pushed to adopt a soda tax. *Tax-News.com*. http://www.tax-news.com/news/Malaysia_Pushed_To_Adopt_A_Soda_Tax____58617.html. Accessed January 22, 2016

The Economic Times. (2014, July 10). Budget 2014: Excise duty hike would lead to slowdown in the sector: Indian Beverage Association. *The Times of India*. New Delhi. http://articles.economictimes.indiatimes.com/2014-07-10/news/51300799_1_duty-hike-excise-duty-pepsico-india. Accessed January 5, 2016

The Telegraph. (2014, July 11). Human and fiscal health pills. *The Telegraph*. Calcutta, India. http://www.telegraphindia.com/1140711/jsp/business/story_18603354.jsp#.VnbNdMArK-w. Accessed December 20, 2015

Thow, A. M., Quested, C., Juventin, L., Kun, R., Khan, A. N., & Swinburn, B. (2011). Taxing soft drinks in the Pacific: Implementation lessons for improving health. *Health Promotion International*, *26*(1), 55–64. doi:10.1093/heapro/daq057

Tourigny, A., Namoro, A., Arsenault, P.-É., Héronneau, C., Guilbert-Couture, A., Anstett, É., et al. (2014). *La prévention en obésité au Québec : un portrait croisé et analytiques des interventions et politiques gouvernementales visant les jeunes*. Québec (Québec): École nationale d'administration publique (ENAP).

United Nations. (2011, September 16). Political declaration of the high-level meeting of the general assembly on the prevention and control of non-communicable diseases. A/66/L.1. http://www.un.org/en/ga/ncdmeeting2011/. Accessed April 25, 2012

Vandevijvere, S., Dominick, C., Devi, A., Swinburn, B., & International Network for Food and Obesity/non-communicable diseases Research, Monitoring and Action Support. (2015). The healthy food environment policy index: Findings of an expert panel in New Zealand. *Bulletin of the World Health Organization*, *93*(5), 294–302. doi:10.2471/BLT.14.145540

Wang, Y. C., Coxson, P., Shen, Y.-M., Goldman, L., & Bibbins-Domingo, K. (2012). A penny-per-ounce tax on sugar-sweetened beverages would cut health and cost burdens of diabetes. *Health Affairs (Project Hope), 31*(1), 199–207. doi:10.1377/hlthaff.2011.0410

World Cancer Research Fund (WCRF) (Ed.). (2015). WCRF international nourishing framework. Use economic tools to address food affordability and purchase incentives. Table last updated 24 November 2015. http://www.wcrf.org/sites/default/files/3_Economic%20Tools_Final.pdf. Accessed December 17, 2015

World Health Organization (WHO). (2011). *Global status report on noncommunicable diseases 2010*. Geneva, Switzerland: World Health Organization.

World Health Organization (WHO). (2014a). *Global status report on noncommunicable diseases 2014: Attaining the nine global noncommunicable diseases targets; a shared responsibility*. Geneva: World Health Organization.

World Health Organization (WHO). (2014b). Noncommunicable Diseases (NCD) Country Profiles: Canada. http://www.who.int/nmh/countries/can_en.pdf?ua=1. Accessed March 1, 2015

World Health Organization (WHO). (2015a). *Guideline: Sugars intake for adults and children*. Geneva: Switzerland.

World Health Organization (WHO). (2015b). *Using price policies to promote healthier diets*. Copenhagen (Denmark): World Health Organization (WHO).

World Health Organization (WHO). (2016). *Report of the commission on ending childhood obesity*. Geneva, Switzerland. http://apps.who.int/iris/bitstream/10665/204176/1/9789241510066_eng.pdf?ua=1. Accessed February 28, 2016

Zikara, M. (2012, February 8). *Les principales dispositions de la loi de finances 2012*. Alger.

Part I
Why: Rationale for Taxing Sugar-Sweetened Beverages

Chapter 2
Reasons for Specifically Targeting Sugar-Sweetened Beverages

Abstract Sugar-sweetened beverages (SSBs) generally designate non-alcoholic beverages containing added sugars. Conclusive epidemiological evidence indicates that overconsuming those beverages places individuals at greater risk of developing overweight, type 2 diabetes, and dental caries. SSB overconsumption is supported by intense marketing practices in terms of offer, distribution, price and promotion. In Canada, sugar-sweetened beverages are a significant dietary source of sugars in some groups of the population, particularly in adolescents. Specific prevention efforts are required at a large scale to tackle SSB consumption trends at population level and to limit SSB contribution to daily energy intakes at individual level.

2.1 Definition of a Sugar-Sweetened Beverage (SSB)

Among a wide array of NCDs' diet-related determinants, the overconsumption of sugar has been singled out as a risk factor for body weight gain, cardiovascular diseases, and dental caries (Te Morenga et al. 2012; Yang et al. 2014; Moynihan and Kelly 2014). Recent results from a systematic review and meta-analyses of 39 randomized controlled trials by Te Morenga et al. (2014) confirm that higher (vs. lower) intakes of dietary sugars negatively impact blood pressure and serum lipids independently of effects on body weight. WHO recently reaffirmed the recommendation to maintain the intake of free sugars[1] to fewer than 10 % of total energy intake (WHO 2015). The report even suggests that halving this limit to 5 % could have additional benefits, which met strong opposition from industry stakeholders (Basu 2015).

Sugar-sweetened beverages (SSBs) represent a significant dietary source of sugars in some groups of the Canadian population (Langlois and Garriguet 2011). It contributes to explaining why the overconsumption of these beverages has raised public health concerns (see Box 2.1).

[1]According to WHO's definition, free sugars include "monosaccharides and disaccharides added to foods and beverages by the manufacturer, cook or consumer, and sugars naturally present in honey, syrups, fruit juices and fruit juice concentrates" (WHO 2015, p. 4).

Y. Le Bodo et al., *Taxing Soda for Public Health*,
DOI 10.1007/978-3-319-33648-0_2

Box 2.1: What is Meant by "Sugar-Sweetened Beverage Overconsumption"?

At individual level, we refer to "sugar-sweetened beverage overconsumption" as the daily consumption of SSB resulting in a daily intake of added sugars that exceeds 10 % of daily energy intake. This threshold is congruent with the WHO recent recommendation to maintain the intake of free sugars to fewer than 10 % of total energy intake (WHO 2015). In average, an individual serving of SSB (size varying between 200 and 600 ml) contains between 20 and 60 g of added sugars (see Table 2.1), which represents between 4 and 12 % of daily energy intake (when considering a normal-weight person consuming 2000 kcal daily). As a result, the daily consumption of one serving of SSB per day could already be considered "overconsumption", depending on the sugar content of this beverage and the intake of added sugars from other foods.

The definition of SSB varies in the literature, but generally converges to designate non-alcoholic beverages containing added sugars. According to the Canadian Food and Drug Regulations, the term "sugars" encompasses all monosaccharides and disaccharides (including glucose, fructose, sucrose) and the term "sugar" specifically refers to sucrose.[2] Consequently, beverages containing "added sugars" designate beverages whose sugars content has been increased through the addition of sugars, or through the addition of ingredients containing added sugars (Government of Canada 2015).[3]

In this book, SSBs include carbonated drinks, fruit drinks, flavoured waters, energy drinks and sports drinks, as long as these beverages contain added sugars. Consequently, our definition excludes beverages containing no added sugars, such as bottled water, 100 % fruit juice, milk and non-calorically sweetened beverages (NCSBs). We also propose to exclude from our definition all forms of sugar-sweetened milk (e.g. chocolate milks and flavoured milk): although these beverages may contain as much added sugars as the aforementioned SSBs, it also brings essential micronutrients (especially calcium for youth) and does not appear to reach high intakes levels in Canada (MC Paquette, personal communication, unpublished supermarket sales data). Finally, we propose to exclude from our definition hot chocolate, hot coffee- and hot tea-based beverages for three main reasons: (1) the intake levels of these beverages in the Canadian population and their contribution to energy intakes remain weakly documented (Agriculture and

---

[2]Combination of the monosaccharides glucose and fructose.

[3]According to Agriculture and Agri-Food Canada (2015), corn sugar (mainly imported from the US) is much more commonly used than cane or beet sugar by the Canadian soft drink industry.

Table 2.1 Sugar content of various sugary beverages, measured in teaspoons (tsp)[a] or grams (g) per 250 ml and per individual serving size (as sold)

	250 ml portion	Individual serving size
*Beverages containing added sugars **included** in our SSB definition*		
Energy drink (Full Throttle citrus ®)	8 tsp	473 ml 14 tsp (58 g)
Cola flavoured soda (Coca-Cola ®)	7 tsp	355 ml 10 tsp (39 g)
Fruit drink (Grape cocktail Rougemont ®)	6 tsp	200 ml 5 tsp (21 g)
Fruit punch (Fruité ®)	6 tsp	300 ml 8 tsp (30 g)
Sports drink (Gatorade Perform Cool Blue ®)	4 tsp	591 ml 9 tsp (35 g)
Vitamin water (Glaceau Vitaminwater watermelon pink grapefruit ®)	3 tsp	591 ml 8 tsp (31 g)
*Beverages containing added sugars **excluded** from our SSB definition*		
Chocolate milk base beverage (Hershey ® milkshake creamy chocolate)	9 tsp	355 ml 13 tsp (51 g)
Yogurt drink (Strawberry Yop ®)	7 tsp	200 ml 6 tsp (22 g)
Chocolate milk (Beatrice ®)	7 tsp	200 ml 6 tsp (22 g)
Cold specialty coffee (Iced capp ®)	6 tsp	648 ml (large) 16 tsp (62 g)
Hot specialty coffee (Starbucks ® Caramel macchiato)	4 tsp	591 ml (venti) 10 tsp (40 g)
Beverages containing no added sugars		
100% apple juice (tradition ®)	6 tsp	300 ml 8 tsp (30 g)
Plain milk (Québon ®)	3 tsp	237 ml 3 tsp (11 g)
Coffee	0 tsp	0 g
Water	0 tsp	0 g

[a]1 teaspoon of white sugar eq. 4 g of sugar
Sources companies' websites

Agri-food Canada 2010; Coffee Association of Canada 2011; Cardwell 2015)[4]; (2) although these beverages may contain as much added sugars as SSBs, this content may vary importantly since the beverage is prepared at the point of

[4]In the future, if the contribution of these beverages to daily energy intakes appears to be significant in some groups of the population, greater consideration should be deserved to their possible inclusion in the SSB definition.

purchase and depends on consumer preferences[5]; (3) current soda taxes implemented at a large scale across the world and related evidence generally do not apply to this type of beverage (see Table 1.1). The sugar content of the above-mentioned beverages and their inclusion (or not) in our SSB definition are summarized in Table 2.1.

It should also be noted that some types of SSBs (particularly energy drinks) have raised other specific health concerns due to their high content in stimulants (caffeine, in particular), considered to enhance mental acuity and physical performance. In 2011, the Canadian Government announced new measures, including capped caffeine content and explicit labelling requirements, in order to reduce the chances of caffeine overconsumption (Health Canada 2011a). Specific taxes on drinks containing stimulants at a high dose have been discussed in other countries (Le Monde 2014). However, only taxation issues related to sugars in non-alcoholic beverages are addressed in this book.

The following sections discuss the reasons why SSBs may be specifically targeted to prevent obesity and related diseases. The health risks that have been associated with SSB intakes are presented, as well as SSB consumption trends and marketing practices in Canada. Finally, additional considerations on other non-alcoholic beverages are discussed.

2.2 SSB and Health Risks

In line with international recommendations (AICR/WCRF 2007; WHO 2015), Canada's food guide recommends limiting the consumption of foods and beverages "high in calories, fat, sugar or salt such as [...] fruit-flavoured drinks, soft drinks, sports and energy drinks, and sweetened hot or cold drinks" (Health Canada 2011b). Indeed, SSB overconsumption has been frequently singled out as an independent risk factor for several health issues in the scientific literature.

The most recent reviews and experimental studies clearly suggest that high SSB intakes are associated with weight gain in children and adults and that limiting SSB intake in youth may contribute to prevent weight gain (de Ruyter et al. 2012; Ebbeling et al. 2012; Malik et al. 2013; Ebbeling 2014).[6] However, it remains

---

[5]In the framework of a food policy targeting specifically the presence (or not) or the amount of added sugars in a beverage, this situation makes technically and administratively burdensome the distinction between on-site prepared drinks containing added sugars or not. The fact that some consumers (and not others) add sweeteners themselves at the point of purchase also raises concerns. Finally, since many of these beverages contain milk, including it in our SSB definition would interfere with the aforementioned exclusion of all forms of sugar-sweetened milk.

[6]Over the last 10 years, nearly 20 literature reviews have been published as regards the relationship between sugar-sweetened beverages and body weight (Massougbodji et al. 2014). Until recently, these reviews offered dissenting conclusions: some of them concluded that the evidence of a causal relation was weak or limited (e.g. Forshee et al. 2008; Mattes et al. 2011) whereas others concluded that the evidence was strong or significant (e.g. Malik et al. 2006;

difficult to precisely estimate the strength of the association because of the diversity of study designs, their respective methodological limitations, as well as because of the difficulty to fully control for a broad range of potential confounders (Sievenpiper and de Souza 2013; Pereira 2014).[7] Attempting to take as much as possible account of these issues, Malik et al.'s recent systematic review and meta-analysis (2013) conclude that, in prospective cohort studies, one daily additional 12-oz serving of SSB is associated with a 0.06 BMI-unit increase[8] among youth and a 0.12–0.22 kg of body weight gain among adults over a one-year period. That is to say that the effect size may be clinically modest and uncertain on the long term, but it may prove significant at the population level, especially in at-risk subgroups (e.g. high SSB consumers, overweight individuals). As regards potential underlying mechanisms, the research is suggestive but not conclusive (Ebbeling 2014). SSB's high contribution to fructose intake has been suggested as a potential explanatory factor, since the latter has been associated with decreased insulin sensitivity, dyslipidemia and visceral adiposity in overweight adults (Welsh et al. 2013). However, as regards body weight, a recent systematic review indicates that the negative impact of a higher (vs. lower) intake of sugars and SSB would be more likely mediated by the increase in energy intake per se than by specific physiological or metabolic effects of sugars (Te Morenga et al. 2012). The lower effect of liquid calories on satiety has proved to cause extra energy intake and weight gain in adults on the short term but, in a longer term, this explanatory hypothesis may not be true (Bachman et al. 2006; Trumbo and Rivers 2014; Allison

(Footnote 6 continued)

Woodward-Lopez et al. 2011). In 2012, the controversy reached a turning point, as shown by the opening debate of the 30th Annual scientific meeting of the Obesity society (Billes 2012) and the subsequent release of new findings from two randomized controlled trials supporting the hypothesis that limiting access to SSBs may contribute to prevent weight gain in youth (Rabin 2012). Subsequent reviews still offer methodological debates and contrasting conclusions, but clearly tend to agree that increasing SSB intake increases the risk of weight gain (Kaiser et al. 2013; Hu 2013). Additionally, two studies have looked at the funding source of reviews on SSB consumption and weight gain (2006–2013) and suggest that it may have introduced a bias in their conclusions, since industry-funded reviews are significantly more likely to suggest weak evidence of an association than other reviews (Bes-Rastrollo et al. 2013; Massougbodji et al. 2014).

[7]In Canada, some cross-sectional and longitudinal studies have analyzed the association between SSB consumption and body weight. We did not particularly emphasize these results in this section since the cause-to-effect relationship between SSB and weight has been largely covered in methodologically more robust systematic reviews and meta-analyses at international level (see the previous footnote). These reviews provide the best available evidence to date. However, on an indicative basis, we refer the reader to some Canadian studies on youth SSB consumption and body weight published over the last years. On the one hand, significant positive associations (with at least one weight-related outcome for the whole population) have been found in several studies based on a cross-sectional design (Gillis and Bar-Or 2003; Reid et al. 2015) or a longitudinal design (Dubois et al. 2007). Absence of any significant positive association was also reported in studies based on a cross-sectional design (Janssen et al. 2005; Vanderlee et al. 2014) or a longitudinal design (Mundt et al. 2006). Finally, mixed results (depending on gender/age group) have been reported in cluster analyses (Danyliw et al. 2012).

[8]For a definition of body mass index (BMI), see Chap. 1, footnote 2.

2014). Finally, research suggests that genetic factors (some of which associated with ethnicity), adiposity and glucose intolerance may influence the susceptibility to SSB consumption's effects on metabolism and body weight. Further research is required in this area (Wang et al. 2013; Ebbeling 2014).

A high SSB intake has also been associated with increased risks of type 2 diabetes, cardiovascular disorders and metabolic syndrome, partially beyond the interaction with obesity (Malik et al. 2010; Ambrosini et al. 2013; Huang et al. 2014; Keller et al. 2015; Greenwood et al. 2014). For instance, results from a recent systematic review and meta-analysis of 17 prospective cohort studies following adults initially without diabetes revealed that the intake of one SSB serving/day was associated with an 18 % greater relative risk of type 2 diabetes. From a public health perspective, if such association would correspond to causation, the authors estimate that SSB consumption may be responsible for 4–13 % of type 2 diabetes new cases in the USA between 2010 and 2020 (Imamura et al. 2015). A high consumption of SSB has also been associated with a higher risk of kidney stone formation in large cohort studies (Ferraro et al. 2013). Finally, the quantity and frequency of added sugars consumption have been recognized as an important risk factor in the development of dental caries in both children and adults. This risk is shown to be cumulative over the life course, which indicates that reducing added sugars consumption during childhood can have a protective effect for dental health in the long term (WHO 2003, 2015).

All in all, considering well-powered prospective cohort studies alongside short-term randomized controlled trials, Hu (2013, p. 1) concludes that "a significant association, established temporality and direct dose–response relationship between SSB consumption and long-term weight gain and risk of type 2 diabetes" have been demonstrated. He adds that "taken together, the evidence that decreasing SSBs will decrease the risk of obesity and related diseases such as type 2 diabetes is compelling" (Hu 2013, p. 1). As Hu suggests, we consider that "we should avoid the trap of waiting for absolute proof before allowing public health action to be taken" (2013, p. 1).

Other relevant reasons for specifically focusing public health action on SSBs are that these beverages have no or low added value for health (they are often energy-dense beverages with few micronutrients) and that healthier substitutes are commonly available, starting with water (Popkin et al. 2006). In sum, although SSB overconsumption is not the only dietary risk factor linked to obesity and NCDs (Sievenpiper and de Souza 2013), it clearly seems to be a relevant target for public health action, particularly in heavy consumers (including youth), patients at health risk, and on territories where average consumption is high (Ebbeling 2014; Le Bodo et al. 2015).

2.3 SSB Marketing and Sales Trends

In their recent analyses of SSB sales trends around the world,[9] Popkin and Hawkes (2016) indicate that the four regions with the highest consumption[10] are as follows: (1) North America, (2) Latin America, (3) Australasia and (4) Western Europe. With the exception of Latin America, average SSB intake has been declining in these four regions over the last years. In North America, this decline has been mainly driven by a decrease in the sales of caloric soft drinks, despite the fact that the sales of sports and energy drinks increased significantly.[11] All in all, in 2014, North America still ranked first (by far) for total sales of SSBs, and Canada ranked 14th out of 54 countries[12] for which data have been analysed (Popkin and Hawkes 2016).

A closer analysis of Canadian Market data tends to confirm these trends: the consumption of carbonated beverages (excluding sports drinks and energy drinks) increased from 55 to 117 l/capita between 1972 and 1998, before declining to 85 l/capita in 2009, which still represents a very high consumption level (Statistics Canada 2010). Overall, in a study based on soft drink sales data from the Canadian Socioeconomic Information Management System (Statistics Canada),[13] Merchant et al. (2010) demonstrate that the average available energy from SSB[14] per capita has increased of 37 kcal per day between 1976 and 2007, which represents 23 % of the increase in total available energy per capita per day over the same period. Interestingly, yearly averages show that the higher is the contribution of soft drinks to total available energy, the higher tends to be the total available energy per capita per day. Whereas the ecological study design precludes assuming cause-to-effect, this analysis suggests that a greater SSB intake may have been both a contributor to and a marker of increasing energy intakes in the diet of the Canadian population over the last 30 years[15](Merchant et al. 2010).

---

[9]Popkin and Hawkes' calculations are based on the Euromonitor Passport International database, for the period 2009–2014.

[10]Expressed in Kcal per person per day.

[11]Popkin and Hawkes' analyses are based on three SSB categories: caloric soft drinks (carbonated and non-carbonated); fruit drinks (sweetened beverages of diluted fruit juice and often other caloric sweeteners and flavourings); and sports and energy drinks (including sugar-sweetened waters).

[12]Before Canada ranked Chile, Mexico, USA, Argentina, Saudi Arabia, Germany, Netherlands, Slovakia, Austria, Brazil, Belgium, Israel and Ireland.

[13]Provided by the Canadian Soft Drink Association.

[14]Excluding juices.

[15]Indeed, the fact that SSB intake has been frequently associated with overall energy intake could be partly related to the fact that SSB consumption is a marker of food patterns involving other dietary factors favouring excessive energy intake (Woodward-Lopez et al. 2011). Anyway, considering the evidence presented overall in this section, it seems plausible that the increase in SSB consumption over the last decades has somewhat contributed in itself to increase energy intakes in some groups of the Canadian population.

Although SSB manufacturers and retailers have made voluntary efforts over the last years, e.g. to reduce SSB availability in schools or to reduce SSB advertising to children (Canadian Beverage Association 2015), heavy marketing is still practised in terms of product diversification, availability, prices and promotion (Welsh et al. 2013). Here below are summarized some key facts and figures.

An attractive and diversified choice: in 2009, in Canada, carbonated soft drinks represented the second non-alcoholic beverage volume market share (16.3 %), just after coffee (16.6 %) and before tea (12.9 %), milk (12.7 %), bottled water (10.6 %) and non-carbonated fruit beverages (including juices) (8.7 %) (Agriculture and Agri-Food Canada 2015). The non-alcoholic beverage industry offers an increasing variety of products to adjust to consumers' preferences. It includes the creation and development of juices, fruit-flavoured beverages, dairy-based beverages, iced tea, vitamin waters, sports drinks and energy drinks. In Canada, the largest SSB companies propose more than 30 different brands and a diversity of packages and innovative flavours (Agriculture and Agri-Food Canada 2015). Over the last decades, higher SSB intakes have been prompted by steady increases in portion sizes. Between 1977 and 1996, US consumption surveys show that the average SSB portion size consumed by all age groups increased from 402 to 621 ml, whereas milk beverages' average portion size decreased from 346 to 322 ml (Nielsen and Popkin 2004). Some US data also indicate a two- to threefold increase in the size of soda poured from fountains in two large fast-food chains between their time of introduction (in the 1950s) and 2002 (Young and Nestle 2003).

A widespread distribution: according to Agriculture and Agri-Food Canada (2015), two-thirds of SSB sales are not for on-site consumption, which reflects important takeaway sales from fast-food restaurants but also a particularly large diversity of easy-to-access points of sales such as vending machines, convenience stores and gas stations. Although representing a small market segment, SSB presence in schools has been used as a way to position SSB brands and to encourage out-of-school consumption across the world (Hawkes 2010). The situation in Canada does not seem to make exception. A study relying on data from the British Columbia Adolescent Health Survey (n = 174 schools/10,879 grades 7–12 students, 2007–2008) found SSB availability at school to be associated with a higher SSB intake[16] (Mâsse et al. 2014). As a result, many governments have taken action. In Quebec, for instance, policy recommendations have been issued to limit SSB and NCSB availability in schools[17] and childcare centres (Ministère de

[16]Independently of neighbourhood-level post-secondary education and school setting (urban/suburban/rural) at school level, and independently of age and sex at student level.

[17]This "Policy framework for a health turn at school" (*Politique cadre pour un virage santé à l'école*) has been set up in 2007. A process evaluation has been published in 2009. For the food environment, this evaluation is based on data collected from 720 schools. The sample was further weighted to make it representative of all Quebec schools (according to type, network and region). Results indicate that 94 % of primary schools (eq. to elementary schools) and 59 % of secondary schools (eq. to middle and high schools) report having removed SSBs and NCSBs from their main

l'Éducation, du Loisir et du Sport 2009; Martin et al. 2014). On the other hand, Canadian manufacturers have collectively issued voluntary-based guidelines for the sale of beverages in schools (Canadian Beverage Association [CBA] 2012).[18] These efforts should be further encouraged and evaluated. A recent research conducted in the province of Quebec (Sherbrooke area, $n = 39$ schools/7099 students, 2007–2008) indicates that beyond individual characteristics and the school's socioeconomic status, a significant difference of daily SSB consumption was observed between schools. It was further estimated that a student moving from a school with a lower rate of SSB consumption to a school with a higher rate of SSB consumption would typically increase his risk of becoming a daily SSB consumer by 52 %.[19] Beside the school's socioeconomic status, students in schools located in a higher urban density area had a significantly higher risk of SSB consumption (Lebel et al. under review). These results suggest that beyond SSB availability at school, the school vicinity may play an important role in favouring SSB consumption in youth.

A very affordable option: in high-income countries, an increasing affordability gap is generally observed between low nutritive energy-dense foods and more nutritive choices (e.g. meat, fruit and vegetables). For example, in Canada, a study conducted on the price of a diversity of food items sold in Edmonton supermarkets illustrated that the price per calorie of "processed sweets" was 7 times cheaper than the price per calorie of "fish and poultry" (Cash and Lacanilao 2007). Among these low nutritive energy-dense foods, SSBs are considered to be particularly and increasingly affordable. Agriculture and Agri-food Canada (2015) indicates that this is partly due to a highly competitive environment: "the [soft drink] industry has

(Footnote 17 continued)

food service. Very few primary schools offer vending machines (2 %) but among secondary schools, 47 % report having removed these beverages from vending machines (Ministère de l'Éducation, du Loisir et du Sport 2012).

[18]These guidelines prone the removal of soft drinks from elementary and middle schools and displacement by water, 100 % juice and milk. In high schools, the guidelines indicate that "at least 50 % of the beverages offered [...] should be water and low- or no-calorie options" and also encourage capped caloric content for SSBs, capped portion sizes for SSBs, and non-availability of energy drinks. According to the Canadian beverage association, its members have "completed [the] implementation of the guidelines during the 2009–2010 school year" (CBA 2012, §1).

[19]The multilevel logistic modelling procedures used in this study made possible to appreciate the school built environment effect on daily SSB consumption independently of other variables that may influence consumption at individual level (gender, academic cycle, cultural origin, participation in organized physical activities) and independently of the school socioeconomic status. Concerning the school built environment's vicinity, it is noteworthy that none of the contextual characteristics measured (number of fast-food restaurants, number of convenience stores, walkability, degree of vegetation cover, distance to the closest fast-food restaurant, distance to the closest convenient store) had a significant association with SSB consumption when taken separately. However, the combination of these characteristics showed an important difference in students' SSB consumption between schools. This observation confirms the complex aetiology of eating habits and the most probable influence of a pattern of contextual characteristics rather than a significant influence of one risk factor considered in isolation.

experienced intense price competition with the expansion of private label sales and decreasing consumer demand for carbonated soft drinks. Price reductions have been an important element to enable the industry to maintain its dominant market share in a beverage market where the choice of products is increasing". Additionally, the organization suggests that SSBs are frequently used as a drawing card: "During the recent difficult economic climate, price reductions are still occurring, especially in the retail grocery sector where chains have reduced prices to encourage consumers to visit their establishments". Indeed, SSBs have been shown to be highly subject to specific pricing strategies such as rebates, bulk offers and free refills in fast-food restaurants (Quebec Weight Coalition [CQPP] 2012). An analysis of food and beverages purchases in supermarkets by a large representative panel of UK households (n = 30,000) over two years shows that price promotions on carbonated drinks are heavily practised (PHE 2015). Unsurprisingly, a very large proportion of consumers are sensitive to price promotions on those drinks (the so-called promotionally sensitive shoppers)[20] (PHE 2015, p. 19). On average, overtime, it was estimated that price promotions increase total purchases of regular carbonated drinks by 31 % in comparison with a hypothetical situation where no price promotion was practised at all (PHE 2015, p. 28). Accounting for compensation effects, it was also estimated that the total volume of sugars included in purchased food and beverages could be 0.7 % lower if no price promotions were practised on regular carbonated drinks[21] (PHE 2015, p. 34). These data (although specific to the UK) clearly suggest that SSBs are a relevant starting point for public health efforts aimed to tackle price promotions on sugary food and drinks.

A strong promotion, especially towards youth: in Canada as elsewhere, brand names, advertising and promotion are essential dimensions of the competition in the soft drink sector. Placement in the store and on the shelf is deemed particularly critical to stimulate consumer purchases (Agriculture and Agri-Food Canada 2015). Evidence indicates that advertisements are usually appreciated by youth and influence their dietary behaviours (Hastings et al. 2007). Accordingly, in 2009 in the USA, carbonated beverage companies' marketing expenditures targeting youth (2–17 years of age) reached US$395 millions (i.e. 22 % of all food categories). It was the #1 food category in terms of teen-directed marketing expenditures. Traditional media (TV, radio, print advertising) are complemented by a diversity of marketing techniques including appealing packages, point-of-purchase displays, movie cross-promotions, sponsorship of sports events, advertisements using celebrities (Federal Trade Commission [FTC] 2012). Over the last years, facing international pressure (WHO 2010), the food and beverage industry has accelerated its large-scale voluntary commitments to limit marketing to children, as shown in

[20]In this study, three types of price promotions were considered: "temporary price reductions", "multibuy" offers (e.g. "buy 2 for 5$" or "buy one and get one free") and "extra free" offers (i.e. portions larger than usual for the same price). The analysts emphasize that food price promotions are heavily practised in the UK and may represent a proportion of purchases that is twice as much elevated than in other European countries (i.e. ≈40 % vs. ≈20 %).

[21]With equal percentage found for chocolate and cakes.

North America by the US Children's Food and Beverage Advertising Initiative (Council of Better Business Bureaus 2006) or the Canadian Children's Food and Beverage Advertising Initiative (CAI) (Advertising Standards Canada 2007). Overall, the evaluations of such "pledges" have shown mixed results (sometimes in the same country and for the same intervention) because of different sampling methods and heterogeneous criteria used to measure success (Hawkes 2013). The US Federal Trade Commission did observe an overall decrease[22] in marketing expenditures on non-alcoholic beverages towards young people between 2006 and 2009 and note "small but positive changes in the nutritional profile of drinks marketed to youth"[23] between 2006 and 2009 (FTC 2012, ES-5). However, the FTC also highlights that "almost all of the sugar in drinks marketed to both children and teens was added sugar, rather than naturally occurring sugar from fruit or fruit juice" (FTC 2012, ES-5). Another concern is the significant increase in youth-directed new media expenditures[24] on food and beverages between 2006 and 2009,[25] with carbonated beverages ranking #1 among all food and beverage categories in 2009.[26] The report adds that "carbonated beverages marketed to teens averaged 10 % more calories and added sugars in new media than in traditional measured media" in 2009 (FTC 2012, p. ES-5). Uncertainties as regards the effectiveness of self-regulation were also expressed on a Canadian ground. A comprehensive assessment of food and beverage TV advertising in Toronto and Vancouver between 2006 and 2011[27] shows that the total number of soft drinks spots broadcasted on children's speciality stations actually decreased.[28] As a result, the average number of soft drinks spots actually viewed by children (aged 2–11) on these channels effectively decreased. However, when considering all stations, the total number of soft drinks spots broadcasted did not decrease over the same period. As a consequence, the authors note that the decrease in soft drinks TV ads viewed by children between 2006 and 2009 was mitigated in Toronto and fully compensated in Vancouver when all stations were considered (Potvin Kent et Wanless

[22]59 % decrease (adjusted for inflation) in child-directed marketing of carbonated beverages (from US$96 million to $US42 million) and 29 % decrease (adjusted for inflation) in teen-directed marketing of carbonated beverages (from US$508 million to US$382 million).

[23]The US Federal Trade Commission (2012, p. ES-5) observes that drinks marketed to youth (2–17 years old) "averaged 20 fewer calories per serving in 2009 than in 2006".

[24]Consisting of "company-sponsored websites, Internet, digital, word-of-mouth, and viral marketing" (FTC 2012, p. 9).

[25]50.5 % increase (adjusted for inflation) for all food categories (from US$76.6 million in 2006 to US$122.5 million in 2009).

[26]Moreover, between 2006 and 2009, the relative increase in child-directed new media expenditures focused on juices and non-carbonated beverages has been substantial (+617 %, adjusted for inflation) (FTC 2012).

[27]Using Nielsen Media Research data specific to Vancouver and Toronto cities.

[28]Considering the evolution of TV advertisements targeting kids in Toronto between 2006 and 2011, Potvin Kent et al. (2014) note that this decrease was particularly due to the fact that companies adhering to CAI had eliminated soft drinks advertisements on these stations, which was not the case of non-CAI members.

2014). Research also indicates that the TV marketing techniques used to target children and teens in 2011 have been more intense than in 2006. Another concern appears to be that the promotion of "less healthy" items by companies adhering to CAI has increased in children and teens over the same period (Potvin Kent et al. 2014). Finally, another study suggests that Canadian self-regulatory measures imperfectly protect children from Internet marketing of energy-dense food and beverages (Potvin Kent et al. 2013).

Altogether, this overview shows that SSB consumption is substantially supported by marketing practices. In the next section, we will review consumption trends at the individual level.

2.4 SSB Consumption Trends

In 2004, according to the Canadian Community Health Survey—Nutrition (CCHS, 24-h dietary recall), average dietary sugar intakes (including added and naturally occurring sugars) represented 21 and 25 % of total daily calories in adults (aged 19 + years) and youth (aged 9–18), respectively. Regular soft drinks represented the second dietary source of sugar intake in adults and the first one in youth aged 9–18 before milk, fruit and confectionery (Langlois and Garriguet 2011).

According to CCHS-2004 data, Brisbois et al. (2014) estimate that on average, sweetened drinks[29] contribute 3.3 % to daily calories in Canadian adults. According to Garriguet (2008a, same data source), the highest proportion of adults reporting having consumed regular soft drinks the day before the survey was found in adult men aged 19–30 (47 %), for whom the average daily consumption reached 649 g.[30] The lowest proportion was found in Canadian women aged 71 or older (10 %), for whom the average daily consumption reached 305 g.

According to CCHS-2004 data, average consumption of sweetened drinks in Canadian youth is similar across provinces, with slightly higher values in New Brunswick, Newfoundland and Labrador, and slightly lower values in British Columbia. The contribution of milk and 100 % fruit juice to daily calories decreases with age, whereas the contribution of sweetened drinks increases with age (see Fig. 2.1) (Garriguet 2008b).

In teens aged 14–18, average daily intake of regular soft drinks[31] reached 179 and 376 g in girls and boys, respectively. For this age group, the average contribution of sweetened drinks[32] to daily calories was 7 % in girls and 8 % in boys (see

[29]Including regular soft drinks and fruit drinks.

[30]Regular soft drinks volumes expressed in grams (g) or in millilitres (ml) are generally close. For instance, according to the Canadian Nutrient File (CNF), 100 ml of cola-type soft drinks is equivalent to 104.2 g (Source: Health Canada, file consulted in March 2012 from: http://webprod3.hc-sc.gc.ca/cnf-fce/index-fra.jsp).

[31]Excluding fruit beverages and fruit juices.

[32]Including regular soft drinks and fruit drinks.

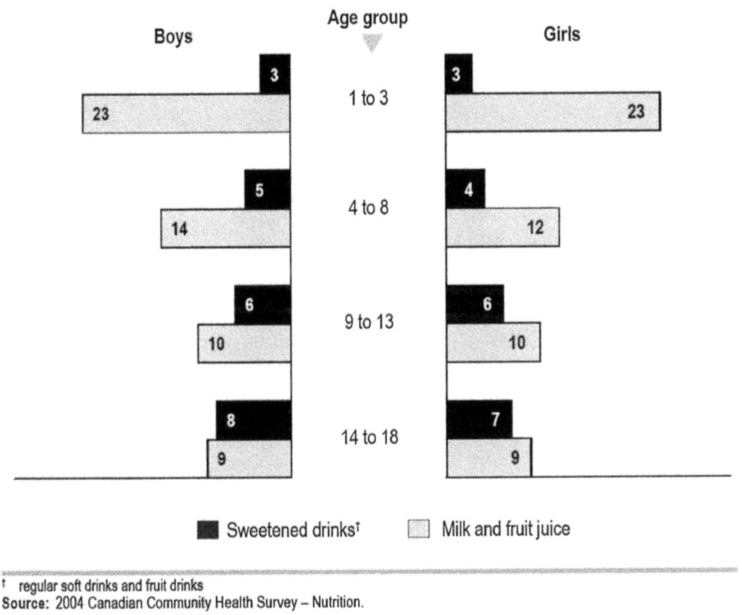

Fig. 2.1 Percentage of daily calories derived from sweetened drinks and from milk and fruit juice, by gender and age groups, household population aged 1–18, Canada, excluding territories, 2004. *Source* Garriguet (2008b). Beverage consumption of children and teens. Health Matters. Statistics Canada, Catalogue no. 82-003-XPE. Health Reports, Vol. 19, no. 4, December 2008 (Reproduced with permission)

Fig. 2.1) (Garriguet 2008b). In comparison, average SSB contribution to total energy intake has been estimated to be 13 % in US teens aged 1–19 and 5 % in European teens aged 12.5–17.5 (Le Bodo et al. 2015). These numbers represent averages of total population daily intake of SSB, including people who did not consume any. According to CCHS-2004 data, among Canadian adolescents (aged 14–18) reporting having consumed sweetened drinks the day before the survey (i.e. 35 % of girls and 53 % of boys), this consumption reached 514 g/day in girls and 715 g/day in boys (Garriguet 2008b). The CCHS—Nutrition has been repeated in 2015. The release of the data in 2017 will provide an updated description of the situation (Statistics Canada 2014).

Meanwhile, more recent data from provincial surveys tend to confirm that SSBs are still consumed in high proportions by Canadian youths. In British Columbia (BC), for example, data from the BC Adolescent Health Survey (2007–08) indicate that grades 7–12 students (N = 11,385) were 32 and 10 % to report having consumed one and at least two SSB(s) the previous day, respectively[33] (Mâsse et al. 2014). In the province of Quebec, the large survey on the health of secondary

[33]As for all self-reported data, measurement errors may occur (e.g. underreporting of SSB consumption) and may influence the results.

school children (2010–2011, $N = 63{,}196$) indicates that 25 % of students report consuming SSB at least once a day (28 % of boys vs. 21 % in girls)[34] (Camirand et al. 2012). In another Quebec survey concerning students from all the 42 public primary and secondary schools of the Sherbrooke area (2008, $N = 13{,}500$), data confirm that the proportion of weekly SSB consumers is sizeable and increases with age, fruit drinks being the most popular beverages (from 34 % of consumers at age 5, to 65 % at age 17), followed by soft drinks (12 % at age 5, to 52 % at age 17) (Morin 2010; Morin et al. 2013). Recent data from other provinces confirm and even exceed these trends. According to a study including data from 10,188 youth (aged 13–18) from Hamilton and Thunder Bay (Ontario) and the province of Prince Edward Island (PEI) in 2009–10, 36 % of the students reported having consumed one to two SSB(s)[35] the day preceding the survey and 44 % of the students reported having consumed three or more SSBs. The proportion of daily SSB consumers in the PEI subgroup was statistically lower than in the Ontarian ones (Vanderlee et al. 2014).

A growing body of evidence also suggests that SSB consumption is associated with a less healthy dietary pattern (e.g. frequent fast-food meals, lower intakes of healthy beverages, breakfast skipping) and TV watching among children and adolescents (Mathias et al. 2013; Welsh et al. 2013). In the province of Quebec, the aforementioned survey on the health of more than 60,000 secondary school children indicates that carbonated drinks consumption is correlated to "junk food"[36] consumption in restaurants (see Fig. 2.2) (Camirand et al. 2012). The same survey indicates that students having at least 6 portions of fruit and vegetable per day are less likely to be daily SSB consumers[37] compared to those having less fruit and vegetable (22 % vs. 25 %) (Camirand et al. 2012, p. 87). Daily SSB consumers are also slightly[38] less likely to report consumption of at least 4 glasses of water per day (38 %) and slightly more likely to report consumption of fewer than 1 glass of water per day (9 %) compared to those having fewer than one SSB per day (40 and 6 %, respectively) (Camirand et al. 2012, p. 84). Although statistically significant, these differences (2 or 3 % points) remain small in magnitude.

In other provinces, the aforementioned study conduced in Ontario and PEI also shows that a higher SSB consumption in youth is slightly correlated with a higher

---

[34]In this survey, beyond the proportion of daily consumers, the proportion of occasional SSB consumers is sizeable. For instance, 36 % of students report consuming fruit-based drinks from one to six times per week (38 % for boys vs. 33 % for girls) and 36 % of students report consuming carbonated drinks between once to six times per week (43 % for boys vs. 29 % for girls).

[35]In addition to fruit-flavoured drinks, regular soft drinks, sports drinks and high-energy drinks, SSBs here also include hot chocolate, cappuccino or frappuccino, slurpees, and shakes.

[36]In this survey, the term "junk food" encompasses foods that should only be consumed exceptionally according to official Quebec guidelines, such as French fries, hamburgers, pizza, chicken wings, fried chicken, hot dogs.

[37]i.e. having SSB at least once a day.

[38]Difference statistically significant ($p \leq 0.05$).

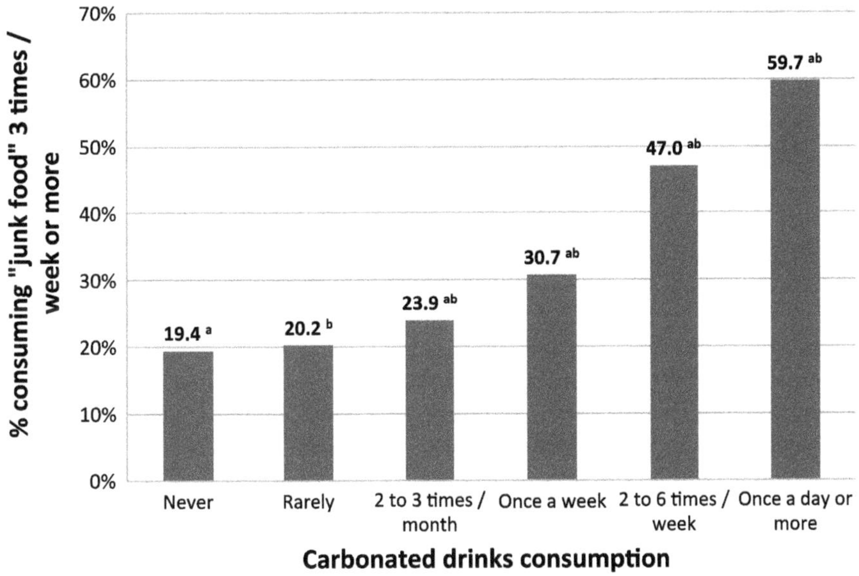

Fig. 2.2 "Junk food" consumption 3 times per week or more in a restaurant or a café during the school week, according to carbonated drinks consumption, secondary school children, Quebec, 2010–2011. Same exponent (a or b) means a statistical difference between proportions ($p \leq 0.05$). *Source* Camirand et al. (2012) *Habitudes alimentaires*, dans L'Enquête québécoise sur la santé des jeunes du secondaire 2010–2011. Le visage des jeunes d'aujourd'hui: leur santé physique et leurs habitudes de vie, Tome 1, Québec, Institut de la statistique du Québec, p. 89

consumption of 100 % fruit juice and milk. However, the authors note that a reverse correlation between SSB consumption and milk consumption has been more frequently documented elsewhere in the literature, which raises calcium displacement concerns (Vanderlee et al. 2014).

Using CCHS-2004 data and distinguishing sources of naturally occurring and added sugars in the diet, Brisbois et al. (2014) estimate that the average contribution of added sugars to daily calories in Canadians is 9.9 % in adults, 10.4 % in children (aged 1–8) and 14.1 % in adolescents (aged 9 to 18). These average values for all Canadians suggest how even more difficult it may be for SSB consumers (youth in particular) to limit the daily energy intake from free sugars below 10 % of total daily energy intake, as recommended by the World Health Organization (WHO 2015).

The evidence presented in this section confirms that specific efforts to prevent SSB overconsumption are relevant in Canada. Our conclusions are consistent with Saunders' ones, who considered large pieces of evidence to rank potential contributors to the Canadian paediatric obesity epidemic. He concludes that "although influenced by numerous factors, available evidence suggests that the Canadian childhood obesity epidemic is most closely related to deleterious changes in sugar-sweetened beverage intake, sedentary behaviour, reduced sleep, and adult obesity [i.e. parental obesity and gestational weight gain]" (Saunders 2011, p. 6).

2.5 Additional Considerations on Other Non-alcoholic Beverages

Beyond the fact that SSBs are generally the core target of taxation proposals, the need to prevent excessive intake of other non-alcoholic caloric beverages sometimes comes into the debate.

The relevance to include 100 % fruit juice is sometimes questioned. The calorie and sugar content of these beverages may be similar or even higher than for SSBs (see Table 2.1). Furthermore, 100 % fruit juice intakes have increased in some groups of the Canadian population over the last years. For example, in Quebec adults, fruit juice[39] intake has increased by 80 % between 1990 and 2004[40] (Blanchet et al. 2009). However, 100 % fruit juice's contribution to micronutrients recommended daily intakes makes them "sparingly" drinkable on a daily basis, although whole fruit consumption should be preferred (Popkin et al. 2006). This last consideration makes these beverages generally exempt from soda taxes currently implemented across the world (see Table 1.1), even though they are included in the WHO recommendations as a source of free sugar (WHO 2015).

More controversial is the case of non-calorically sweetened beverages (NCSBs or "diet drinks"), whose consumption has been correlated to increased appetence for sweetness, risks of metabolic syndrome and type 2 diabetes in several clinical and observational studies (Popkin et al. 2006; Nettleton et al. 2009; Fagherazzi et al. 2013). However, to date, evidence supporting a cause-to-effect relationship is limited: no clear mechanism has been demonstrated and conclusions from epidemiological studies are inconsistent (de Koning et al. 2011; Greenwood et al. 2014; Imamura et al. 2015). According to Pereira's reviews (2013, 2014), the more rigorous prospective studies controlling for many potential confounders are not conclusive or suggest a reverse causality bias (i.e. higher NCSB consumption in overweight persons attempting to lose weight or to prevent further weight gain by substituting SSBs for non-caloric beverages). Indeed, consistent evidence from experimental and cohort studies suggest that substituting NCSBs for SSBs can help control total energy intake and prevent weight gain in the long term (de Ruyter et al. 2012; Zheng et al. 2015). Consequently, to date, NCSBs may remain an acceptable alternative to SSBs, although water should be explicitly promoted as the healthiest option (Popkin et al. 2006; Pereira 2013; de Ruyter et al. 2014). Excluding diet drinks from the scope of soda taxes has been recommended in the literature (Brownell et al. 2009; Mytton 2015). Generally, with some exceptions (e.g. the

---

[39]In the CCHS-Nutrition survey, the fruit juice category corresponds to 100 % pure juice. It includes the "juice portion of alcoholic beverages and juice recipes (concentrate and water)". Fruit drinks correspond to "beverages that contain less than 100 % fruit juice" (Garriguet 2008b, p. 27).

[40]Difference statistically significant ($p \leq 0.05$).

French and Belgian soda taxes),[41] "diet drinks" are exempt from soda taxes currently implemented across the world (see Table 1.1).

These considerations will have important implications when we analyse if exempting beverages deemed healthier than SSBs from the tax may encourage higher consumption of the former and lower consumption of the latter (see Sect. 7.3). Altogether, this will also have important implications when considering which beverages should be included (or not) in the scope of a soda tax (see Sect. 12.4.1).

Key Messages

- Most recent systematic reviews, meta-analyses and experimental studies confirm that SSB overconsumption increases the risk of weight gain, type 2 diabetes and dental caries.
- SSB consumption is stimulated by an attractive and diversified offer, a widespread distribution, low prices and a strong promotion (especially towards youth).
- SSB consumption represents a significant proportion of sugar intake and daily energy intake in some groups of the Canadian population, in particular youth.
- From a public health perspective, preventing SSB overconsumption in Canada requires specific efforts at a large scale.

References

Advertising Standards Canada. (2007). Children's food and beverage advertising initiative. http://www.adstandards.com/en/childrensinitiative/default.htm. Accessed December 15, 2015

Agriculture and Agri-Food Canada. (2010). The Canadian coffee industry. Sub-Sector Profile. October 2010. http://www.agr.gc.ca/eng/industry-markets-and-trade/statistics-and-market-information/by-product-sector/processed-food-and-beverages/the-canadian-coffee-industry/?id=1172237152079. Accessed March 1, 2015

Agriculture and Agri-Food Canada. (2015). The Canadian soft drink industry. http://www.agr.gc.ca/eng/industry-markets-and-trade/statistics-and-market-information/by-product-sector/processed-food-and-beverages/the-canadian-soft-drink-industry/?id=1172167862291. Accessed March 1, 2015

[41] Actually, the inclusion of NCSBs in the French soda tax was not part of the initial bill. It was even strongly fought by the government during parliamentary sessions, who considered that it was jeopardizing the public health credibility of the tax. The extension of the SSB tax scope to NCSBs rather holds to budgetary considerations defended by members of the parliament whose intent was to raise greater fiscal revenues in order to fund a reduction of wage costs in the farming sector. However, once the extension of the tax scope was voted at the parliament, the French government also invoked NCSBs' potential harmful nutritional effects to justify this political move (French Constitutional Council 2011) (see also Sect. 13.6).

Allison, D. B. (2014). Liquid calories, energy compensation and weight: What we know and what we still need to learn. *The British Journal of Nutrition, 111*(3), 384–386. doi:10.1017/S0007114513003309.

Ambrosini, G. L., Oddy, W. H., Huang, R. C., Mori, T. A., Beilin, L. J., & Jebb, S. A. (2013). Prospective associations between sugar-sweetened beverage intakes and cardiometabolic risk factors in adolescents. *The American Journal of Clinical Nutrition, 98*(2), 327–334. doi:10.3945/ajcn.112.051383.

American Institute for Cancer Research (AICR), & World Cancer Research Fund (WCRF) (Eds.). (2007). *Food, nutrition, physical activity and the prevention of cancer: A global perspective: A project of World Cancer Research Fund International*. Washington, D.C: American Institute for Cancer Research.

Bachman, C. M., Baranowski, T., & Nicklas, T. A. (2006). Is there an association between sweetened beverages and adiposity? *Nutrition Reviews, 64*(4), 153–174.

Basu, J. (2015, March 5). Academics and industry clash over WHO sugar advice. FOODNavigator.com. http://www.foodnavigator.com/Policy/WHO-releases-final-sugar-advice-for-children-and-adults. Accessed March 5, 2015

Bes-Rastrollo, M., Schulze, M. B., Ruiz-Canela, M., & Martinez-Gonzalez, M. A. (2013). Financial conflicts of interest and reporting bias regarding the association between sugar-sweetened beverages and weight gain: A systematic review of systematic reviews. *PLoS Medicine, 10*(12), e1001578; dicsussion e1001578. doi:10.1371/journal.pmed.1001578

Billes, S. K. (2012, September 21). Obesity 2012 opens with keynote debate. In *TOS Times, OBESITY 2012 30th Annual scientific meeting*, San Antonio, Texas, September 20–24.

Blanchet, C., Lamontagne, P., Rochette, L., Plante, C., Institut national de santé publique du Québec, & Direction recherche, formation et développement. (2009). *La consommation alimentaire et les apports nutritionnels des adultes québécois*. Montréal, Québec: Institut national de santé publique du Québec. http://site.ebrary.com/id/10350771. Accessed February 25, 2016

Brisbois, T., Marsden, S., Anderson, G., & Sievenpiper, J. (2014). Estimated intakes and sources of total and added sugars in the Canadian diet. *Nutrients, 6*(5), 1899–1912. doi:10.3390/nu6051899.

Brownell, K. D., Farley, T., Willett, W. C., Popkin, B. M., Chaloupka, F. J., Thompson, J. W., & Ludwig, D. S. (2009). The public health and economic benefits of taxing sugar-sweetened beverages. *New England Journal of Medicine, 361*(16), 1599–1605. doi:10.1056/NEJMhpr0905723.

Camirand, H., Blanchet, C., & Pica, L. A. (2012). Habitudes alimentaires. In *L'enquête québécoise sur la santé des jeunes du secondaire, 2010–2011. Le visage des jeunes d'aujourd'hui : leur santé physique et leurs habitudes de vie.* (Vol. Tome 1, pp. 71–96). Québec: Institut de la statistique du Québec.

Canadian Beverage Association (CAB). (2012, September). Industry Guidelines for sale of beverages in schools. http://www.canadianbeverage.ca/wp-content/uploads/2013/09/School-guidelines-CBA-2012-FINAL-ENG-as-ratified-by-Bd-20120918.pdf. Accessed October 1, 2015

Canadian Beverage Association (CAB). (2015). Guidelines. http://www.canadianbeverage.ca/nutrition/guidelines/. Accessed March 1, 2015

Cardwell, M. (2015, March 9). Coffee consumption habits continue to shift in Canada. Grocers stand to profit from the popularity of in-home, single-serve coffee machines. *MarketingMag.ca*. http://www.marketingmag.ca/consumer/coffee-consumption-habits-continue-to-shift-in-canada-139728. Accessed December 14, 2015

Cash, S., & Lacanilao, R. (2007). Taxing food to improve health: Economic evidence and arguments. *Agricultural and Resource Economics Review, 36*(2)

Coffee Association of Canada. (2011). Canadian coffee drinking study—2011 highights. http://www.coffeeassoc.com/coffee-in-canada/canadian-coffee-drinking-study-2011-highights/. Accessed December 14, 2015

Council of Better Business Bureaus. (2006). Children's food and beverage advertising initiative. https://www.bbb.org/council/the-national-partner-program/national-advertising-review-services/childrens-food-and-beverage-advertising-initiative/. Accessed December 15, 2015

Danyliw, A. D., Vatanparast, H., Nikpartow, N., & Whiting, S. J. (2012). Beverage patterns among Canadian children and relationship to overweight and obesity. *Applied Physiology, Nutrition and Metabolism, 37*(5), 900–906. doi:10.1139/h2012-074.

de Koning, L., Malik, V. S., Rimm, E. B., Willett, W. C., & Hu, F. B. (2011). Sugar-sweetened and artificially sweetened beverage consumption and risk of type 2 diabetes in men. *American Journal of Clinical Nutrition, 93*(6), 1321–1327. doi:10.3945/ajcn.110.007922.

de Ruyter, J. C., Katan, M. B., Kas, R., & Olthof, M. R. (2014). Can children discriminate sugar-sweetened from non-nutritively sweetened beverages and how do they like them? *PLoS One, 9*(12), e115113. doi:10.1371/journal.pone.0115113.

de Ruyter, J. C., Olthof, M. R., Seidell, J. C., & Katan, M. B. (2012). A trial of sugar-free or sugar-sweetened beverages and body weight in children. *New England Journal of Medicine, 367*(15), 1397–1406. doi:10.1056/NEJMoa1203034.

Dubois, L., Farmer, A., Girard, M., & Peterson, K. (2007). Regular sugar-sweetened beverage consumption between meals increases risk of overweight among preschool-aged children. *Journal of the American Dietetic Association, 107*(6), 924–934; discussion 934–935. doi:10.1016/j.jada.2007.03.004

Ebbeling, C. B. (2014). Sugar-sweetened beverages and body weight. *Current Opinion in Lipidology, 25*(1), 1–7. doi:10.1097/MOL.0000000000000035.

Ebbeling, C. B., Feldman, H. A., Chomitz, V. R., Antonelli, T. A., Gortmaker, S. L., Osganian, S. K., & Ludwig, D. S. (2012). A randomized trial of sugar-sweetened beverages and adolescent body Weight. *New England Journal of Medicine, 367*(15), 1407–1416. doi:10.1056/NEJMoa1203388.

Fagherazzi, G., Vilier, A., Saes Sartorelli, D., Lajous, M., Balkau, B., & Clavel-Chapelon, F. (2013). Consumption of artificially and sugar-sweetened beverages and incident type 2 diabetes in the Etude Epidemiologique aupres des femmes de la Mutuelle Generale de l'Education Nationale-European Prospective Investigation into Cancer and Nutrition cohort. *American Journal of Clinical Nutrition, 97*(3), 517–523. doi:10.3945/ajcn.112.050997.

Federal Trade Commission (FTC). (2012). *A review of food marketing to children and adolescents: Follow-up report.* https://www.ftc.gov/sites/default/files/documents/reports/review-food-marketing-children-and-adolescents-follow-report/121221foodmarketingreport.pdf. Accessed February 25, 2016

Ferraro, P. M., Taylor, E. N., Gambaro, G., & Curhan, G. C. (2013). Soda and other beverages and the risk of kidney stones. *Clinical Journal of the American Society of Nephrology, 8*(8), 1389–1395. doi:10.2215/CJN.11661112.

Forshee, R. A., Anderson, P. A., & Storey, M. L. (2008). Sugar-sweetened beverages and body mass index in children and adolescents: A meta-analysis. *The American Journal of Clinical Nutrition, 87*(6), 1662–1671.

French Constitutional Council. (2011). *Décision No2011-644DC. Article, consolidation, travaux parlementaires.* http://www.conseil-constitutionnel.fr/conseil-constitutionnel/francais/les-decisions/acces-par-date/decisions-depuis-1959/2011/2011-644-dc/decision-n-2011-644-dc-du-28-decembre-2011.104235.html. Accessed March 1, 2012

Garriguet, D. (2008a). Beverage consumption of Canadian adults. Health matters. *Health Reports, 19*(4), 23.

Garriguet, D. (2008b). Beverage consumption of children and teens. Health matters. *Health Reports, 19*(4), 17.

Gillis, L. J., & Bar-Or, O. (2003). Food away from home, sugar-sweetened drink consumption and juvenile obesity. *Journal of the American College of Nutrition, 22*(6), 539–545.

Government of Canada. (2015). Food and drug regulations. C.R.C., c. 870. FOOD AND DRUGS ACT. http://laws-lois.justice.gc.ca/eng/regulations/C.R.C.%2C_c._870/FullText.html. Accessed September 22, 2015

Greenwood, D. C., Threapleton, D. E., Evans, C. E. L., Cleghorn, C. L., Nykjaer, C., Woodhead, C., & Burley, V. J. (2014). Association between sugar-sweetened and artificially sweetened soft drinks and type 2 diabetes: Systematic review and dose-response meta-analysis of prospective studies. *The British Journal of Nutrition, 112*(5), 725–734. doi:10.1017/S0007114514001329.

Hastings, G., McDermott, L., Angus, K., Stead, M., & Thomson, S. (2007). *The extend nature and effect of food promotion to children: a review of the evidence* (No. Technical paper prepared for the World Health Organization). Geneva, Switzerland: Institute for Social Marketing, University of Stirling & The Open University, United Kingdom. http://www.who.int/dietphysicalactivity/publications/Hastings_paper_marketing.pdf. Accessed February 25, 2016

Hawkes, C. (2010). The worldwide battle against soft drinks in schools. *American Journal of Preventive Medicine, 38*(4), 457–461. doi:10.1016/j.amepre.2010.01.011.

Hawkes, C. (2013). *Promoting healthy diets through nutrition education and changes in the food environment: an international review of actions and their effectiveness*. Rome (Italy): Nutrition Education and Consumer Awareness Group, Food and Agriculture Organization of the United Nations. http://www.fao.org/docrep/017/i3235e/i3235e.pdf. Accessed February 25, 2016

Health Canada. (2011a). Caffeinated energy drinks. http://www.hc-sc.gc.ca/fn-an/prodnatur/caf-drink-boissons-eng.php. Accessed January 14, 2016

Health Canada. (2011b). Canada's food guide. http://www.hc-sc.gc.ca/fn-an/food-guide-aliment/index-eng.php. Accessed September 23, 2015

Huang, C., Huang, J., Tian, Y., Yang, X., & Gu, D. (2014). Sugar sweetened beverages consumption and risk of coronary heart disease: A meta-analysis of prospective studies. *Atherosclerosis, 234*(1), 11–16. doi:10.1016/j.atherosclerosis.2014.01.037.

Hu, F. B. (2013). Resolved: There is sufficient scientific evidence that decreasing sugar-sweetened beverage consumption will reduce the prevalence of obesity and obesity-related diseases: Sugar-sweetened beverages and risk of obesity. *Obesity Reviews, 14*(8), 606–619. doi:10.1111/obr.12040.

Imamura, F., O'Connor, L., Ye, Z., Mursu, J., Hayashino, Y., Bhupathiraju, S. N., & Forouhi, N. G. (2015). Consumption of sugar sweetened beverages, artificially sweetened beverages, and fruit juice and incidence of type 2 diabetes: Systematic review, meta-analysis, and estimation of population attributable fraction. *BMJ (Clinical Research Ed.), 351*, h3576.

Janssen, I., Katzmarzyk, P. T., Boyce, W. F., Vereecken, C., Mulvihill, C., Roberts, C., et al. (2005). Comparison of overweight and obesity prevalence in school-aged youth from 34 countries and their relationships with physical activity and dietary patterns. *Obesity Reviews: An Official Journal of the International Association for the Study of Obesity, 6*(2), 123–132. doi:10.1111/j.1467-789X.2005.00176.x.

Kaiser, K. A., Shikany, J. M., Keating, K. D., & Allison, D. B. (2013). Will reducing sugar-sweetened beverage consumption reduce obesity? Evidence supporting conjecture is strong, but evidence when testing effect is weak. *Obesity Reviews: An Official Journal of the International Association for the Study of Obesity, 14*(8), 620–633. doi:10.1111/obr.12048.

Keller, A., Heitmann, B. L., & Olsen, N. (2015). Sugar-sweetened beverages, vascular risk factors and events: A systematic literature review. *Public Health Nutrition, 18*(7), 1145–1154. doi:10.1017/S1368980014002122.

Langlois, K., & Garriguet, D. (2011). Sugar consumption among Canadians of all ages. *Health Reports, 22*(3). 23. http://www.statcan.gc.ca/pub/82-003-x/2011003/article/11540-eng.pdf. Accessed February 28, 2016

Le Bodo, Y., Paquette, M.-C., Vallières, M., & Alméras, N. (2015). Is sugar the new tobacco? Insights from laboratory studies, consumer surveys and public health. *Current Obesity Reports, 4*(1), 111–121. doi:10.1007/s13679-015-0141-3.

Le Monde. (2014, September 19). Le Conseil constitutionnel censure de nouveau la taxe sur les boissons énergisantes. *LeMonde.fr*. http://www.lemonde.fr/sante/article/2014/09/19/le-conseil-constitutionnel-censure-de-nouveau-la-taxe-sur-les-boissons-energisantes_4490601_1651302.html. Accessed March 1, 2015

Malik, V. S., Pan, A., Willett, W. C., & Hu, F. B. (2013). Sugar-sweetened beverages and weight gain in children and adults: A systematic review and meta-analysis. *The American Journal of Clinical Nutrition, 98*(4), 1084–1102. doi:10.3945/ajcn.113.058362.

Malik, V. S., Popkin, B. M., Bray, G. A., Després, J.-P., Willett, W. C., & Hu, F. B. (2010). Sugar-sweetened beverages and risk of metabolic syndrome and type 2 diabetes: A meta-analysis. *Diabetes Care, 33*(11), 2477–2483. doi:10.2337/dc10-1079.

Malik, V. S., Schulze, M. B., & Hu, F. B. (2006). Intake of sugar-sweetened beverages and weight gain: A systematic review. *The American Journal of Clinical Nutrition, 84*(2), 274–288.

Martin, V., Parent, M.-P., Blouin, M., & Durand, C. (2014). *Gazelle et Potiron: cadre de référence pour créer des environnements favorables à la saine alimentation, au jeu actif et au développement moteur en services de garde éducatifs à l'enfance*. Québec, Canada: Ministère de la Famille du Québec. http://www.uqtr.ca/biblio/notice/document/30823897D.pdf. Accessed February 28, 2016

Mâsse, L. C., de Niet-Fitzgerald, J. E., Watts, A. W., Naylor, P.-J., & Saewyc, E. M. (2014). Associations between the school food environment, student consumption and body mass index of Canadian adolescents. *The International Journal of Behavioral Nutrition and Physical Activity, 11*(1), 29. doi:10.1186/1479-5868-11-29.

Massougbodji, J., Le Bodo, Y., Fratu, R., & De Wals, P. (2014). Reviews examining sugar-sweetened beverages and body weight: Correlates of their quality and conclusions. *The American Journal of Clinical Nutrition, 99*(5), 1096–1104. doi:10.3945/ajcn.113.063776.

Mathias, K. C., Slining, M. M., & Popkin, B. M. (2013). Foods and beverages associated with higher intake of sugar-sweetened beverages. *American Journal of Preventive Medicine, 44*(4), 351–357. doi:10.1016/j.amepre.2012.11.036.

Mattes, R. D., Shikany, J. M., Kaiser, K. A., & Allison, D. B. (2011). Nutritively sweetened beverage consumption and body weight: A systematic review and meta-analysis of randomized experiments. *Obesity Reviews: An Official Journal of the International Association for the Study of Obesity, 12*(5), 346–365. doi:10.1111/j.1467-789X.2010.00755.x.

Merchant, A. T., Tripathi, A., & Pervaiz, F. (2010). Available energy from soft drinks: More than the sum of its parts. *Public Health Nutrition, 13*(12), 1997–1999. doi:10.1017/S136898001000128X.

Ministère de l'Éducation, du Loisir et du Sport. (2009). *Framework policy on healthy eating and active living—Going the healthy route at school. Foods and beverages containing sweeteners.* Québec, Canada. http://www.education.gouv.qc.ca/fileadmin/site_web/documents/dpse/adaptation_serv_compl/VirageSanteEcole_FicheInfo_Edulcorant_a.pdf. Accessed February 28, 2016

Ministère de l'Éducation, du Loisir et du Sport. (2012). *Bilan de la mise en œuvre de la politique-cadre pour une saine alimentation et un mode de vie physiquement actif - Pour un virage santé à l'école, Thème 1 : Environnement scolaire Volet saine alimentation.* Québec, Canada

Morin, P. (Ed.). (2010). *Portrait des jeunes Sherbrookois de 4 à 17 ans en matière d'alimentation et d'activité physique et sportive rapport final.* Québec: Québec en forme. http://public.eblib.com/choice/publicfullrecord.aspx?p=3279285. Accessed February 28, 2016

Morin, P., Turcotte, S., & Perreault, G. (2013). Relationship between eating behaviors and physical activity among primary and secondary school students: Results of a cross-sectional study. *The Journal of School Health, 83*(9), 597–604. doi:10.1111/josh.12071.

Moynihan, P. J., & Kelly, S. A. M. (2014). Effect on caries of restricting sugars intake: Systematic review to inform WHO Guidelines. *Journal of Dental Research, 93*(1), 8–18. doi:10.1177/0022034513508954.

Mundt, C. A., Baxter-Jones, A. D. G., Whiting, S. J., Bailey, D. A., Faulkner, R. A., & Mirwald, R. L. (2006). Relationships of activity and sugar drink intake on fat mass development in youths. *Medicine and Science in Sports and Exercise, 38*(7), 1245–1254. doi:10.1249/01.mss.0000227309.18902.fe.

Mytton, O. (2015). Time for a sugary drinks tax in the UK? *Journal of Public Health, 37*(1), 24–25. doi:10.1093/pubmed/fdu033

Nettleton, J. A., Lutsey, P. L., Wang, Y., Lima, J. A., Michos, E. D., & Jacobs, D. R. (2009). Diet soda intake and risk of incident metabolic syndrome and type 2 diabetes in the multi-ethnic study of atherosclerosis (MESA). *Diabetes Care, 32*(4), 688–694. doi:10.2337/dc08-1799.

Nielsen, S. J., & Popkin, B. M. (2004). Changes in beverage intake between 1977 and 2001. *American Journal of Preventive Medicine, 27*(3), 205–210. doi:10.1016/j.amepre.2004.05.005.

Pereira, M. A. (2013). Diet beverages and the risk of obesity, diabetes, and cardiovascular disease: A review of the evidence. *Nutrition Reviews, 71*(7), 433–440. doi:10.1111/nure.12038.

Pereira, M. A. (2014). Sugar-sweetened and artificially-sweetened beverages in relation to obesity risk. *Advances in Nutrition: An International Review Journal, 5*(6), 797–808. doi:10.3945/an.114.007062.

Popkin, B. M., Armstrong, L. E., Bray, G. M., Caballero, B., Frei, B., & Willett, W. C. (2006). A new proposed guidance system for beverage consumption in the United States. *The American Journal of Clinical Nutrition, 83*(3), 529–542.

Popkin, B. M., & Hawkes, C. (2016). Sweetening of the global diet, particularly beverages: Patterns, trends, and policy responses. *The Lancet Diabetes & Endocrinology, 4*(2), 174–186. doi:10.1016/S2213-8587(15)00419-2.

Potvin Kent, M., Dubois, L., Kent, E. A., & Wanless, A. J. (2013). Internet marketing directed at children on food and restaurant websites in two policy environments. *Obesity (Silver Spring, Md.), 21*(4), 800–807. doi:10.1002/oby.20124

Potvin Kent, M., Martin, C. L., & Kent, E. A. (2014). Changes in the volume, power and nutritional quality of foods marketed to children on television in Canada. *Obesity (Silver Spring, Md.), 22*(9), 2053–2060. doi:10.1002/oby.20826

Potvin Kent, M., & Wanless, A. (2014). The influence of the children's food and beverage advertising initiative: Change in children's exposure to food advertising on television in Canada between 2006–2009. *International Journal of Obesity, 38*(4), 558–562. doi:10.1038/ijo.2014.4.

Public Health England (PHE). (2015). *Sugar reduction: The evidence for action. Annexe 4: An analysis of the role of price promotions on the household purchases of food and drinks high in sugar. A research project for Public Health England conducted by Kantar Worldpanel UK.* London. https://www.gov.uk/government/uploads/system/uploads/attachment_data/file/470175/Annexe_4._Analysis_of_price_promotions.pdf. Accessed January 22, 2016

Quebec Weight Coalition (CQPP). (2012). *Sugar-sweetened beverage marketing unveiled. Volume 2: Price, an argument that pays off.* http://www.cqpp.qc.ca/en/sugar-sweetened-beverages/marketing. Accessed March 1, 2013

Rabin, R. C. (2012, September 21). Avoiding sugared drinks limits weight gain in two studies. *The New York Times.* New York. http://www.nytimes.com/2012/09/22/health/avoiding-sugary-drinks-improves-childrens-weight-in-2-studies.html?_r=1. Accessed October 1, 2015

Reid, J. L., Hammond, D., McCrory, C., Dubin, J. A., & Leatherdale, S. T. (2015). Use of caffeinated energy drinks among secondary school students in Ontario: Prevalence and correlates of using energy drinks and mixing with alcohol. *Canadian Journal of Public Health=Revue Canadienne De Santé Publique, 106*(3), e101–e108.

Saunders, T. (2011). Potential contributors to the Canadian pediatric obesity epidemic. *ISRN Pediatrics, 2011*, 1–10. doi:10.5402/2011/917684.

Sievenpiper, J. L., & de Souza, R. J. (2013). Are sugar-sweetened beverages the whole story? *The American Journal of Clinical Nutrition, 98*(2), 261–263. doi:10.3945/ajcn.113.067215.

Statistics Canada. (2010). *Food statistics 2009* (No. Catalogue no. 21-020-X). http://www.statcan.gc.ca/pub/21-020-x/21-020-x2009001-eng.pdf. Accessed March 1, 2012

Statistics Canada. (2014). Canadian community health survey—nutrition (CCHS). http://www23.statcan.gc.ca/imdb/p2SV.pl?Function=getSurvey&SDDS=5049. Accessed October 29, 2015

Te Morenga, L. A., Howatson, A. J., Jones, R. M., & Mann, J. (2014). Dietary sugars and cardiometabolic risk: Systematic review and meta-analyses of randomized controlled trials of the effects on blood pressure and lipids. *The American Journal of Clinical Nutrition, 100*(1), 65–79. doi:10.3945/ajcn.113.081521.

Te Morenga, L., Mallard, S., & Mann, J. (2012). Dietary sugars and body weight: systematic review and meta-analyses of randomised controlled trials and cohort studies. *BMJ, 346*(3), e7492–e7492. doi:10.1136/bmj.e7492

Trumbo, P. R., & Rivers, C. R. (2014). Systematic review of the evidence for an association between sugar-sweetened beverage consumption and risk of obesity. *Nutrition Reviews, 72*(9), 566–574. doi:10.1111/nure.12128.

Vanderlee, L., Manske, S., Murnaghan, D., Hanning, R., & Hammond, D. (2014). Sugar-sweetened beverage consumption among a subset of Canadian youth. *The Journal of School Health, 84*(3), 168–176. doi:10.1111/josh.12139.

Wang, J. W., Mark, S., Henderson, M., O'Loughlin, J., Tremblay, A., Wortman, J., et al. (2013). Adiposity and glucose intolerance exacerbate components of metabolic syndrome in children consuming sugar-sweetened beverages: QUALITY cohort study. *Pediatric Obesity, 8*(4), 284–293. doi:10.1111/j.2047-6310.2012.00108.x.

Welsh, J. A., Lundeen, E. A., & Stein, A. D. (2013). The sugar-sweetened beverage wars: Public health and the role of the beverage industry. *Current opinion in Endocrinology, Diabetes, and Obesity, 20*(5), 401–406. doi:10.1097/01.med.0000432610.96107.f5.

Woodward-Lopez, G., Kao, J., & Ritchie, L. (2011). To what extent have sweetened beverages contributed to the obesity epidemic? *Public Health Nutrition, 14*(3), 499–509. doi:10.1017/S1368980010002375.

World Health Organization (WHO). (2010). *Set of recommendations on the marketing of foods and non-alcoholic beverages to children*. Geneva, Switzerland: World Health Organization. http://whqlibdoc.who.int/publications/2010/9789241500210_eng.pdf. Accessed February 29, 2016

World Health Organization (WHO). (2015). *Guideline: Sugars intake for adults and children*. Geneva: Switzerland.

World Health Organization (WHO), & Food and Agriculture Organization (FAO) (Eds.). (2003). *Diet, nutrition, and the prevention of chronic diseases: report of a WHO-FAO Expert Consultation on Diet, Nutrition, and the Prevention of Chronic Diseases*. Geneva: World Health Organization.

Yang, Q., Zhang, Z., Gregg, E. W., Flanders, W. D., Merritt, R., & Hu, F. B. (2014). Added sugar intake and cardiovascular diseases mortality among US adults. *JAMA Internal Medicine, 174*(4), 516–524. doi:10.1001/jamainternmed.2013.13563.

Young, L. R., & Nestle, M. (2003). Expanding portion sizes in the US marketplace: Implications for nutrition counseling. *Journal of the American Dietetic Association, 103*(2), 231–234. doi:10.1053/jada.2003.50027.

Zheng, M., Allman-Farinelli, M., Heitmann, B. L., & Rangan, A. (2015). Substitution of sugar-sweetened beverages with other beverage alternatives: A review of long-term health outcomes. *Journal of the Academy of Nutrition and Dietetics, 115*(5), 767–779. doi:10.1016/j.jand.2015.01.006.

Chapter 3
Sugar-Sweetened Beverage Taxation as a Public Health Policy Instrument

Abstract The manufacturing cost of sugar-sweetened beverages (SSBs) is particularly low. It contributes to making these drinks particularly affordable and to encourage consumption. Taxation has been increasingly suggested as a way to increase SSB price and discourage demand, as part of a comprehensive prevention strategy. Lessons from tobacco control confirm that taxation can be a valuable instrument, although the comparison with SSB should be made cautiously. Economic simulations tend to position SSB taxes among the most cost-effective interventions because of their low cost of implementation, the critical mass of the population they may reach, the revenues they generate and their sustainability. Evidence from soda taxes implemented in the "real world" has to be considered to confirm these predictions.

3.1 Taxing Sugar-Sweetened Beverages (SSBs) as Part of a Comprehensive Strategy

Non-communicable diseases (NCDs) are determined by a multiplicity of food-related factors whose specific influences are not easy to disentangle. Therefore, it is widely documented that a combination of incremental interventions and policies affecting many factors at several levels are needed to make a difference (Gortmaker et al. 2011). Such systems thinking perspective may require allowing a certain dose of "uncertainty" and "unpredictability" when considering the potential outcomes of single interventions (Rutter 2011, 2012, p. 657). Indeed, the specific impacts on diet and health of a particular policy, such as a soda tax, are expected to be small in magnitude and difficult to detect (see Chap. 14). As a result, one should not expect specific actions to be magic bullets: taxation, as many other policy options, must be considered as part of a larger societal action.

Y. Le Bodo et al., *Taxing Soda for Public Health*,
DOI 10.1007/978-3-319-33648-0_3

3.1.1 International Evidence

The literature suggests that various interventions at individual and environmental levels should be simultaneously pursued to reduce SSB consumption in the population (IOM 2012; Hsiao and Wang 2013; Yon and Johnson 2014). School-based policies limiting the availability of SSBs in cafeterias and vending machines have shown effectiveness in reducing SSB consumption at school, although it may not necessarily translate into lower SSB intake overall. Compensatory behaviours at home may contribute to explaining this result (Levy et al. 2011; Taber et al. 2012; Avery et al. 2015; WCRF 2015). Some studies have also demonstrated that a higher availability of safe drinking water at school combined with educational activities promoting healthy beverages could contribute to increasing daily water consumption and decrease SSB consumption at school (Centers for Disease Control and Prevention [CDC] 2010; Armstrong et al. 2012). Mass communication campaigns using a diversity of channels (TV, radio, the Internet and information posting in multiple places) also have the potential to contribute influencing SSB-related knowledge, attitudes, intentions and eventually behaviours in the population (Jordan et al. 2012; Boles et al. 2014; WCRF 2015).

On the market side, examples of collaborating initiatives with retailers have demonstrated effectiveness to offer and promote healthier beverages and influence consumer behaviours (WCRF 2015). Commitments made by manufacturers to reduce the sugar content of their beverages also deserve attention (Liu et al. 2014; British Broadcasting Corporation 2015). From a more coercive standpoint, a controlled experiment in the corner store of a large US hospital led to encouraging results as to the potential effects of warning labels on "less healthy" beverages (Elbel et al. 2013). In this regard, lessons from the tobacco case brings additional arguments (Capewell 2014). Furthermore, whereas evidence on the behavioural effects of policies aimed at regulating SSB marketing to children is still scarce, simulations and recommendations abound in their favour (WHO 2010; WCRF 2015; Liu et al. 2014). Simulations on SSB portion-size caps in various points of sale also predict encouraging results, but raise feasibility and acceptability concerns (Wang and Vine 2013; Roberto and Pomeranz 2015). Banning free or unlimited SSB refills in restaurants and cafeterias may be an alternative. For example, this measure has been officially adopted in France and will become effective in January 2017 (JORF 2016). Last but not least, taxation likely is the SSB policy option for which the scientific literature is the more abundant but also one of the more controversial (Hsiao and Wang 2013; Le Bodo et al. 2015). All these options have been documented: encouraging evidence supports several of them, while others remain debated. On-going efforts should be evaluated as a whole to better demonstrate what works (or not), for whom, how, why and in combination with what (Studdert et al. 2015).

New York City's fight against SSB overconsumption is an interesting illustration of a multipronged strategy based on several of the aforementioned interventions (Farley et al. 2012; Farley and Merrill 2013; Farley and Dowell 2014). Considering that educational strategies should be complemented by interventions in the environment, between 2003 and 2011, New York's Department of Health eliminated sugary drinks from beverage vending machines and made incremental improvements in the nutritive value of school meals served in public schools. In 2008, the energy content and portion sizes of beverages available and the quality of meals and snacks served by city agencies (2008) have also been regulated. These efforts were complemented by the "Pouring on the Pounds" mass communication campaign launched in 2009. Furthermore, in 2010–2011, the content of beverages sold in elementary and middle schools has been capped to fewer than 10 calories per 8 ounces. The Department of Health also supported a potential penny-per-ounce excise tax on sugary drinks, beverage policy changes in the Supplemental Nutrition Assistance Program, and a cap on the portion size of sugary drinks served at restaurants and other places with food to go. Although these proposals did not succeed, it generated media attention and raised awareness in the population. Even if causal inferences cannot be proved, surveillance data are encouraging. From 2007 to 2013, the number of New Yorkers reporting drinking one or more sugary drink(s)[1] per day decreased from 36 to 23 % in adults and from 57 to 42 % in public high school students (Kansagra et al. 2015).

In sum, as any of these interventions, SSB taxation should not be considered as a "holy grail" or a "panacea" but as a contributing instrument within a comprehensive intersectoral strategy aimed to decrease SSB overconsumption and, ultimately, to contribute preventing obesity and NCDs (Block and Willett 2013, p. 183; Briggs et al. 2013, p. 7; Mytton 2015; WHO 2015).

3.1.2 *The Canadian Situation*

Such a comprehensive approach is already in progress across Canada. Many interventions involving a diversity of stakeholders have been implemented at all levels to promote healthy eating and drinking behaviours over the last years. Overarching policies and frameworks provide a backbone for such developments. At federal level, it includes the Pan-Canadian Declaration on Prevention and Promotion (PHAC 2010), the Curbing Childhood Obesity Framework (PHAC 2011), the Centre for Chronic Disease Prevention's Strategic Plan 2016–2019 (PHAC 2015), Nutrition North Canada (Government of Canada 2014) and the Canada's Food Guide (Health Canada 2011b). Provincial governments, whose mandate for health promotion predominates (see Sect. 12.2), have also provided comprehensive public health nutrition strategies and action plans supporting the implementation of educational initiatives and environmental changes aimed to

[1]Including carbonated beverages, fruit drinks, sports and energy drinks, and other drinks with added sugars.

increase the consumption of healthy beverages instead of sugary drinks.[2] It includes school nutrition policies and guidelines—sometimes mandatory, often voluntary—that have been developed across all Canadian provinces and territories.[3] Similar initiatives also exist in governmental and healthcare facilities as well as in municipal buildings, day care centres, parks, recreational facilities, community centres and other settings (CQPP 2012b, pp. 37–9). The implementation of these frameworks, policies and guidelines involves a broad diversity of public, not-for-profit and private stakeholders at federal, provincial/territorial and local levels. Large communication and educational campaigns have also been developed specifically to promote healthy drinking behaviours in the population such as "Sip Smart! BC" in British Columbia (Healthy schools BC 2012), "New Drive" and "Liquid sugars—NOT every day!" in Quebec (Réseau du sport étudiant du Québec [RESQ] 2011; CQPP 2015b). Supporting and evaluating all these efforts is critical. In Nova Scotia and Alberta, comprehensive school-based policies and interventions promoting healthy eating proved to be very promising (including reducing SSB consumption) (Veugelers and Fitzgerald 2005; Fung et al. 2012, 2013). Larger community-based interventions may require intense efforts sustained over a longer term to be successful (Raine et al. 2013b).

---

[2]Including Ontario's *Healthy kids strategy* (Ontario Ministry of health and long term care 2014), Quebec's *Plan d'action gouvernemental de promotion des saines habitudes de vie et de prévention des problèmes reliés au poids* 2006–2012 (Lachance et al. 2006) and *Québec en forme* movement (2016), *Healthy eating Nova Scotia strategy* (Government of Nova Scotia 2011), New Brunswick's *Wellness strategy 2014–2021* (Government of New Brunswick 2014), Manitoba's *Healthy eating guidelines, services and programmes* (Government of Manitoba 2016), *Healthy families BC* (Government of British Columbia 2012), Prince Edward Island's *Healthy Eating Strategy 2011–2015* (Healthy eating alliance 2011b), Saskatchewan's *Nourishing minds* (Saskatchewan Ministry of education 2012) and *Northern Healthy Communities Partnership* (NHCP 2013), Alberta's *Healthy eating starts here* (Alberta health services 2016), Newfoundland and Labrador's *Healthy People, Healthy Families, Healthy Communities* (Newfoundland and Labrador Department of health and community services 2015), Northwest territories' *Community wellness Initiatives* (Northwest territories health and social services 2015), Yukon's *Health investment fund* (Yukon Health and social services 2015) and Nunavut's *Nutrition framework for action* (Nunavut Department of health and social services 2007).

[3]Including Ontario's *School food and beverage policy* (Ontario Ministry of education 2010), Quebec's *Framework Policy on Healthy Eating and Active Living—Going the Healthy Route at School* (Ministère de l'Éducation, du Loisir et du Sport 2009), *Food and Nutrition Policy for Nova Scotia Public Schools* (Government of Nova Scotia 2006), New Brunswick's *Policy 711: Healthier Foods and Nutrition in Public Schools* (New Brunswick Department of education 2008), Manitoba's *Moving Forward with School Nutrition Guidelines* (Government of Manitoba 2014), *Guidelines for food and beverage sales in B.C. schools* (British Columbia's Ministry of health 2014), Prince Edward Island's *School nutrition policies* (Healthy eating alliance 2011a), *Healthy Eating, Nutrition, and Food Safety Guidelines for Saskatchewan Schools* (Saskatchewan Ministry of education 2012), Alberta's *Nutrition Guidelines for Children and Youth* (Alberta government 2012), Newfoundland and Labrador's *School food guidelines* (Government of Newfoundland and Labrador 2006), Northwest territories' *Guidelines for Healthy Foods in Facilities* (Northwest territories health and social services 2006), Yukon's *School nutrition policy* (Yukon Education 2008) and Nunavut's *Food guide educator's handbook* (Nunavut health and social services 2012).

As indicated before, benefits would likely be obtained if the aforementioned initiatives were complemented by policies targeting the market environment. Until now, few mandatory nutrition policies have been implemented in Canada and most apply to various foods and beverages (possibly including SSBs) rather than focusing on particular ingredients (with the exception of *trans* fats or caffeine) (Ries and von Tigerstrom 2011; Health Canada 2011a). For example, a nutrition facts table on food packages is mandatory in Canada, although it does not distinguish naturally occurring and added sugars (Canadian Food Inspection Agency 2014). This situation could change as Health Canada recently proposed to single out added sugars in the nutrition facts table of processed foods (Health Canada 2015). On a similar ground, in 2015, Ontario was the first province to adopt the "Healthy Menu Choices Act" making calorie posting mandatory in chain restaurants with 20 locations or more (Legislative Assembly of Ontario 2015). As regards food promotion, the Quebec Consumer Protection Act (1980) includes pioneering articles (Sect. 248 and 249) prohibiting commercial advertising directed at children aged less than 13 (Office de la protection du consommateur 2012). Some complaints filed by the Quebec Weight Coalition over the last years have led to condemnations by the Quebec consumer protection bureau (CQPP 2015a), including a recent 27 K $CAD fine requested against a major SSB manufacturer for child-targeted advertising in a Montreal amusement park (Blais 2015). Evaluations of this advertising ban conducted in 2009 suggest that Quebec French-speaking children (aged 10–12) were not exposed to less food and beverages TV ads than Ontarian and Quebec English-speaking children. However, the formers were less intensively targeted than the latter, and the nutritional quality of the items advertised was "marginally healthier" in Quebec French-speaking children. Nonetheless, the authors note that the overall nutritional quality of the items marketed was not optimal (Potvin-Kent et al. 2011, 2012). Another study suggests that the Quebec ban does not preclude Internet food and beverage marketing to kids (Potvin Kent et al. 2013). Calls have been made to reinforce this law in Quebec and to enact a similar policy in other Canadian jurisdictions (Ries and von Tigerstrom 2011; CQPP 2012c, pp. 75–9; Raine et al. 2013a). Over the last years, recommendations and bills have proposed further regulations on SSB labelling, reformulation, availability, taxation and promotion (Ries and von Tigertsrom 2011; Faulkner et al. 2011; Buhler et al. 2013; Heart and stroke foundation of Canada [HSF] 2014; CQPP 2016). Most have not been enforced, and among others, soda taxation remains particularly debated across the country (see Chap. 13).

Yet, taxation is a policy deserving particular attention: (1) it aims to tackle a key determinant of food choice: price; (2) it has already proved beneficial as part of strategies applied to other public health domains, in particular for tobacco control; (3) simulations present it as one of the most cost-effective interventions as part of comprehensive obesity and NCD prevention strategies. These aspects will be covered in the next three sections.

3.2 SSB, An Affordable Choice

In high-income countries, food and beverages rich in added sugars, such as SSBs, are increasingly affordable in comparison with other products. An analysis of Canadian consumer price indexes (Statistics Canada) conducted by the Quebec Weight Coalition (CQPP 2012a, pp. 14–15) shows that between 1995 and 2009, price increases[4] in some staple foods (e.g. 98 % for bread, 53 % for butter) have been significantly higher than those in cola-type soft drinks (11 %) and lemon-lime-type soft drinks (6 %). Such evolution has been correlated with increased SSB consumption (Brownell and Frieden 2009; Étilé 2012). This is not surprising, since food price has proved to be among the most significant determinants of food choice (after taste preferences), especially in youth and lower-income population groups (Glanz et al. 1998; IOM 2006). Special offers on SSBs (discount, coupons, bulk offers) are likely to accentuate the phenomenon (CQPP 2012a; Public Health England (PHE) 2015).

So, how did sugary food and drinks become so affordable? This has to be related to advances in agriculture and food technologies as well as food policies having jointly contributed to lowering the production costs of added sweeteners, making these ingredients a convenient and low-cost option for manufacturers (Drewnowski et Darmon 2005). According to Drewnowski (2007), the relative lower cost of these ingredients and others (including refined grains and added fats) has progressively made processed and palatable energy-dense foods very affordable, which has significantly contributed to increasing dietary energy intakes in the population. Drewnowski also postulates that this evolution has been a significant driver of weight gain at the population level and contributes to explaining why lower-income consumers tend to consume those foods in higher proportions.

In such circumstances, taxes aimed at increasing the price of energy-dense food and drinks have been increasingly suggested at international level. Sodas have raised particular attention because of the health risks related to their overconsumption, the high intakes often observed in youth and the aggressive marketing practices of manufacturers (see Chap. 2). In the next section, we will argue that interest in soda taxes also builds on successful tax uses in other public health domains, particularly tobacco control.

3.3 Learning from Tobacco Taxation

Fiscal policies have already proved to be successful in other public health domains, such as tobacco control (Chaloupka et al. 2012). In high-income countries, the International Agency for Research on Cancer (IARC) considers that price elasticities of demand for cigarettes average −0.4, which indicates that for every 10 %

[4]Versus "overall inflation in Canada".

increase in cigarettes price, we tend to observe a 4 % decrease in purchases.[5] In adults, consequently, increases in tobacco prices and taxes have shown an independent effect on smoking prevalence, intensity (number of cigarettes smoked) as well as on quitting interest, quitting attempts and successful cessation: for every 10 % increase in cigarettes price, we tend to observe a 2–6 % decrease in smoking by reducing both smoking prevalence and smoking intensity. Young people tend to be more tobacco price sensitive than adults: in most high-income countries, price elasticities of demand for cigarettes in this group range between −0.5 and −1.2. In this case, longitudinal studies indicate that tobacco taxation positively impacts smoking initiation, intensity and cessation (IARC 2011). According to a recent systematic review on the impact of tobacco control interventions, tobacco taxes show a greater independent effect than any other policies, such as smoking bans or mass media campaigns (Wilson et al. 2012a).

In Canada, tobacco price elasticities have also been estimated with confidence. Using approaches aimed at neutralizing smuggling-induced biases in estimations,[6] Gruber et al. (2003) have measured the price elasticity of tobacco demand in Canada between −0.45 and −0.47, which is similar to the price elasticities reported in international reviews. Additionally, lower-income smokers appeared to be much more sensitive to price hikes, which indicates that tobacco taxes may result in relatively greater health benefits for this population subgroup. As regards teens, Sen and Wirjanto (2010) have considered the exceptional tobacco tax rollbacks implemented in February 1994 in Canada (40–60 % decrease in cigarette excise taxes) as well as data from a diversity of Canadian cross-provincial tobacco consumption surveys to estimate price elasticities of smoking participation, initiation and persistence in youth. They estimate price elasticities of daily and occasional smoking between −0.10 and −0.14. They also used a longitudinal data set from the Waterloo Smoking Prevention Project (WSPP) to estimate the price elasticity of smoking initiation and persistence among teens between −0.2 and −0.5. They consider that these estimates are "considerably below traditional estimates of the efficacy of cigarette taxes, but quite credible given the significant variation in cigarette taxes" observed during this period (Sen and Wirjanto 2010, p. 1266). These results suggest that consumers' sensitivity to price change may not be linear and involve threshold and saturation effects. It also reminds us that taxation should rather be used in combination with other interventions to gain effectiveness (National Cancer Institute 2006).

For instance, in New York City, research indicates that between 2002 and 2003, the 32 % tax-induced cigarettes' price increase appears to have been the largest policy contributor (vs. smoking bans on the workplace and tobacco cessation programs) to the 11 % decrease in smoking prevalence among adults. Nonetheless,

---

[5]For price elasticity terms of reference, see Chap. 6, Box 6.1.

[6]The authors experimented two approaches, based (1) on legal sales data excluding the regions and years where the smuggling problem was the worst and (2) micro-data on consumer cigarette expenditures. Their estimations of price elasticities using these two different methods were "strikingly similar", which increases confidence in their conclusions (Gruber et al. 2003, p. 840).

the analysis of various combinations of policies across time and geographical areas as well as comparison with other studies led the authors to suggest that beyond the effect of each policy intervention per se, synergies among all preventive actions likely provided additional benefits (Frieden et al. 2005).

However, tobacco and SSB taxes must be carefully compared in many respects. In particular, SSB taxes raise much more substitution concerns than tobacco taxes (i.e. compensatory intakes of untaxed calorie-dense food and beverages) (Le Bodo et al. 2015) (see Chap. 7). Notwithstanding, SSB taxation is considered to be potentially among the most cost-effective options to improve eating behaviours and population health.

3.4 Taxation, A Highly Cost-Effective and Potentially Cost-Saving Option

So far, most nutritional policies have focused on better informing consumer choices (e.g. via mass informational campaigns, nutritional labelling), with a modest impact on consumers' diet and health. Beyond the beverage industry's voluntary commitments, scientific evidence suggests that further efforts could be put on more coercive policies aimed at changing the market environment including regulations on the content of some foods, prescriptive labelling, stronger advertising restrictions and additional taxes on unhealthy foods (Cecchini et al. 2010; Ries and von Tigertstrom 2011; Mozzafarian et al. 2012; Réquillart and Soler 2014; Ananthapavan et al. 2014). Such efforts have been urged at international level[7] and are increasingly experimented by some governments. According to economic simulations, taxes could be among the most cost-saving and health-improving obesity interventions because of their low cost of implementation (centrally administered), the critical mass of the population they may reach, the revenues they generate and their sustainability (Cecchini et al. 2010; Lehnert et al. 2012; Ananthapavan et al. 2014).

According to the modelling works by the Organization for Economic Cooperation and Development and the World Health Organization (Sassi 2010, p. 196), fiscal policies could even be the interventions most likely to "pay for

[7]Such recommendations have also been suggested by various Canadian organizations (Ries and von Tigerstrom 2011). At the federal level, this includes the parliamentary standing committee on health (Merrifield 2007), the advisor on healthy children and youth (Health Canada 2007), the Canadian partnership against cancer (CPAC 2010), the Curbing Childhood Obesity Framework (PHAC 2011) or the McMaster Health Forum (Wilson et al. 2012b). In provinces, recent examples of key recommendations include those issued by the Select Standing Committee on Health in British Columbia (Sultan 2006), the Select Standing Committee on Wellness in New Brunswick (Collins et al. 2008), the Quebec Common vision on environments in favour of healthy eating, active lifestyles and the prevention of weight-related issues (Ministère de la Santé et des Services sociaux 2012) and the Healthy kids Panel in Ontario (2013).

[themselves], i.e. the only one[s] which generate larger savings in health expenditure than costs of delivery". Rather than taxing large food groups (e.g. tax on a broad range of non-alcoholic beverages), scientific evidence suggests focusing on intracategory taxation (e.g. a specific tax on sugar-sweetened beverages) since, in this case, consumers' substitution for untaxed healthier alternatives (e.g. diet drinks, milk, 100 % fruit juice) is encouraged, and manufacturers may have more incentives to improve the nutritional quality of their portfolio (Hawkes 2012; Nghiem et al. 2013; Réquillart and Soler 2014; WHO 2015). As a result, since SSB overconsumption has been recognized as a significant public health issue (see Chap. 2), a specific tax on SSBs may be legitimately considered as a good place to start. However, evidence from soda taxes implemented in the "real world" has to be considered to confirm these predictions. This will be a central topic in the second part of this book.

Key Messages

- Rather than questioning whether SSB taxation alone is effective in preventing obesity and related chronic diseases, it is more realistic to question whether it can bring a valuable contribution to the issue, as part of a large and comprehensive prevention strategy.
- Many educational initiatives and environmental changes in public places already contribute to preventing SSB overconsumption across Canada. International evidence and recommendations suggest that policies targeting the market environment could bring additional benefits.
- Considering that sugar-sweetened beverages are particularly affordable, taxation has been increasingly suggested as a way to increase price and discourage demand.
- The success of taxes applied to other public health domains (e.g. tobacco control), as well as their potential cost-effectiveness, make SSB taxation a policy instrument deserving particular attention.

References

Alberta government. (2012). *Alberta nutrition guidelines for children and youth. A childcare, school and recreation/community centre resource manual.* http://www.healthyalberta.com/NutritionGuidelines-Sept2012.pdf. Accessed Feb 16, 2016

Alberta health services. (2016). *Healthy eating starts here. Healthy eating starts where you live, work, learn and play.* http://www.albertahealthservices.ca/nutrition/Page12598.aspx. Accessed Feb 16, 2016

Ananthapavan, J., Sacks, G., Moodie, M., & Carter, R. (2014). Economics of obesity–Learning from the past to contribute to a better future. *International Journal of Environmental Research and Public Health, 11*(4), 4007–4025. doi:10.3390/ijerph110404007.

Armstrong, L. E., Barquera, S., Duhamel, J.-F., Hardinsyah, R., Haslam, D., & Lafontan, M. (2012). Recommendations for healthier hydration: addressing the public health issues of obesity and type 2 diabetes: Recommendations for healthier hydration. *Clinical Obesity, 2*(5–6), 115–124. doi:10.1111/cob.12006.

Avery, A., Bostock, L., & McCullough, F. (2015). A systematic review investigating interventions that can help reduce consumption of sugar-sweetened beverages in children leading to changes in body fatness. *Journal of Human Nutrition and Dietetics: The Official Journal of the British Dietetic Association, 28*(Suppl 1), 52–64. doi:10.1111/jhn.12267.

Blais, A. (2015, October 17). *Coca-Cola: amende pour publicité illégale à La Ronde. LaPresse.Ca.* http://www.lapresse.ca/actualites/montreal/201510/17/01-4911032-coca-cola-amende-pour-publicite-illegale-a-la-ronde.php. Accessed Feb 16, 2016

Block, J. P., & Willett, W. C. (2013). Taxing sugar-sweetened beverages: Not a "Holy Grail" but a cup at least halfcomment on "Food Taxes: A new holy grail?". *International Journal of Health Policy and Management, 1*(2), 183–185. doi:10.15171/ijhpm.2013.33.

Boles, M., Adams, A., Gredler, A., & Manhas, S. (2014). Ability of a mass media campaign to influence knowledge, attitudes, and behaviors about sugary drinks and obesity. *Preventive Medicine, 67*, S40–S45. doi:10.1016/j.ypmed.2014.07.023.

Briggs, A. D. M., Mytton, O. T., Kehlbacher, A., Tiffin, R., Rayner, M., & Scarborough, P. (2013). Overall and income specific effect on prevalence of overweight and obesity of 20 % sugar sweetened drink tax in UK: econometric and comparative risk assessment modelling study. *BMJ, 347*(oct31 4), f6189–f6189. doi:10.1136/bmj.f6189

British Broadcast Corporation (BBC). (2015, May 22). *Tesco to cut soft drinks sugar content. BBC NEWS.* http://www.bbc.com/news/health-32835245. Accessed Jan 22, 2016

British Columbia Ministry of Health, Population and Public Health Division, & British Columbia Ministry of Education. (2014). *Guidelines for food & beverage sales in BC schools.* http://site.ebrary.com/id/10863168. Accessed Feb 16, 2016

Brownell, K. D., & Frieden, T. R. (2009). Ounces of prevention—The public policy case for taxes on sugared beverages. *New England Journal of Medicine, 360*(18), 1805–1808. doi:10.1056/NEJMp0902392.

Buhler, S., Raine, K. D., Arango, M., Pellerin, S., & Neary, N. E. (2013). Building a strategy for obesity prevention one piece at a time: The case of sugar-sweetened beverage taxation. *Canadian Journal of Diabetes, 37*(2), 97–102. doi:10.1016/j.jcjd.2013.03.025.

Canadian Food Inspection Agency. (2014). *Information within the nutrition facts table. Mandatory information and serving size. Government of Canada.* http://www.inspection.gc.ca/food/labelling/food-labelling-for-industry/nutrition-labelling/information-within-the-nutrition-facts-table/eng/1389198568400/1389198597278?chap=1. Accessed Mar 1, 2015

Canadian Partnership Against Cancer. (2010). *Canadian priorities for addressing obesity as a cancer and chronic disease risk factor.*

Capewell, S. (2014). Sugar sweetened drinks should carry obesity warnings. *BMJ, 348*(may27 10), g3428–g3428. doi:10.1136/bmj.g3428

Cecchini, M., Sassi, F., Lauer, J. A., Lee, Y. Y., Guajardo-Barron, V., & Chisholm, D. (2010). Tackling of unhealthy diets, physical inactivity, and obesity: Health effects and cost-effectiveness. *The Lancet, 376*(9754), 1775–1784. doi:10.1016/S0140-6736(10)61514-0.

Centers for Disease Control and Prevention (CDC). (2010). *The CDC guide to STRATEGIES FOR REDUCING THE CONSUMPTION OF SUGAR-SWEETENED BEVERAGES* (p. 41). http://www.cdph.ca.gov/SiteCollectionDocuments/StratstoReduce_Sugar_Sweetened_Bevs.pdf. Accessed Feb 24, 2016

Chaloupka, F. J., Yurekli, A., & Fong, G. T. (2012). Tobacco taxes as a tobacco control strategy. *Tobacco Control, 21*(2), 172–180. doi:10.1136/tobaccocontrol-2011-050417.

Collins, C., Lavoie, C., Leblanc, B., Miles, R., & MacAlpine-Stiles, J. (2008). *Wellness—we each have a role to play individuals, communities, stakeholders and government: final report of the select committee on wellness to the second session of the 56th legislative assembly of New Brunswick.* [Fredericton]: Select Committee on Wellness.

Drewnowski, A. (2007). The real contribution of added sugars and fats to obesity. *Epidemiologic Reviews, 29*(1), 160–171. doi:10.1093/epirev/mxm011.

Drewnowski, A., & Darmon, N. (2005). The economics of obesity: Dietary energy density and energy cost. *The American Journal of Clinical Nutrition, 82*(1 Suppl), 265S–273S.

Elbel, B., Taksler, G. B., Mijanovich, T., Abrams, C. B., & Dixon, L. B. (2013). Promotion of Healthy Eating Through Public Policy. *American Journal of Preventive Medicine, 45*(1), 49–55. doi:10.1016/j.amepre.2013.02.023.

Étilé, F. (2012). La taxation nutritionnelle comme outil de santé publique: Justifications et effets attendus. *Cahiers de Nutrition et de Diététique, 47*(1), 25–34. doi:10.1016/j.cnd.2011.12.002.

Farley, T. A., & Dowell, D. (2014). Preventing childhood obesity: What are we doing right? *American Journal of Public Health, 104*(9), 1579–1583. doi:10.2105/AJPH.2014.302015.

Farley, T. A., & Merrill, T. (2013). "Sugary drinks" and obesity. *The New England Journal of Medicine, 369*(7), 680–681. doi:10.1056/NEJMc1307331#SA1.

Farley, T., Just, D. R., & Wansink, B. (2012). Clinical decisions. Regulation of sugar-sweetened beverages. *The New England Journal of Medicine, 367*(15), 1464–1466. doi:10.1056/NEJMclde1210278.

Faulkner, G. E., Grootendorst, P., Nguyen, V., Andreyeva, T., Arbour-Nicitopoulos, K., Auld, M. C., et al. (2011). Economic instruments for obesity prevention: results of a scoping review and modified Delphi survey. *International Journal of Behavioral Nutrition and Physical Activity, 8* (1), 109. doi:10.1186/1479-5868-8-109.

Frieden, T. R., Mostashari, F., Kerker, B. D., Miller, N., Hajat, A., & Frankel, M. (2005). Adult tobacco use levels after intensive tobacco control measures: New York City, 2002-2003. *American Journal of Public Health, 95*(6), 1016–1023. doi:10.2105/AJPH.2004.058164.

Fung, C., Kuhle, S., Lu, C., Purcell, M., Schwartz, M., Storey, K., & Veugelers, P. J. (2012). From "best practice" to "next practice": The effectiveness of school-based health promotion in improving healthy eating and physical activity and preventing childhood obesity. *The International Journal of Behavioral Nutrition and Physical Activity, 9*, 27. doi:10.1186/1479-5868-9-27.

Fung, C., McIsaac, J.-L. D., Kuhle, S., Kirk, S. F. L., & Veugelers, P. J. (2013). The impact of a population-level school food and nutrition policy on dietary intake and body weights of Canadian children. *Preventive Medicine, 57*(6), 934–940. doi:10.1016/j.ypmed.2013.07.016.

Glanz, K., Basil, M., Maibach, E., Goldberg, J., & Snyder, D. (1998). Why Americans eat what they do: Taste, nutrition, cost, convenience, and weight control concerns as influences on food consumption. *Journal of the American Dietetic Association, 98*(10), 1118–1126. doi:10.1016/S0002-8223(98)00260-0.

Gortmaker, S. L., Swinburn, B. A., Levy, D., Carter, R., Mabry, P. L., Finegood, D. T., et al. (2011). Changing the future of obesity: science, policy, and action. *Lancet (London, England), 378*(9793), 838–847. doi:10.1016/S0140-6736(11)60815-5

Government of British Columbia. (2012). *Welcome to HealthyFamilies BC!* https://www.healthyfamiliesbc.ca/home/about-us. Accessed Feb 16, 2016

Government of Canada. (2014). *How nutrition North Canada works.* http://www.nutritionnorthcanada.gc.ca/eng/1415538638170/1415538670874. Accessed Feb 16, 2016

Government of Manitoba. (2014). *Healthy kids, healthy schools, healthy communities. Moving Forward with School Nutrition Guidelines. Revised October 2014.* http://www.gov.mb.ca/healthyschools/foodinschools/documents/mfsng/mfsng.pdf. Accessed Feb 16, 2016

Government of Manitoba. (2016). *Healthy living and seniors. Healthy living and healthy populations. Healthy eating.* http://www.gov.mb.ca/cgi-bin/print_hit_bold.pl/healthyliving/hlp/nutrition/index.html. Accessed Feb 16, 2016

Government of New Brunswick. (2014). *New Brunswick's wellness strategy 2014-2021: The heart of our future.* http://www.deslibris.ca/ID/244266. Accessed Feb 28, 2016

Government of Newfoundland and Labrador. (2006). *School food guidelines. For administrators and caterers. Healthy students healthy schools.* http://www.ed.gov.nl.ca/edu/publications/k12/school_food_guide.pdf. Accessed Feb 16, 2016

Government of Nova Scotia. (2006). *Food and nutrition policy for nova scotia public schools. Food and beverage standards for nova scotia public schools.* https://www.ednet.ns.ca/sites/default/files/pubdocs-pdf/foodnutritionfoodbeveragestandard.pdf. Accessed Feb 16, 2016

Government of Nova Scotia. (2011). *Evaluation of the healthy eating nova scotia strategy. Executive summary.* http://novascotia.ca/dhw/healthy-communities/documents/Evaluation-of-the-Healthy-Eating-Nova-Scotia-Strategy-Executive-Summary.pdf. Accessed Feb 16, 2016

Gruber, J., Sen, A., & Stabile, M. (2003). Estimating price elasticities when there is smuggling: The sensitivity of smoking to price in Canada. *Journal of Health Economics, 22*(5), 821–842. doi:10.1016/S0167-6296(03)00058-4.

Hawkes, C. (2012). Food policies for healthy populations and healthy economies. *BMJ (Clinical research ed.), 344*, e2801.

Health Canada. (2007). *Reaching for the top: A report by the advisor on healthy children & youth.* http://www.hc-sc.gc.ca/hl-vs/pubs/child-enfant/advisor-conseillere/index-eng.php. Accessed Mar 1, 2015

Health Canada. (2011a). Caffeinated energy drinks. http://www.hc-sc.gc.ca/fn-an/prodnatur/caf-drink-boissons-eng.php. Accessed Jan 14, 2016

Health Canada. (2011b). Canada's food guide. http://www.hc-sc.gc.ca/fn-an/food-guide-aliment/index-eng.php. Accessed Sept 23, 2015

Health Canada. (2015). *Health Canada's proposed changes to the core nutrients declared in the Canadian nutrition facts table.* http://www.hc-sc.gc.ca/fn-an/consult/2014-core-nutrients-principaux-nutriments/document-consultation-eng.php#a71. Accessed Feb 15, 2016

Healthy Eating Alliance. (2011a). School nutrition policies. http://www.healthyeatingpei.ca/school-nutrition-policies/. Accessed Feb 16, 2016

Healthy Eating Alliance. (2011b). *Prince Edward island's healthy eating strategy 2011–2015.* http://www.healthyeatingpei.ca/kitchen/wp-content/uploads/2014/10/Healthy-Eating-Strategy-2011-15_April-final-draft-2011.pdf. Accessed Feb 16, 2016

Healthy Kids Panel. (2013). *No time to wait: The healthy kids strategy.* http://www.health.gov.on.ca/en/common/ministry/publications/reports/healthy_kids/healthy_kids.aspx. Accessed March 1, 2015

Healthy schools BC. (2012). *Healthy schools BC programs & supports.* Sip Smart! BC. http://www.healthyschoolsbc.ca/program/298/sip-smart-bc. Accessed Feb 16, 2016

Heart and Stroke Foundation of Canada (HSF). (2014). *Sugar, heart disease and stroke. Position statement.* http://www.heartandstroke.com/atf/cf/%7B99452D8B-E7F1-4BD6-A57D-B136CE6C95BF%7D/Sugar-Eng.pdf. Accessed Feb 16, 2016

Hsiao, A., & Wang, Y. C. (2013). Reducing sugar-sweetened beverage consumption: Evidence, policies, and economics. *Current Obesity Reports, 2*(3), 191–199. doi:10.1007/s13679-013-0065-8.

Institute of Medicine (U.S.), Glickman, D., Committee on Accelerating Progress in Obesity Prevention, & Food and Nutrition Board (Eds.). (2012). *Accelerating progress in obesity prevention: solving the weight of the nation.* Washington, DC: National Academies Press.

Institute of Medicine (U.S.), McGinnis, J. M., Gootman, J. A., & Kraak, V. I. (Eds.). (2006). *Food marketing to children and youth: threat or opportunity?* Washington, D. C: National Academies Press.

International Agency for Research on Cancer (IARC) (Ed.). (2011). *IARC handbooks of cancer prevention, tobacco control, volume 14, effectiveness of tax and price policies for tobacco control.* Lyon: IARC.

Jordan, A., Taylor Piotrowski, J., Bleakley, A., & Mallya, G. (2012). Developing media interventions to reduce household sugar-sweetened beverage consumption. *The ANNALS of the American Academy of Political and Social Science, 640*(1), 118–135. doi:10.1177/0002716211425656.

Journal Officiel de la République Française. (JORF). (2016). *JORF n°0022 du 27 janvier 2016 texte n° 1. LOI n° 2016-41 du 26 janvier 2016 de modernisation de notre système de santé (1). Article 16.* http://www.legifrance.gouv.fr/affichTexte.do?cidTexte=JORFTEXT000031912641&dateTexte=&categorieLien=id. Accessed Nov 29, 2015

Kansagra, S. M., Kennelly, M. O., Nonas, C. A., Curtis, C. J., Van Wye, G., Goodman, A., & Farley, T. A. (2015). Reducing sugary drink consumption: New York City's approach. *American Journal of Public Health, 105*(4), e61–e64. doi:10.2105/AJPH.2014.302497.

Lachance, B., Bertrand, L., Drouin, D., Pageau, M., Poirier, A., Roy, S., & Ministère de la Santé et des Services sociaux du Québec. (2006). *Investir pour l'avenir plan d'action gouvernemental de promotion des saines habitudes de vie et de prévention des problèmes reliés au poids 2006-2012.* [Québec, Qué.]: Santé et services sociaux Québec. http://site.ebrary.com/id/10350793. Accessed Feb 28, 2016

Le Bodo, Y., Paquette, M.-C., Vallières, M., & Alméras, N. (2015). Is sugar the new tobacco? Insights from laboratory studies, consumer surveys and public health. *Current Obesity Reports, 4*(1), 111–121. doi:10.1007/s13679-015-0141-3.

Legislative Assembly of Ontario. *An act to enhance public health by enacting the HEALTHY MENU CHOICES ACT, 2015 and the electronic cigarettes act, 2015 and by amending the Smoke-Free Ontario Act.*, Pub. L. No. 45 (Chapter 7 Statutes of Ontario, 2015) (2015). http://www.ontla.on.ca/bills/bills-files/41_Parliament/Session1/b045ra.pdf. Accessed Feb 28, 2015

Lehnert, T., Sonntag, D., Konnopka, A., Riedel-Heller, S., & König, H.-H. (2012). The long-term cost-effectiveness of obesity prevention interventions: systematic literature review. *Obesity Reviews: An Official Journal of the International Association for the Study of Obesity, 13*(6), 537–553. doi:10.1111/j.1467-789X.2011.00980.x.

Levy, D. T., Friend, K. B., & Wang, Y. C. (2011). A review of the literature on policies directed at the youth consumption of sugar sweetened beverages. *Advances in Nutrition: An International Review Journal, 2*(2), 182S–200S. http://doi.org/10.3945/an.111.000356.

Liu, Y., Lopez, R. A., & Zhu, C. (2014). The impact of four alternative policies to decrease soda consumption. *Agricultural and Resource Economics Review, 43*(1), 53–68.

Merrifield, R. (2007). *Healthy weights for healthy kids: Report of the standing committee on health* (39th Parliament, 1st Session). *House of commons*, Canada. http://www.parl.gc.ca/content/hoc/Committee/391/HESA/Reports/RP2795145/hesarp07/hesarp07-e.pdf. Accessed Feb 25, 2016

Ministère de la Santé et des Services sociaux. (2012). *Pour une vision commune des environnements favorables à la saine alimentation, à un mode de vie physiquement actif et à la prévention des problèmes reliés au poids* (Document rédigé en collaboration avec Québec en Forme et l'Institut national de santé publique du Québec) (p. 24). Québec, Canada: Gouvernement du Québec. http://publications.msss.gouv.qc.ca/acrobat/f/documentation/2012/12-289-03.pdf. Accessed Nov 18, 2015

Ministère de l'Éducation, du Loisir et du Sport. (2009). *Framework policy on healthy eating and active living—Going the healthy route at school. Foods and beverages containing sweeteners.* Québec, Canada. http://www.education.gouv.qc.ca/fileadmin/site_web/documents/dpse/adaptation_serv_compl/VirageSanteEcole_FicheInfo_Edulcorant_a.pdf. Accessed Feb 28, 2016

Mozaffarian, D., Afshin, A., Benowitz, N. L., Bittner, V., Daniels, S. R., Franch, H. A., et al. (2012). Population approaches to improve diet, physical activity, and smoking habits: A scientific statement from the American Heart Association. *Circulation, 126*(12), 1514–1563. doi:10.1161/CIR.0b013e318260a20b.

Mytton, O. (2015). Time for a sugary drinks tax in the UK? *Journal of Public Health (Oxford, England), 37*(1), 24–25. doi:10.1093/pubmed/fdu033

National Cancer Institute (NCI). (2006). *Evaluating ASSIST: A blueprint for understanding state-level tobacco control* (No. Tobacco Control Monograph No. 17). Bethesda, MD: US Department of Health and Human Services, National Institutes of Health, National Cancer Institute. NIH.

New Brunswick Department of education. (2008). *Healthier eating and nutrition in public schools: A handbook for policy 711.* http://www2.gnb.ca/content/dam/gnb/Departments/ed/pdf/K12/policies-politiques/e/711A-Handbook.pdf. Accessed Feb 16, 2016

Newfoundland and Labrador Department of health and community services. (2015). Healthy people, healthy families, healthy communities: A primary health care framework for newfoundland and labrador 2015–2025. http://www.health.gov.nl.ca/health/publications/phc_framework.pdf. Accessed Feb 16, 2016

Nghiem, N., Wilson, N., Genç, M., & Blakely, T. (2013). Understanding price elasticities to inform public health research and intervention studies: Key issues. *American Journal of Public Health, 103*(11), 1954–1961. doi:10.2105/AJPH.2013.301337.

Northern Healthy Communities Partnership (NHCP). (2013). Drop the pop. http://www.nhcp.ca/healthy_eating_team/drop_the_pop.html. Accessed Feb 16, 2016

Northwest territories health and social services. (2006). Healthy foods in facilities. food and beverage guidelines for health and social services. http://www.hss.gov.nt.ca/sites/default/files/healthy_foods_in_facilities.pdf. Accessed Feb 16, 2016

Northwest Territories health and social services. (2015). *Northwest territories community wellness initiatives. Annual report 2013–2014. Best health. Best care. A better future.* http://www.hss.gov.nt.ca/sites/default/files/nwt-community-wellness-initiatives-report.pdf. Accessed Feb 16, 2016

Nunavut Department of health and social services. (2007). *Nutrition in nunavut. A framework for action.* http://www.gov.nu.ca/sites/default/files/files/Nutrition%20Framework.pdf. Accessed Feb 16, 2016

Nunavut Health and Social Services. (2012). *Nunavut's food guide educator's handbook.* http://www.livehealthy.gov.nu.ca/sites/default/files/GN_3564_Educators_Handbook_eng-small%20res.pdf. Accessed Feb 16, 2016

Office de la protection du consommateur (OPC). (2012). *Advertising directed at children under 13 Years of age. Guide to the application of sections 248 and 249. Consumer protection act.* https://www.opc.gouv.qc.ca/fileadmin/media/documents/consommateur/sujet/publicite-pratique-illegale/EN_Guide_publicite_moins_de_13_ans_vf.pdf. Accessed Feb 16, 2016

Ontario Ministry of Education. (2010). *Policy/program memorandum No. 150. Ontario's school food and beverage policy.* http://www.edu.gov.on.ca/extra/eng/ppm/150.html. Accessed Feb 16, 2016

Ontario Ministry of Health and Long Term Care. (2014). *Ontario's healthy kids strategy.* https://news.ontario.ca/mohltc/en/2014/02/ontarios-healthy-kids-strategy-1.html?_ga=1.131925839.1984439746.1455638462. Accessed Feb 16, 2016

Potvin Kent, M., Dubois, L., Kent, E. A., & Wanless, A. J. (2013). Internet marketing directed at children on food and restaurant websites in two policy environments. *Obesity (Silver Spring, Md.), 21*(4), 800–807. doi:10.1002/oby.20124

Potvin Kent, M., Dubois, L., & Wanless, A. (2011). Food marketing on children's television in two different policy environments. *International journal of pediatric obesity: IJPO: an official journal of the International Association for the Study of Obesity, 6*(2–2), e433–441. doi:10.3109/17477166.2010.526222.

Potvin Kent, M., Dubois, L., & Wanless, A. (2012). A nutritional comparison of foods and beverages marketed to children in two advertising policy environments. *Obesity, 20*(9), 1829–1837. doi:10.1038/oby.2011.161.

Public Health Agency of Canada (PHAC). (2010). *Creating a healthier Canada: Making prevention a priority. A declaration on prevention and promotion from Canada's ministers of health and health promotion/healthy living.* http://www.phac-aspc.gc.ca/hp-ps/hl-mvs/declaration/pdf/dpp-eng.pdf. Accessed Feb 16, 2016

Public Health Agency of Canada (PHAC). (2011). *Curbing childhood obesity: A federal, provincial and territorial framework for action to promote healthy weights.* http://www.phac-aspc.gc.ca/hp-ps/hl-mvs/framework-cadre/pdf/ccofw-eng.pdf. Accessed Mar 1, 2015

Public Health Agency of Canada (PHAC). (2015). *Centre for chronic disease prevention: Strategic plan 2016–2019. Improving health outcomes—A paradigm shift.* http://www.phac-aspc.gc.ca/cd-mc/assets/pdf/ccdp-strategic-plan-2016-2019-plan-strategique-cpmc-eng.pdf. Accessed Feb 16, 2016

Public Health England (PHE). (2015). *Sugar reduction: The evidence for action. Annexe 4: An analysis of the role of price promotions on the household purchases of food and drinks high in sugar. A research project for Public Health England conducted by Kantar Worldpanel UK.*

London. https://www.gov.uk/government/uploads/system/uploads/attachment_data/file/470175/Annexe_4._Analysis_of_price_promotions.pdf. Accessed Jan 22, 2016

Québec en Forme. (2016). Pour que les jeunes boivent davantage d'eau, de lait et de substituts de lait, et réduisent leur consommation de boissons sucrées et énergisantes. http://www.quebecenforme.org/section-outils/pour-que-les-jeunes-boivent-davantage-d-eau-de-lait-et-de-substituts-de-lait-et-re%CC%81duisent-leur-consommation-de-boissons-sucre%CC%81es-et-e%CC%81nergisantes.aspx. Accessed Feb 16, 2016

Quebec Weight Coalition (CQPP). (2012a). *Sugar-sweetened beverage marketing unveiled. Volume 2: Price, an Argument That Pays Off.* http://www.cqpp.qc.ca/en/sugar-sweetened-beverages/marketing. Accessed Mar 1, 2013

Quebec Weight Coalition (CQPP). (2012b). *Sugar-sweetened beverage marketing unveiled. Volume 3: Place, A Sugar-sweetened beverage always at your fingertips.* http://www.cqpp.qc.ca/en/sugar-sweetened-beverages/marketing. Accessed Mar 1, 2013

Quebec Weight Coalition (CQPP). (2012c). *Sugar-sweetened beverage marketing unveiled. Volume 4: Promotion, buying the approval of young people.* http://www.cqpp.qc.ca/en/sugar-sweetened-beverages/marketing. Accessed March 1, 2013

Quebec Weight Coalition (CQPP). (2015a). *Complaints lodged.* http://www.cqpp.qc.ca/en/complaints-lodged. Accessed Feb 16, 2016

Quebec Weight Coalition (CQPP). (2015b). *Liquid sugars—NOT every day!* http://liquidsugar.ca/. Accessed Feb 16, 2016

Quebec Weight Coalition (CQPP). (2016). Boissons sucrées. Solutions prometteuses. http://www.cqpp.qc.ca/fr/dossiers/boissons-sucrees/solutions-prometteuses#. Accessed Feb 16, 2016

Raine, K. D., Lobstein, T., Landon, J., Kent, M. P., Pellerin, S., Caulfield, T., et al. (2013a). Restricting marketing to children: Consensus on policy interventions to address obesity. *Journal of Public Health Policy, 34*(2), 239–253. doi:10.1057/jphp.2013.9.

Raine, K. D., Plotnikoff, R., Schopflocher, D., Lytvyak, E., Nykiforuk, C. I. J., Storey, K., et al. (2013b). Healthy Alberta communities: Impact of a three-year community-based obesity and chronic disease prevention intervention. *Preventive Medicine, 57*(6), 955–962. doi:10.1016/j.ypmed.2013.08.024.

Réquillart, V., & Soler, L.-G. (2014). Is the reduction of chronic diseases related to food consumption in the hands of the food industry? *European Review of Agricultural Economics, 41*(3), 375–403. doi:10.1093/erae/jbu010.

Réseau du sport étudiant du Québec (RESQ). (2011). New drive—What's that? http://www.moncarburant.ca/en/texte/17/new-drive-whats-that. Accessed Feb 16, 2016

Ries, N. M., & von Tigerstrom, B. (2011). Legal interventions to address obesity: Assessing the State of the Law in Canada. *The University of British Columbia Law Review 361*, *43*(2).

Roberto, C. A., & Pomeranz, J. L. (2015). Public health and legal arguments in favor of a policy to cap the portion sizes of sugar-sweetened beverages. *American Journal of Public Health, 105* (11), 2183–2190. doi:10.2105/AJPH.2015.302862.

Rutter, H. (2011). Where next for obesity? *Lancet (London, England), 378*(9793), 746–747. doi:10.1016/S0140-6736(11)61272-5

Rutter, H. (2012). The single most important intervention to tackle obesity…. *International Journal of Public Health, 57*(4), 657–658. doi:10.1007/s00038-012-0385-6.

Saskatchewan Ministry of Education. (2012). *Nourishing minds. Eat well, learn well, live well. Towards Comprehensive School Community Health: Nutrition Policy Development in Saskatchewan Schools.* http://www.education.gov.sk.ca/nourishing-minds/. Accessed Feb 16, 2016

Sassi, F. (2010). *Obesity and the economics of prevention: Fit not fat* (p. 265). Organisation for Economic Co-operation and Development (OECD).

Sen, A., & Wirjanto, T. (2010). Estimating the impacts of cigarette taxes on youth smoking participation, initiation, and persistence: Empirical evidence from Canada. *Health Economics, 19*(11), 1264–1280. doi:10.1002/hec.1548.

Studdert, D. M., Flanders, J., & Mello, M. M. (2015). Searching for public health law's sweet spot: The regulation of sugar-sweetened beverages. *PLOS Medicine, 12*(7), e1001848. doi:10.1371/journal.pmed.1001848.

Sultan, R. (chair). (2006). *British Columbia, legislative assembly, select standing committee on health, A Strategy for combating childhood obesity and physical inactivity in British Columbia* (BC Report).

Taber, D. R., Chriqui, J. F., Powell, L. M., & Chaloupka, F. J. (2012). Banning all sugar-sweetened beverages in middle schools: Reduction of in-school access and purchasing but not overall consumption. *Archives of Pediatrics and Adolescent Medicine, 166*(3), 256–262. doi:10.1001/archpediatrics.2011.200.

Veugelers, P. J., & Fitzgerald, A. L. (2005). Effectiveness of school programs in preventing childhood obesity: A multilevel comparison. *American Journal of Public Health, 95*(3), 432–435. doi:10.2105/AJPH.2004.045898.

Wang, Y. C., & Vine, S. M. (2013). Caloric effect of a 16-ounce (473-mL) portion-size cap on sugar-sweetened beverages served in restaurants. *The American Journal of Clinical Nutrition, 98*(2), 430–435. doi:10.3945/ajcn.112.054833.

Wilson, L. M., Avila Tang, E., Chander, G., Hutton, H. E., Odelola, O. A., Elf, J. L., et al. (2012a). Impact of tobacco control interventions on smoking initiation, cessation, and prevalence: A systematic review. *Journal of Environmental and Public Health, 2012*, 961724. doi:10.1155/2012/961724.

Wilson, M. G., Guindon, G. E., Baskerville, N. B., & Gauvin, F.-P. (2012b). *Evidence brief: Promoting healthy weights using population-based interventions in Canada.* Hamilton, Canada: McMaster Health Forum.

World Cancer Research Fund (WCRF). (2015). *Curbing global sugar consumption: Effective food policy actions to help promote healthy diets & tackle obesity.* London: World Cancer Research Fund International. http://www.wcrf.org/sites/default/files/Curbing-Global-Sugar-Consumption.pdf. Accessed Feb 28, 2016

World Health Organization (WHO). (2010). *Set of recommendations on the marketing of foods and non-alcoholic beverages to children.* Geneva, Switzerland: World Health Organization. http://whqlibdoc.who.int/publications/2010/9789241500210_eng.pdf. Accessed Feb 29, 2016

World Health Organization (WHO). (2015). *Using price policies to promote healthier diets.* Copenhagen (Denmark): World Health Organization (WHO).

Yon, B. A., & Johnson, R. K. (2014). Dietary patterns and sugar-sweetened beverage consumption among adolescents and adults. *Current Nutrition Reports, 3*(1), 43–50. doi:10.1007/s13668-013-0071-1.

Yukon Education. (2008). *Yukon's School nutrition policy* (No. Policy No. 1025). http://www.education.gov.yk.ca/pdf/policies/school_nutrition_policy.pdf. Accessed Feb 16, 2016

Yukon Health and Social Services. (2015). *Yukon's health investment fund.* http://www.hss.gov.yk.ca/healthinvestment.php. Accessed Feb 16, 2016

Chapter 4
Sugar-Sweetened Beverage Taxation Logics and Ethical Concerns

Abstract Three sugar-sweetened beverage (SSB) taxation logics generally emerge in the literature: (1) increasing SSB prices in order to decrease consumption and impact the health of the population; (2) raising fiscal revenues that may be earmarked for health-related causes; and (3) sending a salient signal to consumers as a way to change social norms, and to manufacturers as a way to encourage product reformulation. Making clear which short-term and long-term logics are pursued is an essential prerequisite to assess the relevancy of an SSB tax proposal in Canada. Furthermore, the economic and ethical justification of SSB taxation remains controversial (e.g. as regards freedom of choice and autonomy) and should be carefully considered before implementation. A democratic dialogue involving all stakeholders (including consumers) may help disentangle this debate.

4.1 Clarifying Sugar-Sweetened Beverage (SSB) Taxation Logics

Literature proposes three intervention logics for SSB taxation: (1) taxing SSBs as a way to increase their price, decrease consumption and impact the health of the population; (2) taxing SSBs as a way to raise fiscal revenues that may be earmarked for health care, health promotion activities or for the reduction of health inequities; and (3) taxing SSBs to send a salient signal to the population and the manufacturers in order to encourage, in a longer term, social norm changes and product reformulation (Bahl et al. 2003; Brownell et al. 2009; Thow et al. 2011; Boizot-Szantai and Étilé 2011; Mytton et al. 2012; Ecorys 2014; WHO 2015). For many years, the first two logics have been extensively used to justify tobacco taxation across the world (WHO 2012).

We have schematized these intervention logics as in Fig. 4.1. This graphical representation, generally called "logic model", is very helpful to articulate each step of the implementation process of an intervention, to clarify expected intermediary effects and to assess whether expected outcomes are realistic, achievable and measurable. Doing so, it represents the "theory" of how SSB taxation should

Y. Le Bodo et al., *Taxing Soda for Public Health*,
DOI 10.1007/978-3-319-33648-0_4

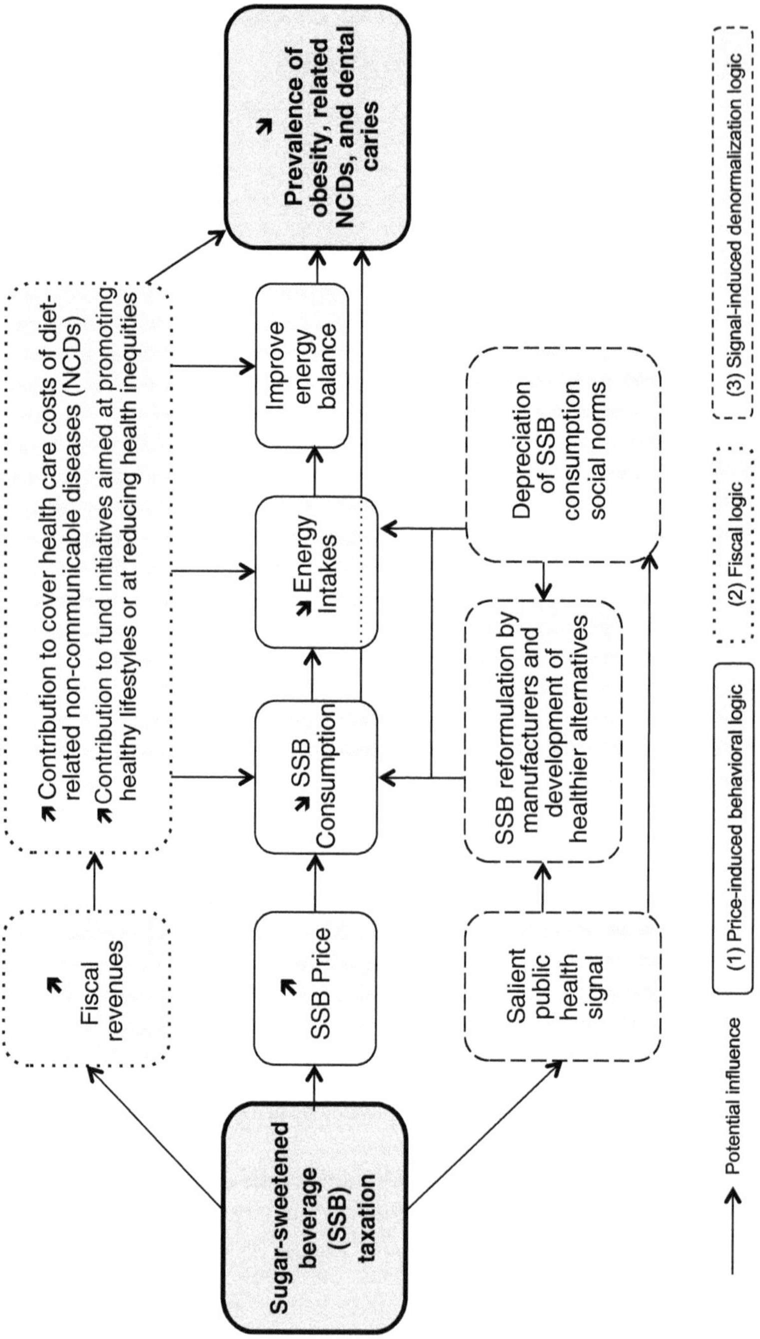

Fig. 4.1 SSB taxation logic model emerging from three different (but not mutually exclusive) public health strategies

produce its intended effects and helps identify the data to be measured and collected at all levels (Morestin et al. 2011; Mytton et al. 2014). The goal of this logic model was not to represent all the factors that influence the issue at stake, but to focus on those targeted by an SSB taxation policy. Interventions pursuing similar goals (e.g. a communication campaign promoting healthy beverages), potentially counteracting forces (e.g. beverage industry's price promotion strategies), potential confounders (e.g. seasonal effects) and unintended effects are generally not represented on logic models, but should be accounted for evaluation purposes (see Chap. 14).

Out of the 21 "real-world" cases of SSB taxation that we have listed in the introductory chapter, the documentation consulted suggests that, in almost all of them, a behavioural logic has been part of the reasons for adoption (see Table 1.1, column 3). In some cases, this was very explicitly stated such as in Denmark (decrease sugar consumption), the Cook Islands and St Helena (increase SSB cost to encourage behaviour change) or Mexico (decrease energy-dense food consumption). In some cases, the fiscal logic was also explicit, such as in French Polynesia (create of a preventive health fund), Hungary (improve the financing of health services), Algeria (create a fund against cancer) or Barbados (finance a national campaign). Finally, a few cases suggest that a "denormalization" logic has been particularly considered, such as in Hungary (promote product reformulation), in the city of Berkeley (discourage SSB distribution) and the Navajo Nation (raise awareness of unhealthy food in the population).

From an *ex-ante* evaluation perspective, the logic model presented in Fig. 4.1 helped define the scope of our review, to orient our documentary search and to structure our analysis (in particular for Part II of the book). All three logics have been considered (Morestin et al. 2011; Chatterji et al. 2014).

Depending on the evidence available and the local context, these different (and not mutually exclusive) intervention logics may be deemed more or less realistic, achievable, acceptable and compatible with ongoing public health actions (Chaloupka et al. 2011; Fletcher et al. 2011). As a result, pursuing one or several of these logics concurrently may influence the assessment of the tax proposal, its eventual implementation and its evaluation. This is why stakeholders (decision-makers in particular) should state clearly what their primary and secondary intentions are. This is generally where economic considerations and ethical concerns come into the debate.

4.2 Debates About the Normative Justification of SSB Taxes

Is there an appropriate rationale for the government to intervene and overtax SSBs? As highlighted by Bahl et al. (2003, p. 514), the "classical view" of "most fiscal economists" generally is that "a good tax policy [aimed to generate revenue] begins with the rule of neutrality; that is, the tax rate and base structure should not interfere with consumption, production or investment choices". Consequently, from this perspective, the broader the tax scope, the better. This is clearly not the case of a tax targeting specifically non-alcoholic beverages containing added sugars: the tax scope is narrow, a single beverage category is explicitly discriminated and, in many cases, the policy is precisely aimed to bring down the production and consumption of the product. Therefore, the legislator may have to demonstrate the negative externality attached to SSB consumption (Preece 2013).

For example, alcohol taxation has been justified as a way to discourage excessive drinking and related car accidents, and/or as a way to raise revenue earmarked for the prevention and treatment of diseases related to alcohol abuse (Bahl et al. 2003). Then, how would negative externalities attached to SSB consumption be formulated? Interestingly, 10 years ago, Bahl et al. (2003, p. 515) observed that concerns related to environmental protection and recycling costs of SSB containers (i.e. made of aluminium and plastic) may have been acceptable but probably not sufficient to single out SSBs among many other goods and services. Since then, however, various public health justifications have been used to make SSB a "good candidate for discriminatory tax treatment" (Bahl et al. 2003, p. 514). These rationales generally refer to "market failures", i.e. imperfect market conditions fostering SSB overconsumption, thus causing additional public health and economic burdens and, consequently, impacting negatively personal and collective welfare. These failures would therefore justify a governmental intervention (Brownell et al. 2009). Concretely, Brownell et al. (2009, p. 1601) illustrate these imperfect market conditions by the facts that: (1) "many persons do not fully appreciate the links between consumption of these beverages and health consequences [and so] make consumption decisions with imperfect information"; (2) that "time-inconsistent preferences (i.e., decisions that provide short-term gratification but long-term harm) […] [are] exacerbated in the case of children and adolescents, who place a higher value on present satisfaction while more heavily discounting future consequences"; and (3) that "consumers do not bear the full costs of their consumption decisions" since SSB overconsumption-related healthcare costs are partially paid via publicly funded programs and services. More broadly, correcting these "externalities" and "market failures" may also be justified from a virtuous systemic behavioural taxing perspective, as Robert H. Frank puts it: "Every dollar raised by taxing harmful activities is one dollar less that we must raise by taxing useful ones" (Frank 2011). Notwithstanding, because of the complex aetiology of obesity and chronic diseases, it remains tricky for the legislator to single out SSB overconsumption as a specific candidate for taxation in comparison with other dietary risk factors. The evidence presented in Chap. 2 of this book supports that SSB is not the only dietary risk factor to focus on, but certainly is a relevant place to start.

On another ground, some US economists have simulated that soda taxes may not be well justified by the economic burden related to obesity and NCDs that is partly supported by the US public health insurance system. They postulate that these costs may actually be offset by a positive "innovation externality" relying on the economic growth related to the development of new pharmaceutical and medical products and services. These gains would be supplemented by welfare gain for patients and consumers who would benefit from such innovations and would be less constrained in terms of self-control and self-protection (Bhattacharya and Packalen 2012). Beyond scientific methodological debates, such an economic perspective clearly raises ethical concerns that should be collectively debated, in particular in terms of governmental action, freedom of choice and autonomy.

4.3 Ethical Concerns Surrounding Freedom of Choice and Autonomy

From an ethical and philosophical standpoint, the debate on SSB taxation essentially goes around the concept of individual food autonomy, itself subdivided into three dimensions highlighted by Barnhill and King (2013b, p. 118): (1) the "liberty/freedom" of choice and action by the consumer; (2) the "understanding of available options" by the consumers and how these may influence his preferences, interests and decisions; (3) the "psychological capacity to make choices". From these dimensions, proponents tend to consider that SSB taxation increases autonomy by mitigating psychological incapacity to make fully informed food choices (in the case of children especially) and, eventually, by increasing the understanding of SSB health-related issues (Brownell et al. 2009; Brownell and Frieden 2009). On the other hand, opponents consider that SSB taxation alters autonomy by restricting freedom of choice and by underestimating the psychological capacity of the consumer to make decision with full knowledge of its consequences (Marlow and Abdukadirov 2012; Lusk 2014; Marlow 2013). As Marlow puts it (2013, p. 16): "(...) People already understand that weight gain results from eating too much and exercising too little. Attempts to replace individual responsibility and motivation with government "incentives" have failed to meet their overstated promises".

In a similar manner as Barnhill and King (2013b) analysed the case of the New York City "big soda ban" proposal (finally ruled unconstitutional), we tend to think that SSB taxation is ethically defensible: it appropriately preserves food autonomy since it offers to the consumer the possibility to "stop and think" about his real preference (and eventually to continue buying SSB), while potentially contributing to health improvements at the individual and population levels in the long term. To some extent, as any SSB public health policy, taxation may also compensate for pre-existing heavy marketing forces already influencing consumers, knowingly or unknowingly. Finally, as underlined by Kass et al. (2014, p. 790), we tend to think that fairness concerns inherent to fiscal policies rather apply to taxes on "necessities such as housing, clothing or food" from which SSB consumption could reasonably be excluded because of their low nutritive value and the existence of a cheap and easily available substitute: tap water. This last point echoes the famous quote addressing sugar necessity from the political economist Adam Smith's in 1776: "Sugar, rum, and tobacco, are commodities which are nowhere necessaries of life, which are become objects of almost universal consumption, and which are therefore extremely proper subjects of taxation" (p. 474).

Notwithstanding, some economists have demonstrated that, from a cost-benefit perspective, SSB taxation may actually generate greater welfare losses than benefits depending on the rationality of the consumers and on the way tax revenues would be used (Étilé 2012; Lusk and Schroeter 2012; Lusk 2014). Additionally, food sociologists and psychologists have extensively demonstrated that the foods and beverages we choose to eat, drink, cook and share convey many important individual and collective values (Poulain 2013). Consequently, considering the risk that some food policy proposals excessively alter these values seems acceptable and even necessary. However, we tend to support the ethical rationale proposed by Resnik (2014, p. 174) and others, according to which altering to some extent this freedom of choice may be justified when the issue at stake is particularly acute, when the concerns are focused on a specific type of food or beverage, when the policy considered is likely to be among the most cost-effective options, when distributive impacts of the policy are equitable and when its implementation is reasonably burdensome. Precisely on all these points, it seems to us that this book gathers sufficient evidence to suggest that, provided several considerations are accounted for, a soda tax could be part of a portfolio of nutrition-enhancing policies in Canada. Some of these aspects will be covered in Parts II and III of this book.

Beyond scientific opinion, the ethical value of food autonomy may still give rise to the pros and cons arguments on SSB taxation. For a comprehensive analysis of ethical concerns surrounding SSB policies, we refer the reader to some key publications (Barnhill and King 2013a, b; Kass et al. 2014; Conly 2013). A democratic dialogue involving all stakeholders may help disentangle this debate on a Canadian ground (Merry 2012; Huang et al. 2015).

Key Messages

- Three SSB taxation intervention logics generally emerge in the literature and could be concurrently pursued: a price-induced behavioural logic; a fiscal logic; and a signal-induced denormalization logic.
- The ethical justification of SSB taxation remains controversial among stakeholders and in the population: this should be carefully considered and debated before implementation on a Canadian ground.
- Several arguments suggest SSB taxation may be ethically defensible: it addresses an acute public health issue (see Chap. 2); it targets a specific category of food and beverages which is not part of basic groceries; taxation is potentially among the most cost-effective options and may not raise insurmountable concerns in terms of equity, undesirable effects and feasibility (see Chap. 3, Part II and III).

References

Bahl, R., Bird, R., & Walker, M. B. (2003). The uneasy case against discriminatory excise taxation: Soft drink taxes in Ireland. *Public Finance Review, 31*(5), 510–533. doi:10.1177/1091142103253753.

Barnhill, A., & King, K. F. (2013a). Evaluating equity critiques in food policy: The case of sugar-sweetened beverages. *The Journal of Law, Medicine & Ethics: A Journal of the American Society of Law, Medicine & Ethics, 41*(1), 301–309. doi:10.1111/jlme.12020.

Barnhill, A., & King, K. F. (2013b). Ethical agreement and disagreement about obesity prevention policy in the United States. *International Journal of Health Policy and Management, 1*(2), 117–120. doi:10.15171/ijhpm.2013.21.

Bhattacharya, J., & Packalen, M. (2012). The other ex ante moral hazard in health. *Journal of Health Economics, 31*(1), 135–146. doi:10.1016/j.jhealeco.2011.09.001.

Boizot-Szantai, C., & Étilé, F. (2011). Taxer les boissons sucrées pour lutter contre l'obésité? Le point de vue de l'économie. *Obésité, 6*(4), 218–225. doi:10.1007/s11690-012-0306-4.

Brownell, K. D., Farley, T., Willett, W. C., Popkin, B. M., Chaloupka, F. J., Thompson, J. W., et al. (2009). The public health and economic benefits of taxing sugar-sweetened beverages. *New England Journal of Medicine, 361*(16), 1599–1605. doi:10.1056/NEJMhpr0905723

Brownell, K. D., & Frieden, T. R. (2009). Ounces of prevention—The public policy case for taxes on sugared beverages. *New England Journal of Medicine, 360*(18), 1805–1808. doi:10.1056/NEJMp0902392.

Chaloupka, F. J., Powell, L. M., & Chriqui, J. F. (2011). Sugar-sweetened beverages and obesity prevention: Policy recommendations. *Journal of Policy Analysis and Management: [The Journal of the Association for Public Policy Analysis and Management], 30*(3), 662–664.

Chatterji, M., Green, L. W., & Kumanyika, S. (2014). L.E.A.D.: A framework for evidence gathering and use for the prevention of obesity and other complex public health problems. *Health Education & Behavior, 41*(1), 85–99. doi:10.1177/1090198113490726.

Conly, S. (2013). Coercive paternalism in health care: Against freedom of choice. *Public Health Ethics, 6*(3), 241–245. doi:10.1093/phe/pht025.

Ecorys. (2014). *Food taxes and their impact on competitiveness in the agri-food sector.* Final report (No. Ref. Ares(2014)2365745–16/07/2014). Rotterdam: Client: DG Enterprise and Industry.

Étilé, F. (2012). La taxation nutritionnelle comme outil de santé publique: justifications et effets attendus. *Cahiers de Nutrition et de Diététique, 47*(1), 25–34. doi:10.1016/j.cnd.2011.12.002.

Fletcher, J. M., Frisvold, D. E., & Tefft, N. (2011). Are soft drink taxes an effective mechanism for reducing obesity? *Journal of Policy Analysis and Management: [The Journal of the Association for Public Policy Analysis and Management], 30*(3), 655–662.

Frank, R. H. (2011). *The Darwin economy: Liberty, competition, and the common good.* Princeton [N.J.]: Princeton University Press.

Huang, T. T.-K., Cawley, J. H., Ashe, M., Costa, S. A., Frerichs, L. M., Zwicker, L., et al. (2015). Mobilisation of public support for policy actions to prevent obesity. *The Lancet, 385*(9985), 2422–2431. doi:10.1016/S0140-6736(14)61743-8.

Kass, N., Hecht, K., Paul, A., & Birnbach, K. (2014). Ethics and obesity prevention: Ethical considerations in 3 approaches to reducing consumption of sugar-sweetened beverages. *American Journal of Public Health, 104*(5), 787–795. doi:10.2105/AJPH.2013.301708.

Lusk, J. L. (2014). Are sugar-sweetened beverage taxes a cost-effective means of reducing weight? *Canadian Journal of Diabetes, 38*(1), 9–10. doi:10.1016/j.jcjd.2013.08.259.

Lusk, J. L., & Schroeter, C. (2012). When do fat taxes increase consumer welfare? *Health Economics, 21*(11), 1367–1374. doi:10.1002/hec.1789.

Marlow, M. L. (2013). Soda regulation is not the solution. *Journal of American Physicians and Surgeons, 18*(1), 14–16.

Marlow, M. L., & Abdukadirov, S. (2012). Can behavioral economics combat obesity? *Regulation Health and Medicine, Summer, 2012*, 14–18.

Merry, M. S. (2012). Paternalism, obesity, and tolerable levels of risk. *Democracy & education, 20* (1), 1–6.

Morestin, F., Gauvin, F.-P., Hogue, M.-C., & Benoit, F. (2011). *Method for synthesizing knowledge about public policies.* [Montréal]: Centre de collaboration nationale sur les politiques publiques et la santé, Institut national de santé publique Québec. http://collections.banq.qc.ca/ark:/52327/2066075. Accessed February 28, 2016.

Mytton, O. T., Clarke, D., & Rayner, M. (2012). Taxing unhealthy food and drinks to improve health. *BMJ (Clinical Research Ed.), 344*, e2931.

Mytton, O. T., Eyles, H., & Ogilvie, D. (2014). Evaluating the health impacts of food and beverage taxes. *Current Obesity Reports, 3*(4), 432–439. doi:10.1007/s13679-014-0123-x.

Poulain, J.-P. (2013). *Sociologies de l'alimentation: les mangeurs et l'espace social alimentaire.* [S.l.]: Presses Universitaires De.

Preece, R. (2013). The effective contribution of excise taxation on non-alcoholic beverages to government revenues and social objectives: A review of the literature. *World Customs Journal, 7*(1), 21–38.

Resnik, D. B. (2014). Paternalistic food and beverage policies: A response to Conly. *Public Health Ethics, 7*(2), 170–177. doi:10.1093/phe/phu014.

Smith, A. (1776). *An inquiry into the nature and causes of the wealth of nations, Book V, Chapter 3* (W. Strahan and t. Cadell.). London.

Thow, A. M., Quested, C., Juventin, L., Kun, R., Khan, A. N., & Swinburn, B. (2011). Taxing soft drinks in the Pacific: Implementation lessons for improving health. *Health Promotion International, 26*(1), 55–64. doi:10.1093/heapro/daq057.

World Health Organization (WHO). (2012). *Tobacco taxation and innovative health-care financing*. New Delhi, India: World Health Organization. http://www.searo.who.int/tobacco/documents/2012-pub1.pdf. Accessed January 22, 2016.

World Health Organization (WHO). (2015). *Using price policies to promote healthier diets.* Copenhagen (Denmark): World Health Organization (WHO).

Part II
What: Impact of Taxing Sugar-Sweetened Beverages

Chapter 5
Effects of Taxation on Sugar-Sweetened Beverage Prices

Abstract When the primary objective of a tax on sugar-sweetened beverages (SSBs) is to discourage consumption on the short term, a key issue is to know whether the tax is actually passed onto retail prices. On the one hand, when the tax is levied on manufacturers, they may decide to under- or overshift the tax. On the other hand, larger contextual factors (e.g. seasonal effects, other policies) may counteract or attenuate tax effects on prices. In this chapter, we present key pieces of evidence from various places where soda taxes have been implemented across the world. Overall, the evidence suggests that such taxes are generally followed by price increases in the short term, sometimes beyond or below the increase theoretically generated by the tax.

5.1 Factors That May Impair Sugar-Sweetened Beverage (SSB) Tax Shift onto Prices

Two main questions arise when considering how an SSB tax would be concretely shifted onto sugar-sweetened beverage (SSB) prices: (1) if the tax is levied at the manufacturer/retailer level, to what extent will they pass the tax onto prices? And, (2) how could larger contextual factors influence and potentially counterbalance SSB tax impact on prices (see Fig. 5.1)? These two concerns are discussed here below.

Out of the 21 places where an SSB tax has been reported across the world (see Table 1.1), taxation is levied on the manufacturer or the importer in 19 cases.[1] In such circumstances, a first risk is that these economic actors incompletely shift the tax onto SSB prices and, consequently, that consumer demand remains fully or

---

[1]Among these 19 places, 15 excise taxes were identified as well as 5 special taxes and 4 import duties. The 2 other places present a somewhat different situation: in Chile, the tax works as a value added tax ultimately reaching the final consumer; in the Navajo Nation, the tax is levied on the retailer directly from beverage sales to the consumer (see Table 1.1, column 5).

Y. Le Bodo et al., *Taxing Soda for Public Health*,
DOI 10.1007/978-3-319-33648-0_5

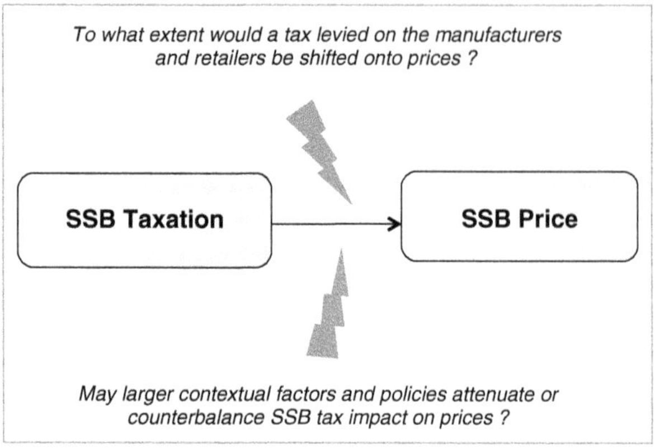

Fig. 5.1 Uncertainties about the propensity of SSB taxation to be passed onto SSB prices

partially unaffected. In other terms, SSB manufacturers, importers and/or retailers may decide to absorb the cost of the tax or to dispatch it onto the price of a wider range of products, including untaxed healthier beverages (e.g. bottled water, non-calorically beverages). This risk may be greater if the tax only applies to a limited territory and a small-size population (e.g. a municipality), in which case retailers may accept to sell at a loss to dissuade consumers from shopping out of the tax zone.

A second risk is the potential interaction of SSB taxation with other policies influencing food price and demand, such as current food taxes and subsidies, trade restrictions and food assistance programs. Some argue that adjusting those existing policies through a health promotion lens should take precedence over the introduction of new taxes (Cash and Lacanilao 2007; Faulkner et al. 2011). Indeed, agriculture and food policy changes may counteract the effect of a new tax on SSB price. For instance, Bonnet and Réquillart (2011, 2012) have simulated the potential impact on SSB price resulting from the European Union's sugar policy reform, which was supposed to lower the price of sugar by 36 % between 2006 and 2010. Their model, accounting for both consumers'[2] and manufacturers' reactions, estimates that the reform could have led to an average 4.2 % decrease in the retail price of SSBs. It also predicts a corresponding 10 % increase in SSB demand. From a public health perspective, these results suggest that such a policy reform may actually counteract or attenuate the expected outcomes of a new SSB taxation policy (Bonnet and Réquillart 2012).

[2]The demand model was based on a representative panel of 19,000 French households providing comprehensive data on "take-home" food purchases between 2003 and 2005.

5.2 Preliminary Evidence from Taxes Implemented Across the World

To appreciate the actual pass-through of SSB taxes onto prices, we have looked at empirical data from countries where SSB taxes have recently been enacted. In some cases, the evidence is limited or even anecdotal, but in others, the evidence becomes sounder as proper evaluations are being conducted (e.g. in Mexico and in the USA). We present here below some key pieces of information from places where taxes were levied on the manufacturers[3]:

- In the Pacific, Thow et al. (2011, p. 58) report that a "casual monitoring" of the price of soft drinks by the Fijian Ministry of Health in 2006 suggests that the new domestic excise tax of 0.05 Fiji dollars (Eq. 0.03 CAD) per litre was fully passed onto prices in the following months. In the Samoan Islands, according to Thow et al. (2011, p. 58), local manufacturers and importers "[…] have reported passing on the tax to consumers" following the increase of the excise tax in 2008 (see Table 1.1). Observations conducted by the Samoa Nutrition Centre (quoted by Thow et al. 2011) suggest that bottled water became cheaper than soft drinks in the following months.[4] Finally, in the island of Nauru, following the introduction of a 30 % "sugar levy" on various sugary beverages (including SSBs) in 2007 (see Table 1.1), Thow et al. (2011, p. 59) report that lower-than-expected price hikes were observed (e.g. 20 % for a 375 ml SSB). The authors suggest that the competition with imported products from Asia may have played a role.
- In Denmark, according to Ecorys data analysis (2014a),[5] successive excise tax changes on colas between 1999 and 2013 indicate that when the tax rate increased, price increased, and when the tax rate decreased, price decreased "out of the ordinary" trend.[6] Two observations, however, indicate that other factors play a role: (1) the price evolution of regular colas and low-calorie colas has been very similar, despite the application of differential tax rates since 2010 (see Table 1.1) and (2) the price of fruit drinks seems to vary independently of tax rate changes (Ecorys 2014a, p. 25, 27). These observations tend to converge

[3]In this section, the actual price changes are generally not adjusted for inflation (unless stated otherwise). In the case of Ecorys data, actual price changes are often compared to the expected price change, calculated as (tax rate) * (tax base) and expressed as a percentage of the pre-tax price (Ecorys 2014a, p. 25).

[4]According to a survey of prices in Apia stores.

[5]In this study, Ecorys predominantly used data from the Euromonitor/Passport system, providing annualized data on the prices and total consumer demand for a diversity of food items in European countries. The authors consider that "(…) in general the datasets of equal quality provide the same conclusions on the trend in the market, although in some cases, other datasets will come to different conclusions in some limited number of instances" (Ecorys 2014b, p. 62).

[6]Over the period considered, the pattern of price changes remains similar after adjustment for inflation.

with Bergman and Hansen's evaluation (2013) of soft drinks' excise tax increases (in 1998 and 2001)[7] and tax cut (in 2003).[8] Their work is based on price data usually aimed to establish the consumer price index in Denmark.[9] The authors' regression analysis is based on price trajectories controlling for inflation, energy costs, rent costs and labour costs. On average, it seems that the two tax hikes have been overshifted to prices and that the tax cut has been undershifted to prices. However, in spite of a model allowing for lagged effects (i.e. in case retailers would shift the tax in the following months), they still observe that between 18 and 41 % of retailers did not particularly change the price of soft drinks in explicit connexion with tax events. According to the authors, it suggests that since retailers appear to update soft drink prices once to twice a year anyway, some of them may just pass the tax onto prices at the most convenient time for them (Bergman and Hansen 2013).

- In Finland, according to Ecorys data analysis (2014a), prices also increased "out of the ordinary" following excise tax changes on soft drinks.[10] In the hypothesis of a 100 % pass-through of the tax, price changes were expected to be +1.5 % in 2011 and +0.9 % in 2013. Instead, Ecorys data analysis indicates that prices increased by 7.3 % in 2011, by 7.3 % in 2012 and by 2.7 % in 2013 (Ecorys 2014a, p. 25, 28). These price changes are not adjusted for inflation. However, further analyses show that, net of inflation, the price changes still exceed the expected price changes due to the tax (Ecorys 2014b, p. 115). These observations suggest the influence of other factors. For instance, it may partly be explained by the fact that both retailers and manufacturers have increased their margins during the period under consideration.
- In France, data analysis based on a large Web-based price comparator[11] suggests that after 6 months, the excise soda tax[12] enacted in January 2012 (eq. to 0.1 CAD/L) had been progressively and fully shifted to the price of carbonated drinks (including "cola, energy, tonic and other soft drinks"). However, it had

[7]From 0.8 DKK (Eq. 0.17 CAD)/ litre to 1 DKK (0.21 CAD) in 1998 and to 1.65 DKK (Eq. 0.34 CAD) in 2001.

[8]Back to 1.15 DKK (Eq. 0.24 CAD)/litre in 2003.

[9]The database provided monthly data from 1997 to 2005 on a broad range of items, including information on product category, product brand, type of retail outlet and its location, and date of purchase. The authors note that the data were not systematically collected in the same stores and products over the period under study.

[10]In Finland (see Table 1.1), the scope of the tax includes bottled water (including flavoured and carbonated water), carbonates (including colas, fruit drinks, low-calorie drinks), juices (including 100 % juices, fruit-flavoured drinks and nectars), ready-to-drink coffee and tea, sports and energy drinks.

[11]The data were collected via Prixing, a digital price comparator that can be used on the Internet or via mobile phone applications. In this study, the authors used a sample of more than 500,000 observations on the prices of 845 products sold in more than 800 supermarkets.

[12]In France, the tax encompasses beverages containing added sugars and non-caloric beverages containing artificial sweeteners (including carbonated drinks, fruit juices, fruit beverages, flavoured milk) (see Table 1.1).

only been partially shifted (≈85 %) to the price of fruit drinks and flavoured waters. In this case, the authors postulate that competition with untaxed substitutes (e.g. unsweetened water, 100 % fruit juice) may have played a role. Variations in the pass-through depending on the time, brands and retailers suggest a diversity of adaptation strategies (Berardi et al. 2012). Ecorys data analysis confirms that the introduction of the tax has been followed by "out of the ordinary" price increases (Ecorys 2014b, p. 132). Following the enactment of the tax, the price of regular colas was expected to increase by 4.5 %, but increased by 5 % in 2012 and 3.1 % in 2013. For low-calorie colas (also included in the scope of the tax), the price was expected to increase by 4.7 % in 2012, but increased by 6 % and 4.6 % in 2012 and 2013, which is partly attributed to increased retail margins. Finally, as regards fruit drinks, the price increased by 5.3 % in 2012 and 3.9 % in 2013 (expected price change from the tax was 6.2 %) (Ecorys 2014a, p. 26). Even if adjusted for inflation, these changes still correspond to "out of the ordinary" price increases and approximate or slightly exceed expected price change due to the tax (Ecorys 2014b, pp. 124–5). A slight price increase was observed for untaxed fruit juices in 2012–2013, but this increase is consistent with the trend observed in the previous years and corresponds to inflation (Ecorys 2014b, p. 128).
- In Hungary, as a result of the 2011 "Public health product tax",[13] according to Ecorys (2014a), the price was expected to climb by 3.1 % in colas and 2.7 % in fruit drinks. Actually, colas' price climbed slightly above expectations (3.4 % in 2011, 1.2 % in 2012 and 0.7 % in 2013), but this increase mainly corresponds to inflation (Ecorys 2014a, p. 26, 2014b, p. 135). As regards fruit drinks, their price climbed slowly and below expectations (0.1 % in 2011, 0.6 % in 2012 and 1.3 % in 2013): in this case, it is difficult to disentangle price changes due to the tax and to other factors, since a similar price development was observed prior to the tax introduction. Moreover, the retail margins on untaxed fruit juices also increased in the same period (Ecorys 2014a, p. 26, 2014b, p. 139).
- In Mexico, preliminary observations had suggested price increases (up to 10 % and beyond, depending on the source) during the first weeks and months following the special soda tax application in January 2014 (Case 2014; Wade 2015). A robust analysis of price changes[14] following the implementation of the tax suggests that it has been fully passed (and even overshifted) to consumers in the case of regular sodas (12 % price increase vs. 9 % anticipated). The pass-through was significant but incomplete in the case of other taxed beverages (i.e. fruit-based beverages), which may be explained by greater competitiveness in this market segment (Grogger 2015). A strength of Grogger's analysis is to have accounted for the national price index monthly evolution between 2005

[13] In Hungary, the tax encompasses sugar-sweetened beverages (including syrups or concentrates for soft drinks) containing more than 8 g of added sugar/100 mL (see Table 1.1).

[14] The analysis was conducted from price data covering a broad range of items and collected monthly by the National Institute for Statistics and Geography in 46 Mexican cities (Grogger 2015).

and 2015: it clearly shows a sharp increase in the price of sodas at the time of the introduction of the tax. This higher price persisted until the end of the time series (March 2015). In the meantime, the price change of most untaxed beverages (including bottled water, 100 % fruit juice, milk) and food (unlikely to be SSB substitutes) did not differ significantly from pre-existing trends. Diet drinks (untaxed), whose price slightly increased, are an exception. The author assumes that SSB substitution towards those drinks may have contributed to explaining this price hike as a result of increased demand. These results clearly suggest a tax-specific effect, to be confirmed in a longer term (Grogger 2015).

- In the US city of Berkeley (California), preliminary observations following the implementation of a 1-cent-per-ounce excise tax on the distribution[15] of SSBs in March 2015 indicate that the tax has been partially shifted to retail prices. Falbe et al. (2015) examined SSB and non-taxed beverage price differences between the pre-tax period (i.e. in fall 2014) and the post-tax period (i.e. in May–June 2015) in a variety of retail stores[16] in Berkeley and two comparison cities (i.e. Oakland and San Francisco).[17] For beverages whose size is lower than 34 oz, they conclude that the relative price increase of soda, fruit-flavoured beverages and SSBs (at large)[18] in Berkeley was, respectively, 0.69, 0.47 and 0.47 cents/ounce higher than in comparison cities.[19] The pass-through was shown to be lower in the case of bulk offers. The fact that the price change of untaxed beverages[20] did not differ significantly between Berkeley and the comparison cities tends to confirm a tax-specific impact on SSB price in the short term. The authors warn that such price changes should be monitored in a longer term, especially since tax shifting onto prices is administratively time-consuming (especially for small stores) and may be achieved at a later stage (Falbe et al. 2015). Another recent study confirmed the partial pass-through of Berkeley's "soda tax" on retail prices. Cawley and Frisvold (2015) used data from a "near census" of convenience stores and supermarkets[21] in the city of Berkeley and

[15]According to the New Berkeley Municipal Code (Chap. 7.72), the tax "[...] shall be levied on the first distributor subject to the jurisdiction of the City". However, there is an exemption for "[...] any SSB distribution to a retailer with less than $100,000 in annual gross receipts" (City of Berkeley 2014, p. 3).

[16]Beverage prices were collected in various stores including "chain supermarkets, drugstores, small grocery stores, liquor stores, and convenience stores", using a random sampling method, and including outlets from "low-income" and "minority" neighbourhoods (Falbe et al. 2015, p. 2196, 2199).

[17]The comparison cities were selected because of their geographical proximity to Berkeley and their quite similar urban profile.

[18]Including all regular soda, energy drinks, sports drinks, sweetened water, presweetened tea, presweetened coffee and fruit-flavoured beverages (not 100 % juice).

[19]These differences were statistically significant.

[20]Including diet versions of taxed beverages, reduced fat milk, water and 100 % orange juice.

[21]Data were manually collected from a sample of grocery stores, supermarkets, convenience stores and pharmacies.

used San Francisco as a control city.[22] The analysis compared the price of SSBs and untaxed beverages before and after the implementation of the tax, i.e. at the end of December 2014 and at the beginning of June 2015. Overall, they estimate[23] that SSB "retail prices rose by less than half of the amount of the tax"[24] (Cawley and Frisvold 2015, p. 2). Their estimates are congruent (although less elevated) with what Falbe et al. (2015) have estimated. It would be interesting to further explore how methodological choices may explain differences in results between the two studies. Cawley and Frisvold (2015) mention several potential explanations for the incomplete pass-through of the tax. First, since cross-border shopping is more conceivable at a municipal level than on large territories, retailers located in Berkeley may have been reluctant to fully pass the tax onto consumers to prevent profit losses. The authors also consider that this may explain why soda taxes seem to have been more significantly shifted to prices when implemented at a country level. Secondly, the authors note that the methods used may have permitted greater control for price trends than former studies and, therefore, may have more strictly delimited the specific contribution of the tax to price changes (Cawley and Frisvold 2015).

Overall, short-term evidence indicates that most of the time, in the months following the enactment of an excise or special tax on SSBs, their prices have increased out of the ordinary. A full pass-through of the tax onto prices was often noted, and in several cases, the price increased beyond or below what could be attributed to the direct effect of the tax.[25] This "over-" or "undershifting" suggests that retailers and/or manufacturers may have increased or reduced their margins and/or that other important factors have played a role in price changes (e.g. inflation, cost of raw materials, compensation of larger taxes on other food items, heterogeneity in the timing and the way retailers pass the tax onto price).[26] In many cases, it is difficult to disentangle precisely tax effects from other contextual factors. However, recent evidence from France, Mexico and Berkeley (California) strongly suggests that new SSB taxes levied on the manufacturers have been partially or

---

[22]For similar reasons to the aforementioned study, San Francisco was used as a control city because of its geographical proximity, and cultural and political similarities with the city of Berkeley. San Francisco was also selected because a soda tax proposal had been included in its November 2014 ballot although it did not gather the required two-thirds of the population vote for adoption (i.e. 56 %). Consequently, the researchers assume that the consumers of the two cities are more likely to have been exposed to similar advocacy efforts in the period preceding the vote. They acknowledge that using only one control city is a methodological limitation.

[23]Controlling for the general time trend in prices for each product.

[24]Across all the models used, average results for the 3 sizes of Coke and Pepsi (i.e. 20-oz bottle, 2-litre bottle and 12-pack of cans) estimate a 21.7 % pass-through of the tax on to retail prices.

[25]Ecorys defines the direct effect of the tax as "the monetary impact of the tax on prices (tax base times tax rate), thereby excluding any additional costs of the tax (administrative costs, product reformulation costs), which might also have an impact on prices" (Ecorys 2014a, p. 23).

[26]This observation compares well to the tobacco case: on most tobacco markets, taxes tend to be overshifted (IARC 2011, p. 350).

fully shifted onto prices. Long-term data monitoring from other large-scale soda taxes currently implemented across the world will bring stronger evidence.

5.3 Additional Proposals to Secure the Pass-Through

Whether the pass-through of excise taxes to SSB prices proved incomplete, complementary strategies have been suggested in the literature. This includes, for example, the enactment of "minimum price laws" calculated per volume of the taxed beverage. It requires "the addition of a minimum percentage mark-up to the base price" of the beverage on store shelves (as already done in some places on tobacco and alcohol) (Pomeranz 2014, p. e13). The applicability of such a coercive policy on SSB will be considered in Chap. 12 of the book.

Despite minimum pricing, another issue that may attenuate tax effects on price and demand is the frequent use of special and promotional offers by retailers (see Sect. 2.3). Such price-discounting strategies have proved to be efficient for sales, especially in individuals who tend to be more sensitive to price (e.g. youth) and/or who may actively look for such offers to circumvent price hikes and keep their consumption unchanged (Pomeranz 2014). Such behaviours have been well documented in the case of tobacco. This is why, beyond taxation and minimum pricing, complementary policies such as prohibitions on coupons and discounting have also been used as part of tobacco control strategies. Potential applications to the case of SSB may be explored (IARC 2011; Pomeranz 2012, 2014).

Key Messages

- Overall, there is convincing evidence that the implementation of a significant SSB tax levied on the manufacturers is generally followed by price increases in the short term, sometimes beyond or below the increase theoretically generated by the tax.
- In many cases, it is difficult to disentangle precisely tax effects from other contextual factors. However, evidence from France, Mexico and Berkeley (California) strongly suggests the existence of tax-specific effects on SSB prices.

References

Berardi, N., Sevestre, P., Tepaut, M., & Vigneron, A. (2012). *The impact of a "Soda Tax" on prices: Evidence from French micro data* (Banque de France Working Paper No. 415) (p. 29). Banque de France Eurosystème. Direction Générale des Études et des Relations internationales. http://ssrn.com/abstract=2192470. Accessed Feb 24, 2016

Bergman, U. M., & Hansen, N. L. (2013). *Are excise taxes on beverages fully passed through to prices? The Danish evidence* (Working paper) (p. 33). http://web.econ.ku.dk/okombe/MBNLH.pdf. Accessed Jan 22, 2016

Bonnet, C., & Réquillart, V. (2011). Does the EU sugar policy reform increase added sugar consumption? An empirical evidence on the soft drink market. *Health Economics, 20*(9), 1012–1024. doi:10.1002/hec.1721.

Bonnet, C., & Réquillart, V. (2012). Les effets de la réforme de la politique sucrière et des politiques de taxation sur le marché des boissons sucrées. *Cahiers de Nutrition et de Diététique, 47*(1), 35–41. doi:10.1016/j.cnd.2011.12.003.

Case, B. (2014, March 27). Soft-drink thirst quenched by Pena Nieto tax: Corporate Mexico. *Bloomberg business*. http://www.bloomberg.com/news/articles/2014-03-27/soft-drink-thirst-quenched-by-pena-nieto-tax-corporate-mexico. Accessed Feb 18, 2015

Cash, S., & Lacanilao, R. (2007). Taxing food to improve health: Economic evidence and arguments. *Agricultural and Resource Economics Review, 36*(2), 174

Cawley, J., & Frisvold, D. (2015). *The incidence of taxes on sugar-sweetened beverages: The case of Berkeley, California* (No. w21465). Cambridge, MA: National Bureau of Economic Research. http://www.nber.org/papers/w21465.pdf. Accessed Feb 25, 2016

City of Berkeley. (2014). Imposing a general tax on the distribution of sugar-sweetened beverage products., Pub. L. No. Measure D § Ballot question.

Ecorys. (2014a). *Food taxes and their impact on competitiveness in the agri-food sector.* Final report (Ref. Ares(2014)2365745—16/07/2014). Rotterdam: Client, DG Enterprise and Industry.

Ecorys. (2014b). *Food taxes and their impact on competitiveness in the agri-food sector.* Annexes to the Main report (No. Ref. Ares(2014)2604304—06/08/2014). Rotterdam: Client, DG Enterprise and Industry.

Falbe, J., Rojas, N., Grummon, A. H., & Madsen, K. A. (2015). Higher retail prices of sugar-sweetened beverages 3 months after implementation of an excise tax in Berkeley, California. *American Journal of Public Health, 105*(11), 2194–2201. doi:10.2105/AJPH.2015.302881.

Faulkner, G. E., Grootendorst, P., Nguyen, V., Andreyeva, T., Arbour-Nicitopoulos, K., Auld, M. C., et al. (2011). Economic instruments for obesity prevention: Results of a scoping review and modified Delphi survey. *International Journal of Behavioral Nutrition and Physical Activity, 8* (1), 109. doi:10.1186/1479-5868-8-109.

Grogger, J. (2015). *Soda taxes and the prices of sodas and other drinks: Evidence from Mexico* (No. w21197). Cambridge, MA: National Bureau of Economic Research. http://www.nber.org/papers/w21197.pdf. Accessed Feb 28, 2016

International Agency for Research on Cancer (IARC) (Ed.). (2011). *IARC handbooks of cancer prevention, tobacco control, volume 14, effectiveness of tax and price policies for tobacco control*. Lyon: IARC.

Pomeranz, J. L. (2012). Advanced policy options to regulate sugar-sweetened beverages to support public health. *Journal of Public Health Policy, 33*(1), 75–88. doi:10.1057/jphp.2011.46.

Pomeranz, J. L. (2014). Sugary beverage tax policy: Lessons learned from tobacco. *American Journal of Public Health, 104*(3), e13–e15. doi:10.2105/AJPH.2013.301800.

Thow, A. M., Quested, C., Juventin, L., Kun, R., Khan, A. N., & Swinburn, B. (2011). Taxing soft drinks in the Pacific: Implementation lessons for improving health. *Health Promotion International, 26*(1), 55–64. doi:10.1093/heapro/daq057.

Wade, L. (2015, July 13). *Mexico's soda tax is working. The US should learn from it. Wired.com.* http://www.wired.com/2015/07/mexicos-soda-tax-working-us-learn/. Accessed Sept 20, 2015

Chapter 6
Effects of Taxation on Sugar-Sweetened Beverage Demand

Abstract Provided a tax on soda/sugar-sweetened beverages (SSBs) is passed onto prices (see Chap. 5), a key issue is to know how the price hike will impact consumer demand. A first drawback would be that soda consumers are not or weakly sensitive to price change. Another drawback would be that other contextual factors counteract price effect on demand (e.g. advertising strategies and meteorological conditions). This chapter reviews evidence from experimental studies, price elasticities estimates from market data and observational studies where soda taxes have been implemented. Overall, the evidence presented confirms that soda consumers are sensitive to price, and that where price hikes followed the enactment of soda taxes, SSB purchases generally decreased. In most cases, it remains difficult to disentangle tax-specific effects from the influence of larger contextual factors. The Mexican case, however, provides stronger evidence. These encouraging trends will have to be confirmed over the long term.

This chapter reviews sugar-sweetened beverage (SSB) taxation effects on consumer demand using various types of evidence. A basic principle of economics is that consumption results from the conjunction of the supply and the demand for a particular good at a set price. According to Whelan and Msefer (1996), the demand is the rate at which consumers want to buy a product and consists of two factors: the taste, which is the desire for a good, and the ability to buy. Both factors are influenced by market price. The demand is an intention, which may translate into behaviours: purchase and consumption. Demand is usually measured by a survey, whereas consumption may be measured by statistics pertaining to sales or purchases. In many studies quoted in this section, the intensity of the demand is assessed by proxies, i.e. the volume of sale or consumption.

First, we mention results from experimental studies. Such studies give an insight on how consumers may react to price hikes on a short-term basis in a controlled setting, but weakly predict what may happen in real-life conditions. Secondly, we actually discuss the factors that may impair tax effects on purchases in the "real world". Thirdly, we present what market data usually tell us in terms of SSB price

Y. Le Bodo et al., *Taxing Soda for Public Health*,
DOI 10.1007/978-3-319-33648-0_6

elasticity of demand (see Box 6.1). Finally, we mention preliminary evidence from taxes implemented across the world.

6.1 Findings from Experimental Studies

This section summarizes experimental evidence from a recent targeted review (Epstein et al. 2012) and from other trials analysing changes in SSB purchases following small-scale pricing interventions in controlled conditions:

- At the level of a hospital cafeteria in the USA (n = 154 customers), a controlled trial imposing a 35 % price increase on SSB showed a 26 % reduction in consumption, and an additional 18 % reduction when the pricing strategy was combined with an educational campaign. In this trial, a partial SSB substitution by non-calorically sweetened drinks has been observed (Block et al. 2010).
- Another study, involving 108 Taiwanese students aged 18–22 years, looked at the effects of price changes combined with the reading of an article containing health claims on the purchase of beverages deemed healthy (e.g. unsweetened green tea) and less healthy (e.g. sweetened iced-tea) in a laboratory randomized controlled trial. They observed higher purchases of healthy beverages when the price of less-healthy beverages was increased. This effect was greater when the price change was combined with the awareness-raising strategy (Yang and Chiou 2010).
- French et al. (2010) studied the effects of food and beverage price changes and availability in 33 vending machines at 4 bus garages (including two control sites) in a US metropolitan area over an 18-month period. They observed that by increasing the availability of healthy items to 50 % of the total offer and by reducing the price of these items by 31 % (on average), sales increased by 10–42 %. In the particular case of cold beverages, healthy choices reached 54 % of total purchases following a 10 % price decrease in intervention sites (vs. 40 % of total purchases in the control sites).
- At supermarket level, in the USA, a controlled trial (n = 113 households) imposing a 10 % tax increase on SSBs showed a decrease in consumption at one month, which was not significant anymore after 3 months and associated with some substitution towards alcoholic beverages (Wansink et al. 2012).
- Another controlled field experiment in a corner store opened on purpose at a large US hospital (n = 2151 transactions over three months) showed that a 30 % tax on less-healthy beverage items generated a 6-percentage-point increase in the purchase of healthier beverages. No additional benefits were observed when the tax and its rationale were made visible on the package or when the tax was combined with a warning label. When considering all foods and beverages purchased by customers within a transaction, it appeared that all conditions (labelling and/or taxation) were significantly associated with fewer calories and sugars purchased (Elbel et al. 2013).

- Finally, in the Netherlands, a randomized controlled trial (n = 102 participants) analysing the effects of a tax increase from 6 to 19 % on SSBs during a single visit at a computer-based virtual supermarket showed a significant decrease in SSB purchases without any significant changes on the purchase of other items (Waterlander et al. 2014).

Overall, these experimental studies tend to confirm that higher SSB prices can reduce SSB purchases. There is also evidence, in some cases, that consumers may become more sensitive to price when the product has a stigmatizing warning label attached to it. Although experimental studies provide encouraging evidence on consumers' short- to medium-term reactions in closed settings, their heterogeneous designs and limited external validity require complementary evidence at the population level. Designing experimental or quasi-experimental studies aimed to better understand the dose–response curve and the shape of the curve (linear, curvilinear or asymptotic) may also help characterize consumers' behaviours and identify potential threshold/saturation effects in relation to SSB price increases.

6.2 Factors that May Impair Tax Effects on Demand in the "Real World"

The aforementioned experimentations are encouraging in terms of SSB price effect on purchases. But in the "real world", larger influences may counteract SSB taxation effect on demand. First, manufacturers may use a diversity of marketing strategies to support sales despite price hikes (e.g. product innovation, upscaled distribution and stronger TV advertising) (Berning 2011). Secondly, beyond price and marketing, larger contextual factors may significantly influence SSB consumption trends (e.g. demographics, purchasing power and weather) (see Fig. 6.1).

To better understand some of these effects, Okrent and MacEwan (2014) have estimated a US non-alcoholic beverage demand system[1] accounting for changes in price, advertising expenditures,[2] income and demographics. Between 1999 and 2008, their results confirm that changes in price and advertising expenditures jointly affected demand in non-alcoholic beverages: for instance, it appears that "changes in advertising expenditures […] reinforced the negative effect of price on demand for milk, coffee and tea", whereas it slightly alleviated the negative effects of price and demographics on demand for regular soda (Okrent and MacEwan 2014, p. 19). For "newer products" categories,[3] their analysis shows that changes in demographic characteristics of the population may have played a larger role than advertising and pricing. Between 2008 and 2010 (period of economic downturn), the demand

[1]Using monthly observations from Nielsen Homescan and Nielsen Fresh Foods panels.

[2]Excluding expenditures on "product placement, movie and video promotions, athletic sponsorships, or celebrity fees" (Okrent and MacEwan 2014, p. 20).

[3]Including bottled water, isotonic and energy drinks, and soya/rice/almond drinks.

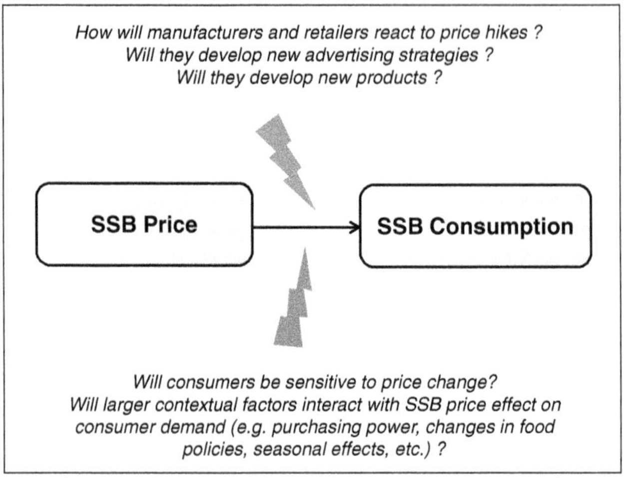

Fig. 6.1 Uncertainties about the propensity of SSB taxation effects on SSB price to impact positively SSB consumption

system shows a different pattern. The estimations rather suggest that the reduction in demand for almost all non-alcoholic beverages was mainly driven by an income-led decrease in households' expenditures. Interestingly, this overall decrease in demand occurred despite the fact that price decrease and advertising efforts were correlated with demand bursts.[4] Such results demonstrate how SSB taxation effects may be associated with or determined, amplified, alleviated or hidden by many other contextual factors. Unfortunately, the inclusion of such variables in SSB tax simulations remains scarce and complex (Berning 2011).

Beyond price and advertising variations, SSB purchases may be influenced by the development of new products. For instance, in France, in 2014, SSB manufacturers have launched new recipes and new flavours to counteract decreasing sales partly due to the soda tax enacted in 2012 (Fédération nationale des boissons [FNB] 2014). Product reformulation (e.g. using cheaper ingredients) may also be used as a way to keep prices unchanged despite taxation. In Mexico, for example, one of the biggest bottlers considered substituting high fructose corn syrup (cheaper) to sugarcane in its recipe, in order to counterbalance potential profit losses related to the new soda tax enacted in 2013. If such a major market change was confirmed, it may significantly alter consumers' taste preferences and, consequently, consumer demand (Flannery 2013).

Altogether, these potential market adaptations are scarcely documented but modelling works by Dharmasena et al. (2014) indicate that a SSB tax may have

[4]This situation was observed for milk, dairy, juices, fruit drinks, iced-tea drinks, regular soda and bottled water, coffee and tea, isotonic and energy drinks. The authors consider that pricing strategies and advertising efforts were used as a way to "abate declining demand for these goods" (Okrent and MacEwan 2014, p. 19).

greater undesirable effects on the "supply side" than on the "demand side". Therefore, not accounting for market supply adaptations could lead to an overestimation of positive SSB tax effects on demand, energy intake and health outcomes. It seems essential to document simultaneously consumers' and firms' reactions to potential SSB taxes in the real world in order to fully appreciate tax impacts "at the market equilibrium" (Réquillart and Soler 2014, p. 390).

Market distortion phenomena driven by consumer demand may also occur, such as cross-border shopping or smuggling practices. Such effects have been documented following taxation of some goods at a higher rate (e.g. tobacco and alcohol). In the case of a specific tax on SSBs, such effects may be less likely to occur because of the low SSB retail price and tax rates generally considered. There is little evidence available on the topic, with the exception of Bergman and Hansen's analysis (2013) of successive tax increases (1998 and 2001)[5] and a tax cut (2003)[6] on soft drinks in Denmark (see also Sect. 5.2). Since soft drinks were known to be cheaper in Germany, observing a lower tax pass-through in Danish outlets located near the German border would have been a clue of cross-border shopping. But their analysis did not confirm this hypothesis (Bergman and Hansen 2013).

Finally, other uncontrollable contextual variables may influence SSB supply and demand and, consequently, counteract or reinforce SSB tax effects. These include variations in purchasing power and economic growth, changes in food policies (e.g. sugar policy reforms), the implementation of other public health interventions (e.g. mass communication campaigns), major special events (e.g. soccer world cup) and weather conditions (Bonnet and Réquillart 2011, 2012; FNB 2014; WCRF 2015). For example, modelling works on large sets of SSB price and demand data over a 20-year period in Ireland shows that, all else being equal, the weather-induced variation in SSB purchases between "good summers" and "bad summers" could reach 2 % (Bahl et al. 2003). Some of these effects may barely be anticipated but should not be ignored, especially for evaluation purposes (see Chap. 14).

6.3 "Real-World" Price Elasticity of SSB Demand

Considering the aforementioned uncertainties (see Sect. 6.2), under the assumption that the tax is fully shifted to prices,[7] what does population-based evidence tell us about the chance that price hikes concretely impact SSB demand? In other terms, will SSB consumers be sensitive to price hikes?

---

[5]From 0.8 DKK (eq. 0.17 CAD)/l to 1 DKK (0.21 CAD) in 1998 and to 1.65 DKK (eq. 0.34 CAD) in 2001.

[6]Back to 1.15 DKK (eq. 0.24 CAD)/l in 2003.

[7]This is a reasonable assumption (see Chap. 5).

Box 6.1: Price elasticity terms of reference, as defined by Nghiem et al. (2013, p. 1956)

Price elasticity of demand: it measures "how much demand for a quantity of good changes in response to changes in prices". Price elasticity is "unit-free". It corresponds to "the percentage change in demand for quantity with respect to a 1 % change in a single price [...], holding all else the same".

Own-price elasticity of demand: it measures "how demand for a quantity of a good responds to a change in [its own] price". Own-price elasticities of demand are "normally negative (i.e. demand decreases as price increases)". Additionally, "if the price elasticity of demand is greater than 1 in absolute value (i.e., the percentage change in the demand for a quantity of a good is greater than the percentage change in its price), then demand is said to be price elastic [...]". Conversely, "if the price elasticity of demand is smaller than 1 in absolute value, demand is said to be price inelastic". For instance, when the demand for a quantity of SSB decreases by 1.5 % when its own price increases by 1 %, then the own-price elasticity of SSB demand is -1.5 and SSB demand is considered to be price elastic.

Cross-price elasticity of demand: it measures how the demand for a quantity of a good (A) responds to a change in the price of another good (B). On the one hand, "if the cross-price elasticity of demand is positive, then good A and good B are substitutes, [meaning that] good B is consumed instead of good A". On the other hand, "if the cross-price elasticity of demand is negative, then good A and good B are complements, [meaning that] good B is consumed together with good A". For instance, considering SSB (good A) and 100 % fruit juice (good B), a cross-price elasticity of demand of 1.2 would indicate that when the price of 100 % fruit juice increases by 1 %, the demand for SSB increases by 1.2 %. In such a case, 100 % fruit juice and SSB are considered to be substitutes.

In usual market conditions, there is strong evidence that SSB demand is particularly elastic to price change (see Box 6.1), in comparison with other food items such as eggs, cheese or oil. This has been well established in the USA (Andreyeva et al. 2010) and subsequently confirmed via empirical market-based data from various countries such as the USA, France, Mexico, Brazil, New Zealand and India. In average, price elasticity of SSB demand fluctuates between −0.8 and −1.3 depending on the model, the SSB category and the country considered. These figures indicate that SSB demand decreases by 0.8 to 1.3 % when its own price increases by 1 % (Andreyeva et al. 2010; Powell et al. 2013; Eyles et al. 2012; Briggs et al. 2013; Cabrera Escobar et al. 2013; Finkelstein et al. 2013; Ni Mhurchu et al. 2013; Basu et al. 2014; Colchero et al. 2015). The elasticity could even prove stronger when the model accounts for drink subcategories, since consumers may switch from a beverage (e.g. regular soda) to another (e.g. diet soda) when their pricing is different. For example, according to Nghiem et al. (2013), the price elasticity of demand for regular soft drinks and diet soft drinks has sometimes proved to climb up to −2.26 and −1.27 (respectively) when drinks subcategories were considered separately.

In their review, Thow et al. (2014, p. 561) analysed the results of 16 simulation studies using price elasticity estimates to investigate potential effects of SSB taxes at rates varying between 5 and 30 %. They observe that "all showed a reduction in consumption of these beverages, ranging from 5 to 48 %, demonstrating overall a response in consumption that was proportional to the taxes applied". Although there is no particular reason to believe that Canada makes exception, price elasticity estimates can be country specific (Nghiem et al. 2013). Thus, it seems important to explore market data and estimate own- and cross-price elasticities of SSB demand in Canada.

6.4 Preliminary Evidence from Taxes Implemented Across the World

Beyond potential effectiveness predicted from price elasticity estimates (see Sect. 6.3), have SSB purchases and consumption concretely decreased in regions and countries that have already enacted a soda tax (see Table 1.1)? On the basis of available evidence, it seems to be the case, Hungary possibly being an exception. Here are some key pieces of evidence:

- Ireland provided a large-scale natural experiment since several changes in the soft drinks excise and value added tax rates occurred over the last decades. Bahl et al. (2003) studied this empirical case over a 21-year period (1975–96) in order to construct a model estimating changes in soft drink consumption in response to changes in the tax rate, accounting for population growth, seasonal effects and price changes in other goods. From this model, they estimate that "a 10 % reduction in the price of soft drinks would lead to an 11 % increase in the number of litres consumed, if all else were held constant" (Bahl et al. 2003, p. 521). This study suggests that, in the "real world" and in a long term, changing SSB tax rates can significantly affect SSB demand.
- In the Samoan Islands, where additional excise taxes on soft drinks have been regularly increased since 1984, survey data quoted by Thow et al. (2011, p. 58) suggest that the number of servings of soda consumed by Samoans "decreased slightly between 1991 and 2003, from around 2.5 to just over two servings per week". Unfortunately, a deeper evaluation of tax effects on demand does not seem to have been conducted.
- In the USA, when matching information on soft drink sales and excise tax applied in numerous states between 1989 and 2006 and data from the National Health and Nutrition Examination Survey (NHANES), researchers estimate that the tax generally applied on soft drink (at a rate of 4–5 %) has been associated with a "moderate reduction in soft drink consumption by children and adolescents" (Fletcher et al. 2010, p. 967). However, they consider the magnitude of

this reduction to be "somewhat modest" (equal to 6 kcal from soda per percentage point of increase in the soft drink tax rate), representing about 5 % of the average daily calorie intakes from those drinks (according to NHANES statistics) (Fletcher et al. 2010, p. 972).

- In Denmark, Ecorys data analysis indicates that the price increase and decrease of regular colas following the introduction and repeal of taxes since 1999 have generally (but not always) been followed by corresponding reductions and increases in demand. Additionally, following the 50 % tax reduction in 2013, a 7.0 and a 4.9 % increase in demand have been observed in regular and low-calorie colas, respectively. We previously mentioned that the price of fruit drinks seems to have evolved independently of tax changes over the last years. Correspondingly, no changes in the intake of these beverages could be explicitly related to tax changes (Ecorys 2014a, p. 36).
- In Finland, excise taxation on SSBs regularly increased in 2011, 2012 and 2014 (see Table 1.1). According to Ecorys data analysis, the yearly cumulated price increase of soft drinks between 2011 and 2013 (combined) reached 17.3 % and was accompanied by a yearly cumulated 4.7 % decrease in consumption. Although soft drinks' demand had been declining since 2007, the decline seemed to intensify following the tax increase in 2011 (Ecorys 2014a, p. 34). The trend is corroborated by other sources: between the 1st trimester of 2013 and the 1st trimester of 2014, statistics compiled by members of the Federation of the Brewing and Soft Drinks Industry indicate a 4.7 % decline in soft drinks sales. The Federation considered that the tax has "accelerated the decline" (Weston 2014, §5).
- In France, the National Beverages Federation (FNB) reports a 2.2 % downward trend in non-alcoholic soft drinks sales volume in 2013, and clearly acknowledges that the tax implemented since 1 January 2012 (≈6 % of average purchase price) is partly responsible for this reduction (FNB 2014). Ecorys data analysis tends to corroborate this observation. In 2012–2013 (combined), it shows that the yearly cumulated price increase for regular colas was 8.1 % and was accompanied by a yearly cumulated 6.7 % decrease in consumption. For low-calorie cola (included in the tax base), the yearly cumulated price increase of 10.6 % was accompanied by a yearly cumulated 6.1 % decrease in consumption. For these two items, the introduction of the tax strikingly coincides with the interruption of a long-term upward trend in demand. For fruit drinks, the yearly-cumulated price increase of 9.2 % was accompanied by a relatively lower decrease in consumption (3.2 %). Additionally, as for the pre-existing upward trend in prices, the demand in those drinks was already declining before the tax introduction (Ecorys 2014a, p. 34, b, p. 135).[8] Stakeholders interviewed in the Ecorys survey mentioned that "meteorological conditions" as well as the

[8]These observations are also congruent with the impact evaluation of the French tax on the price of non-alcoholic beverages during the 6 months following its enactment (see Sect. 5.2). This evaluation suggests that the tax has been overshifted to the price of sodas and incompletely shifted to the price of fruit drinks and flavoured waters (Berardi et al. 2012).

"price of raw materials" might also have impacted SSB consumption (Ecorys 2014b, p. 209).

- In Hungary, a tax on SSBs was newly introduced in 2011 and increased in 2012 (see Table 1.1). According to Ecorys data analysis, the yearly cumulated price increase of colas between 2011 and 2013 (combined) reached 5.3 % and was accompanied by a yearly cumulated 16.2 % decrease in consumption. However, this decline seems to have started prior to the tax introduction. It may have accelerated afterwards as a result of the tax. As regards fruit drinks, we previously mentioned that price changes were similar prior and after the introduction of the tax in 2011. Consequently, it is difficult to correlate the yearly cumulated price increase of 2 % between 2011 and 2013 (combined) and the corresponding 8.4 % decrease in consumption (Ecorys 2014a, p. 34).
- In Mexico, we previously mentioned strong evidence that the SSB tax implemented in January 2014 ($\approx$9 % of soda price) had been overshifted to the price of regular carbonated soft drinks and partially shifted to the price of other SSBs in the following months. Conversely, the price evolution of most untaxed beverages did not differ significantly from pre-existing trends (with the exception of diet drinks) (Grogger 2015). As regards demand, a few months after the soda tax took effect, a sales drop of nearly 3 % has been described by some of the major soda bottlers (Phillips 2014). This trend has been preliminary confirmed by the Mexican Institute of Public Health and the Carolina Population Center, who analysed beverage intakes from a large panel of consumers living in cities of at least 30,000 inhabitants. Preliminary results had indicated "approximately a 10 % decline in purchases of taxed beverages in the 1st quarter of 2014 in comparison with the 1st quarter of 2013" (Instituto Nacional de Salud Pública [INSP] 2014, §2). An observational study later published by authors from the same organizations is based on a panel of more than 6000 households living in Mexican cities counting more than 50,000 inhabitants[9] (Colchero et al. 2016). The authors used SSB purchase data from 2012 to 2014 to compare the after-tax evolution of purchases of SSBs and other drinks to the evolution of these purchases if a tax had not been enacted (on the basis of pre-existing trends). Their empiric model accounts for seasonal trends as well as socio-demographic and other contextual factors across geographic areas (i.e. unemployment rate and minimum salary). It also provides a good temporal resolution (monthly data across 36 months). Their results show that SSB taxation is associated with per capita purchases of taxed beverages that are 6 % lower than what could have been expected in 2014, which the authors translate into an average decrease of seven 600 ml bottles "per average urban Mexican" (Colchero et al. 2016, p. 4). This decrease was mainly driven by a decrease in purchases of non-carbonated SSBs. The analysis also leads to encouraging

[9]Data from Nielsen Mexico's Consumer Panel Services, on the basis of a sample deemed representative of "63 % of the Mexican population and 75 % of the food and beverages expenditures in 2014" (Colchero et al. 2016, p. 2).

results in terms of substitutions (higher purchases of untaxed beverages, mainly bottled plain water) and equity (larger SSB reduction in households with a lower socioeconomic status) (Colchero et al. 2016). These pioneering results show strong evidence of the positive effects of a moderate SSB tax on purchases at a large scale and in the short term. Similar evaluations would be most informative in other countries where such taxes have been enacted.

- Finally, in Berkeley (California), we previously mentioned that the SSB tax implemented in March 2015 had been followed by a significant but incomplete shift to SSB prices (Falbe et al. 2015; Cawley and Frisvold 2015). Preliminary analyses also suggest that this price change was accompanied by a significant reduction in SSB consumption in the year following the implementation of the tax (Obesity Week 2015).

Overall, these observations and results strongly suggest that, where SSB taxation generated price hikes, SSB demand has been negatively affected. In several cases (e.g. in France, Denmark, Finland and Hungary), the observations made from market data without proper evaluation designs make difficult to disentangle tax-specific effects from other factors. The Mexican case, however, provides strong evidence of the positive effects of a moderate SSB tax on purchases. These encouraging trends will have to be confirmed over the long term.

Key messages

- Experimental studies in controlled conditions indicate that SSB consumers are sensitive to price change and that a tax has the potential to favour SSB substitution towards healthier beverages.
- In usual market conditions, data analyses from several countries estimate that a 1 % SSB price increase is generally accompanied by a 0.8–1.3 % decrease in SSB demand. Exploring Canadian data is necessary.
- In places where a 10–20 % price hike followed the enactment of soda taxes, SSB demand generally decreased. In most cases, it remains difficult to isolate precisely tax-specific effects from other factors.
- An evaluation of the Mexican soda tax, however, provides sound evidence that an average 9 % price increase was associated with SSB purchases that were 6 % lower than if no tax had been imposed.
- These encouraging trends will have to be confirmed over the long term.

References

Andreyeva, T., Long, M. W., & Brownell, K. D. (2010). The impact of food prices on consumption: A systematic review of research on the price elasticity of demand for food. *American Journal of Public Health, 100*(2), 216–222. doi:10.2105/AJPH.2008.151415.

Bahl, R., Bird, R., & Walker, M. B. (2003). The uneasy case against discriminatory excise taxation: Soft drink taxes in Ireland. *Public Finance Review, 31*(5), 510–533. doi:10.1177/1091142103253753.

Basu, S., Vellakkal, S., Agrawal, S., Stuckler, D., Popkin, B., & Ebrahim, S. (2014). Averting obesity and type 2 diabetes in India through sugar-sweetened beverage taxation: An economic-epidemiologic modeling study. *PLoS Medicine, 11*(1), e1001582. doi:10.1371/journal.pmed.1001582.

Berardi, N., Sevestre, P., Tepaut, M., & Vigneron, A. (2012). *The impact of a "soda tax" on prices: Evidence from French micro data* (Banque de France Working Paper No. 415) (p. 29). Banque de France Eurosystème. Direction Générale des Études et des Relations internationales. http://ssrn.com/abstract=2192470. Accessed February 24, 2016.

Bergman, U. M., & Hansen, N. L. (2013). *Are excise taxes on beverages fully passed through to prices? The Danish evidence* (Working paper) (p. 33). http://web.econ.ku.dk/okombe/MBNLH.pdf. Accessed January 22, 2016.

Berning, J. (2011). Evaluating excise taxes: The need to consider brand advertising. *Choices 3rd Quarter 2011, 26*(3).

Block, J. P., Chandra, A., McManus, K. D., & Willett, W. C. (2010). Point-of-purchase price and education intervention to reduce consumption of sugary soft drinks. *American Journal of Public Health, 100*(8), 1427–1433. doi:10.2105/AJPH.2009.175687

Bonnet, C., & Réquillart, V. (2011). Does the EU sugar policy reform increase added sugar consumption? An empirical evidence on the soft drink market. *Health Economics, 20*(9), 1012–1024. doi:10.1002/hec.1721

Bonnet, C., & Réquillart, V. (2012). Les effets de la réforme de la politique sucrière et des politiques de taxation sur le marché des boissons sucrées. *Cahiers de Nutrition et de Diététique, 47*(1), 35–41. doi:10.1016/j.cnd.2011.12.003

Briggs, A. D. M., Mytton, O. T., Kehlbacher, A., Tiffin, R., Rayner, M., & Scarborough, P. (2013). Overall and income specific effect on prevalence of overweight and obesity of 20 % sugar sweetened drink tax in UK: Econometric and comparative risk assessment modelling study. *BMJ, 347*, f6189. doi:10.1136/bmj.f6189

Cabrera Escobar, M. A., Veerman, J., Tollman, S. M., Bertram, M. Y., & Hofman, K. J. (2013). Evidence that a tax on sugar sweetened beverages reduces the obesity rate: A meta-analysis. *BMC Public Health, 13*(1), 1072. doi:10.1186/1471-2458-13-1072

Cawley, J., & Frisvold, D. (2015). *The incidence of taxes on sugar-sweetened beverages: The case of Berkeley, California* (No. w21465). Cambridge, MA: National Bureau of Economic Research. http://www.nber.org/papers/w21465.pdf. Accessed February 25, 2016.

Colchero, M. A., Popkin, B. M., Rivera, J. A., & Ng, S. W. (2016). Beverage purchases from stores in Mexico under the excise tax on sugar sweetened beverages: Observational study. *BMJ*, h6704. doi:10.1136/bmj.h6704

Colchero, M. A., Salgado, J. C., Unar-Munguía, M., Hernández-Ávila, M., & Rivera-Dommarco, J. A. (2015). Price elasticity of the demand for sugar sweetened beverages and soft drinks in Mexico. *Economics & Human Biology, 19*, 129–137. doi:10.1016/j.ehb.2015.08.007

Dharmasena, S., Davis, G. C., & Capps, O. (2014). Partial versus general equilibrium calorie and revenue effects associated with a sugar-sweetened beverage tax. *Journal of Agricultural and Resource Economics, 39*(2), 157–173.

Ecorys. (2014a). *Food taxes and their impact on competitiveness in the agri-food sector. Final report* (No. Ref. Ares(2014)2365745—16/07/2014). Rotterdam: Client: DG Enterprise and Industry.

Ecorys. (2014b). *Food taxes and their impact on competitiveness in the agri-food sector. Annexes to the Main report* (No. Ref. Ares(2014)2604304—06/08/2014). Rotterdam: Client: DG Enterprise and Industry.

Elbel, B., Taksler, G. B., Mijanovich, T., Abrams, C. B., & Dixon, L. B. (2013). Promotion of healthy eating through public policy. *American Journal of Preventive Medicine, 45*(1), 49–55. doi:10.1016/j.amepre.2013.02.023

Epstein, L. H., Jankowiak, N., Nederkoorn, C., Raynor, H. A., French, S. A., & Finkelstein, E. (2012). Experimental research on the relation between food price changes and food-purchasing patterns: A targeted review. *American Journal of Clinical Nutrition, 95*(4), 789–809. doi:10.3945/ajcn.111.024380

Eyles, H., Ni Mhurchu, C., Nghiem, N., & Blakely, T. (2012). Food pricing strategies, population diets, and non-communicable disease: A systematic review of simulation studies. *PLoS Medicine, 9*(12), e1001353. doi:10.1371/journal.pmed.1001353

Falbe, J., Rojas, N., Grummon, A. H., & Madsen, K. A. (2015). Higher retail prices of sugar-sweetened beverages 3 months after implementation of an excise tax in Berkeley, California. *American Journal of Public Health, 105*(11), 2194–2201. doi:10.2105/AJPH.2015.302881

Fédération Nationale des Boissons (FNB). (2014). Les BRSA et la période estivale, une belle histoire ! *E-Mag. L'actualité des entrepositaires grossistes en boissons, 26*(Édition estivale). http://www.fnb-info.fr/fr/Entrez-dans-la-tendance-620.html?mbID=2668. Accessed February 25, 2016.

Finkelstein, E. A., Zhen, C., Bilger, M., Nonnemaker, J., Farooqui, A. M., & Todd, J. E. (2013). Implications of a sugar-sweetened beverage (SSB) tax when substitutions to non-beverage items are considered. *Journal of Health Economics, 32*(1), 219–239. doi:10.1016/j.jhealeco.2012.10.005

Flannery, N. P. (2013, November 7). Mexico may make coca-cola give up cane sugar. *Forbes.* http://www.forbes.com/sites/nathanielparishflannery/2013/11/07/in-response-to-new-soda-tax-coca-cola-bottler-considers-switch-from-sugar-to-high-fructose-corn-syrup-in-mexico/#233d63e46ca1. Accessed March 1, 2015.

Fletcher, J. M., Frisvold, D. E., & Tefft, N. (2010). The effects of soft drink taxes on child and adolescent consumption and weight outcomes. *Journal of Public Economics, 94*(11–12), 967–974. doi:10.1016/j.jpubeco.2010.09.005

French, S. A., Hannan, P. J., Harnack, L. J., Mitchell, N. R., Toomey, T. L., & Gerlach, A. (2010). Pricing and availability intervention in vending machines at four bus garages. *Journal of Occupational and Environmental Medicine/American College of Occupational and Environmental Medicine, 52*(Suppl 1), S29–S33. doi:10.1097/JOM.0b013e3181c5c476

Grogger, J. (2015). *Soda taxes and the prices of sodas and other drinks: Evidence from Mexico* (No. w21197). Cambridge, MA: National Bureau of Economic Research. http://www.nber.org/papers/w21197.pdf. Accessed February 28, 2016.

Instituto Nacional de Salud Pública (INSP). (2014). Resultados preliminares sobre los efectos del impuesto de un peso a bebidas azucaradas en México. www.insp.mx/epppo/blog/preliminares-bebidas-azucaradas.html. Accessed February 18, 2015.

Nghiem, N., Wilson, N., Genç, M., & Blakely, T. (2013). Understanding price elasticities to inform public health research and intervention studies: Key issues. *American Journal of Public Health, 103*(11), 1954–1961. doi:10.2105/AJPH.2013.301337

Ni Mhurchu, C., Eyles, H., Schilling, C., Yang, Q., Kaye-Blake, W., Genç, M., & Blakely, T. (2013). Food prices and consumer demand: Differences across income levels and ethnic groups. *PLoS ONE, 8*(10), e75934. doi:10.1371/journal.pone.0075934

Obesity Week. (2015, November 6). California bubbly: The consequences and public debate around soda taxes. In *Symposium of oral presentations*, Track 6, Los Angeles, California.

Okrent, A. M., & MacEwan, J. P. (2014). The effects of prices, advertising, expenditures, and demographics on demand for nonalcoholic beverages. *Agricultural and Resource Economics Review*, *43*(1), Prepublication copy.

Phillips, M. (2014, July 24). Mexico's completely sensible anti-soda tax is working. *Quartz*. http://qz.com/239478/mexicos-completely-sensible-anti-soda-tax-is-working. Accessed February 18, 2015.

Powell, L. M., Chriqui, J. F., Khan, T., Wada, R., & Chaloupka, F. J. (2013). Assessing the potential effectiveness of food and beverage taxes and subsidies for improving public health: A systematic review of prices, demand and body weight outcomes: Review of prices, demand and weight. *Obesity Reviews, 14*(2), 110–128. doi:10.1111/obr.12002

Réquillart, V., & Soler, L.-G. (2014). Is the reduction of chronic diseases related to food consumption in the hands of the food industry? *European Review of Agricultural Economics, 41*(3), 375–403. doi:10.1093/erae/jbu010

Thow, A. M., Downs, S., & Jan, S. (2014). A systematic review of the effectiveness of food taxes and subsidies to improve diets: Understanding the recent evidence. *Nutrition Reviews, 72*(9), 551–565. doi:10.1111/nure.12123

Thow, A. M., Quested, C., Juventin, L., Kun, R., Khan, A. N., & Swinburn, B. (2011). Taxing soft drinks in the Pacific: Implementation lessons for improving health. *Health Promotion International, 26*(1), 55–64. doi:10.1093/heapro/daq057

Wansink, B., Hanks, D., Just, D. R., Cawley, J., Sobal, J., Wethington, E., et al. (2012). From coke to coors: A field study of a sugar-sweetened beverage tax and its unintended consequences. *SSRN Electronic Journal,*. doi:10.2139/ssrn.2079840

Waterlander, W. E., Ni Mhurchu, C., & Steenhuis, I. H. M. (2014). Effects of a price increase on purchases of sugar sweetened beverages. Results from a randomized controlled trial. *Appetite, 78*, 32–39. doi:10.1016/j.appet.2014.03.012

Weston, S. (2014, June 2). Tax increases affect drinks sales in Finland. *FoodBev.com.* http://www.foodbev.com/news/tax-increases-affect-drinks-sales-in-fin/#.VTAntWTBwXB. Accessed March 1, 2015.

Whelan, J., & Msefer, K. (1996). *Economic supply and demand. Prepared for the MIT system dynamics in education project under the supervision of Professor Jay W. Forrester* (No. D-4388). MIT. http://ocw.mit.edu/courses/sloan-school-of-management/15-988-system-dynamics-self-study-fall-1998-spring-1999/readings/economics.pdf. Accessed February 11, 2016.

World Cancer Research Fund (WCRF). (2015). Curbing global sugar consumption: Effective food policy actions to help promote healthy diets & tackle obesity. London: World Cancer Research Fund International. http://www.wcrf.org/sites/default/files/Curbing-Global-Sugar-Consumption.pdf. Accessed February 28 2016

Yang, C.-C., & Chiou, W.-B. (2010). Substitution of healthy for unhealthy beverages among college students. A health-concerns and behavioral-economics perspective. *Appetite, 54*(3), 512–516. doi:10.1016/j.appet.2010.02.004

Chapter 7
Effects of Sugar-Sweetened Beverage Taxation on Energy Intakes and Population Health

Abstract Provided a tax on sugar-sweetened beverages (SSBs) is effectively passed onto prices and significantly decreases SSB demand (see Chaps. 5 and 6), a question frequently raised is to know whether, overall, this reduction can have a meaningful impact on energy balance and population health. In this chapter, reviewed evidence suggests that a lower SSB consumption may be partially compensated by a higher consumption of other caloric foods and drinks (e.g. untaxed fruit juice and snack foods). Yet, studies simulating a 20 % price increase on SSBs accounting for substitution effects generally predict a positive—although modest—impact on energy balance. These studies are based on fragile assumptions. Substitution concerns should be further explored from soda taxes actually implemented across the world.

On the basis of the evidence presented in Chaps. 5 and 6, it seems reasonable to expect that a significant tax on sugar-sweetened beverages (SSBs) would be passed onto prices, and that a 10 to 20 % SSB price increase could lead to a decrease in SSB demand although uncertainties persist regarding the magnitude and sustainability of these effects. In this chapter, we will review evidence on the propensity of these SSB taxation effects on prices and demand to impact positively energy balance and population health (see Fig. 7.1).

7.1 Overview of Substitution Concerns

Provided SSB demand decreases as a result of price hikes, substitution effects may neutralize positive effects on energy balance. In other terms, consumers may totally or partially compensate for their lower SSB consumption by an increase in the consumption of other caloric foods and/or beverages. Cross-price elasticity estimates from the USA and Mexico actually indicate that 100 % fruit juice or milk can be substitutes for SSBs; that is, SSB price increases could cause an increase in the demand for these beverages (Cabrera-Escobar et al. 2013; Colchero et al. 2015).

Y. Le Bodo et al., *Taxing Soda for Public Health*,
DOI 10.1007/978-3-319-33648-0_7

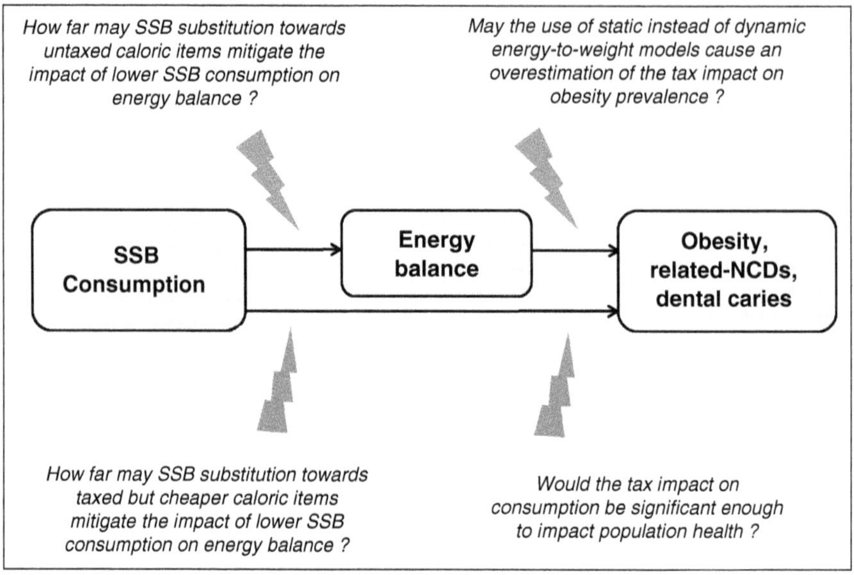

Fig. 7.1 Uncertainties about the propensity of SSB taxation effects on SSB consumption to impact positively energy balance and population health

From a nutritional perspective, it may be seen as an improvement (see Sect. 2.5) but the drawback is that depending on the extent of these substitution effects, the tax-induced reduction in SSB calories may be partially or fully compensated. One could tend to think that exempting non-calorically sweetened beverages (NCSBs) from the tax may instead encourage the consumption of low- or 0-calorie substitutes. However, cross-price elasticity data indicate that NCSBs sometimes appear to be either SSB substitutes or complements,[1] depending on the country (UK, USA and Mexico) or the level of intake[2] (Cabrera-Escobar et al. 2013; Briggs et al. 2013). This observation confirms the need to obtain country-specific cross-price elasticity estimates to reflect on the optimal scope of sweetened beverages (caloric or not) to be taxed. Another potential risk is that some consumers substitute their usual SSB brand for a similar but cheaper SSB product (e.g. store brand), letting their SSB consumption unchanged overall. To date, the extent to which this "intra-category" substitution could happen remains weakly documented (Hawkes 2012). Last but not least, one could hope taxing soda may encourage water consumption. Unfortunately, until recently, observational studies usually did not consider the potential substitution for water (Fletcher et al. 2010; Cabrera-Escobar

[1]i.e. Increase in SSB prices tends to cause a decrease in the demand for both SSBs and NCSBs.
[2]For example, according to a British modelling study, diet cola appears to be a substitute to regular cola in high consumers, but a complement to regular soda in moderate consumers (Tiffin et al. 2015, p. 8).

et al. 2013; Finkelstein et al. 2013). Analysing soda cross-price elasticity of demand from a consumer panel representative of the French population, Boizot-Szantai and Étilé (2011) estimate that soda price changes do not significantly influence still and carbonated water purchases. However, a recent observational study on the Mexican SSB taxation case did show that the tax was associated with higher-than-expected purchases of bottled plain water (see Sect. 7.3.1) (Colchero et al. 2016). In order to promote healthy substitutions, Fletcher et al. (2013) suggest a comprehensive approach that may require combining an SSB tax with subsidies on healthier beverages (e.g. bottled water) or an extension of the SSB tax on a broader range of processed sugary foods.

This brief overview of substitution concerns highlights that, in all cases, substitution effects should be carefully monitored and anticipated, as they may significantly impact SSB taxation effects on energy balance. In this regard, lessons from two types of simulation studies and from real-world SSB taxation cases are presented in the forthcoming two sections.

7.2 Results from Simulations

Methods used in simulations are very diverse. For the purpose of this section, we build on Étilé's categorization (2012) to propose a distinction between two main types of studies:

(1) **Individual-oriented studies**: these studies use large-scale consumer panels to estimate price elasticities of demand, predict potential tax effects on energy balance, and then simulate potential weight-related health outcomes at individual level;
(2) **Population-oriented studies**: these studies use large-scale consumer panels to estimate price elasticities (or use estimations from type-1 studies), combine it with data from national health surveys to simulate potential tax effects on SSB consumption and energy balance, and finally use health models to simulate corresponding health outcomes at the population level.

7.2.1 Individual-Oriented Simulations

The first type of study, often published in economics journals, generally uses a comprehensive consumption model based on data collected from large and representative consumer panels. Such database provides disaggregated and objective information both on a broad range of food and beverages purchases (date, quantity, price, category, packaging, point of sale characteristics, etc.) and consumers (age, income, baseline SSB consumption, eating habits, health status, body mass index, etc.). This material often makes possible to build an "almost ideal demand system" accounting for own and cross-price elasticities[3] of all items considered, to predict directly the impact of an SSB tax-induced price increase on the consumption of these items and, finally, to predict changes in overall energy intakes. A distinctive feature of this type of study is the matching of price and demand data for a broad range of items with consumers' individual characteristics. From there, a health model based on epidemiological evidence may be used to predict how variations in SSB consumption and other components of the diet may impact energy balance, body weight[4] and, eventually, other individual health outcomes. Results from two examples of these studies and a review are summarized below:

- A US simulation based on a Nielsen Homescan panel[5] predicts that a tax increasing the price of SSBs by 20 % would result in a decrease in "store-bought" energy of 24.3 kcal per person per day, accounting for potential substitutions/complementarity between SSB and 16 other food and beverage categories (Finkelstein et al. 2013). According to the demand model, most of this decrease in energy intake comes from a lower intake of fruit drinks (−7 kcal), regular soda (−5 kcal), but also salty snacks (−6.3 kcal) and ice cream (−4.8 kcal), which appear to be SSB complements. The model did not reveal substitution concerns with the exception of a slightly higher intake of fruit juices (+2.5 kcal) and canned soup (+1.9 kcal). According to the energy-to-weight model used, the tax would bring about an average individual weight loss of 2.7 kg during the first year and a cumulated weight loss of 4.1 kg in the long term. The authors conclude that these benefits would have to be weighed against potential equity concerns and undesirable effects (Finkelstein et al. 2013).
- In Australia, Sharma et al. (2014) have used a Nielsen Homescan panel of consumers[6] and an energy-to-weight model to predict the effects of a 20 % valoric tax (i.e. based on price) and a 20 cents/litre volumetric tax (i.e. based on volume) on energy intakes and body weight. Their demand calculations account for cross-price elasticities in the demand of 10 types of beverages. The detailed analysis presented for the 20 % valoric tax indicates that it would lead to a

[3]For price elasticity terms of reference, see Chap. 6, Box 6.1.

[4]Some of the "energy-to-weight" models used in SSB taxation simulations (e.g. Smith et al. 2010) rely on a linear static model based on the assumption that "a cumulative energy deficit of 3500 calories is required to lose one pound of body weight" (Lin et al. 2011, p. 338). However, because of the complex and dynamic links between energy intake and body weight, Lin et al. (2011) recommend using a dynamic model accounting for physiological adaptations as body weight changes, in particular because energy expenditures increase (decrease) when body weight increases (decreases). Lin et al.'s comparative analysis based on laboratory data demonstrates that using static instead of dynamic "energy-to-weight" models in SSB taxation simulations can result in a 7-fold overestimation of body weight loss at the end of a 10-year period (Lin et al. 2011, p. 336). Therefore, in this section, we only refer to studies having considered dynamic "energy-to-weight" models.

[5]Weighted according to the US Census 2010.

[6]Sample of demographically representative households from the state of Victoria.

reduced consumption of regular soft drinks (−11.5 %), cordials[7] (−33,2 %) and fruit drink (−25.5 %), whereas the consumption of diet soft drinks would slightly increase (+3.2 %). Overall, such a tax would result in a yearly energy intake reduction of 10678 kJ per person, corresponding approximately to a reduction of 7 kcal per person per day. According to the energy-to-weight model, the valoric and the volumetric taxes would bring about an average individual weight loss of 0.29 kg and 0.41 kg, respectively (Sharma et al. 2014).
- Findings from a recent systematic review of individual-oriented simulations should be highlighted (Eyles et al. 2012). Pooling results from three studies presenting a higher methodological profile, the authors estimate that taxes on carbonated soft drinks may lead to a 0.02 % reduction in energy intake for each 1 % increase in price. Unfortunately, pooled impact on weight and other health outcomes was not feasible.

7.2.2 Population-Oriented Simulations

This second type of study, often published in public health journals, attempts to simulate potential tax effects on the health of a population. To do this, these studies usually combine a beverage consumption model with one or several health model (s). The beverage consumption model uses large-scale consumer panels to estimate price elasticities (or use estimations from type-1 studies, see Sect. 7.2.1) and then combines it with data from national health and nutrition surveys to predict how non-alcoholic beverage consumption could change as a result of an SSB tax-induced price increase. From there, the health model(s) based on epidemiological evidence generally simulate(s) how the predicted variation in non-alcoholic beverage consumption (and eventually other variations in the diet) may impact energy balance, overweight and obesity prevalence, and other population health outcomes (Étilé 2012). Two examples of these studies are described below:

- In the USA, Wang et al. (2012) have used published estimates of price elasticity of demand, data from the National Health and Nutrition Examination Survey (NHANES) and the US Coronary Heart Disease Policy Model to predict the population health impact of a penny-per-ounce (≈34 cents/litre) excise tax on SSBs. Based on the literature, the authors have anticipated that the tax would reduce SSB consumption in adults by 15 % and have considered that 40 % of the reduction in calories from SSBs would be offset by an increased consumption of other caloric drinks (e.g. milk and 100 % fruit juice). All in all, they predict a net reduction in energy intake of about 9 kcal/pers./day. According to their population health model, over 2010–20, they estimate that the tax would "prevent 2.4 million diabetes person-years, 95,000 coronary heart events, 8000

[7]Soft drink concentrates.

strokes, and 26,000 premature deaths, while avoiding more than $17 billion in medical costs" (Wang et al. 2012, p. 199).

- In the UK, Briggs et al. (2013) have estimated price elasticities from the Living Costs and Food Survey, have used data from the National Diet and Nutrition Survey, and relied on a model of dietary changes' impact on chronic diseases and mortality to predict the health impact of a 20 % SSB tax. They estimate that own-price elasticity for SSB is −0.92 and predict that a 20 % tax on SSBs would reduce adult SSB consumption by 15–16 %. Significant substitutes appear to be diet drinks (increased intake between 3.9 and 7.5 %), tea and coffee (+4.1 %), milk (+3.7 %) and fruit juice (+3.1 %). Overall, substitution and complementarity effects led to a mean net reduction in energy intake of 16.7 kJ/person/day (≈4 kcal/pers./day). According to their population health model, the tax "was estimated to reduce the number of obese adults in the UK by 1.3 % […] and the number who are overweight by 0.9 %" (Briggs et al. 2013, p. 1).

7.2.3 *Limitations*

A detailed analysis of the methods used in the aforementioned simulations is beyond the scope of this book, but significant limitations should be highlighted:

First, the characteristics and degree of sophistication of the beverage demand model may affect the robustness of price elasticity estimations and consecutive assumptions as regards tax effects on energy balance and health. We refer the reader to the above-mentioned studies to get a full view of methodological challenges. To give an example, diverging views exist as regards the potential substitution effects towards other food items. In their study, Zhen et al. (2014, pp. 19–22) have used information on 23 packaged foods and beverages and predict that "almost half" of the reduction in energy intake from SSB is offset by increased energy intakes from other foods (in particular frozen dinners, cheese, white bread and candy) when SSB price increases. Consequently, they anticipate much lower weight losses than Finkelstein et al. (2013, results quoted above). Additionally, it should be noted that consumer panels generally do not include all purchases and may only focus on store-bought items. Consequently, accounting for other purchases (e.g. takeaway) may alter the estimations.

Secondly, models assume a linear correlation between price and demand, whereas threshold and saturation effects may occur in the context of a tax. This is an area where further research is needed.

Thirdly, models rarely take into account how the tax may truly impact price and demand in the long term. On the demand side, consumers' lower demand in immediate response to price hike may increase later as they get used to it (as sometimes shown in the case of tobacco taxation). On the supply side, manufacturers may use pricing and advertising strategies as a way to support demand despite taxation.

Finally, as discussed before (see Sects. 5.1 and 6.2), potential interactions with many contextual influences on price and demand are not (and can barely be) taken into account in simulations.

These limitations call for a cautions interpretation of the results of simulation studies. They can help inform decisions and plan evaluation but, per se, cannot support ambitious population health objectives without being exposed to criticism. An important consideration should also be given to observational and evaluation studies focused on SSB taxes recently implemented across the world.

7.3 Preliminary Evidence from Taxes Implemented Across the World

We do not have much evidence on substitution effects associated with "real-world" SSB taxes as these policies have only recently been implemented and suffer from a lack of evaluation. Therefore, it remains difficult to draw firm conclusions. Pieces of evidence from SSB taxes recently implemented across the world (see Table 1.1) are described below. It essentially describes how SSB taxation may have generated substitution towards untaxed beverages and to taxed beverages sold at a lower price (e.g. non-premium SSBs).

7.3.1 Substitution Towards Untaxed Beverages

US observational studies have estimated the impact of state-level SSB sales and excise taxes implemented at a low rate (i.e. ≈5 %) for many years (although not explicitly for health reasons). These studies have only detected a moderate reduction in SSB consumption in children and adolescents and minimal impact on body weight outcomes. Substitution effects towards untaxed high-calorie drinks (whole milk and 100 % fruit juice), the low tax rate practised and the complex aetiology of obesity were presented as potential explanatory factors (Powell et al. 2009; Fletcher et al. 2010; Sturm et al. 2010; Powell et al. 2013).

In France, we previously mentioned that the soda tax (which equally applies to SSBs and NCSBs) had been followed by a decrease in the consumption of taxed carbonated drinks and, to a lower extent, by a decrease in the consumption of other SSBs (Berardi et al. 2012) (see Sect. 6.4). The partial substitution of those beverages by 100 % fruit juices (sometimes richer in calories) may be a drawback. According to Ecorys data analysis, the consumption of 100 % fruit juices increased by 3 % in 2012–2013 (combined). However, this upward trend existed prior to the tax introduction (Ecorys 2014b, p. 128). In order to favour substitution towards lower-calorie options, it has been suggested to maintain a differential tax rate between SSBs and NCSBs (Bonnet and Réquillart 2012).

In Denmark, as previously mentioned, the price increase and decrease of regular colas following the introduction and repeal of taxes since 1999 have generally (but not always) been followed by corresponding reductions and increases in demand (Ecorys 2014a) (see Sect. 6.4). In 2012, a different tax rate was applied on regular and low-calorie colas. According to Ecorys data analyses (2014b, p. 101), the consumption clearly increased for the latter. However, it seems that this upward trend was observable before the application of the new rates, which precludes assuming cause-to-effect. Furthermore, the analysis indicates that the consumption (per capita) of fruit drinks, although taxed at a higher rate, continued to increase after the tax hike.

Methodologically much stronger evidence recently came from an observational study on the Mexican SSB tax that took effect in January 2014 (Colchero et al. 2016). As mentioned before (see Sect. 6.4), in 2014, the authors estimate that SSB taxation has been associated with per capita purchases of taxed beverages that are 6 % lower than what could have been expected on the basis of pre-existing trends. Additionally, their analysis (adjusted for seasonal trends and macroeconomic factors) shows that these lower-than-expected purchases of taxed beverages were accompanied by 4 % higher-than-expected purchases of untaxed beverages (including diet sodas, bottled water, unsweetened dairy beverages and 100 % fruit juices). These positive substitution effects seem to have been mainly driven by increased purchases of bottled plain water. Since only store-bought beverages have been considered, other positive substitution effects may have occurred towards tap water or unsweetened homemade beverages (Colchero et al. 2016).

7.3.2 Substitution Towards Taxed Beverages Sold at a Lower Price

Based on Ecorys data analyses in the cases of France and Denmark, substitution of premium SSBs by store brands and private labels does not seem to be significant, which suggests a strong customer brand loyalty in this sector (Ecorys 2014a). In France, however, results of the Ecorys case study suggest that nectar (i.e. fruit-based beverages containing water and added sugars) has lost market share following the implementation of the tax, which may be related to a lower brand loyalty. In Finland, where soft drinks premium and non-premium brands equally share the market, a decline in the market share of premium brands has been observed over the last years, but this trend existed before successive tax increases in 2011 and 2012 (Ecorys 2014b, p. 117). This is also the case for regular colas in Hungary (and to a lesser extent for low-calorie colas) although the enactment of the tax in 2011 corresponds to an acceleration of the pre-existing progression of non-premium brands (Ecorys 2014b, p. 136).

In any cases, at population level, it should be kept in mind that substitution effects might be influenced by a wide array of historic, sociocultural and economic factors. As a consequence, assumptions as regards the specific effect of a tax should be made with caution and should be carefully evaluated.

7.4 Results from Systematic Reviews

This mixed picture drawn in the previous sections corroborates the conclusions of three recent and comprehensive systematic reviews of the effectiveness of food taxes (including SSBs) on energy intake and body weight outcomes:

- Thow et al.'s review (2014) underlines the heterogeneity of findings across 38 studies analysing the effects of food taxes and subsidies on the diet. They base their analysis on a thorough assessment of the quality of the studies. Four US modelling studies analysing the effects of SSB taxation distinguish themselves in the way substitution concerns were addressed. Although studying different taxation rates (5–20 %), their predictions all confirm that SSB taxation would reduce energy intake from SSBs in adults and children (to various extents). However, their conclusions diverge as regards the changes in the consumption of other beverages (i.e. milk, low-calorie drinks, tea and coffee). Three of these studies conclude on an overall tax-induced reduction in energy intake from all beverages (Dharmasena and Capps 2012; Lin et al. 2011; Zhen et al. 2011) while another concludes on the absence of such reduction overall (Fletcher et al. 2010). In this fourth study, as mentioned before (see Sect. 7.3.1), the tax rate is particularly low (i.e. ≈5 %), which may contribute to explaining this outcome. The conclusions of the review suggest two conditions that may contribute to a positive tax impact on energy intake: (1) a 10–20 % price increase; (2) exemption of close non-caloric SSB substitutes from the tax scope (e.g. diet soft drinks). The authors indicate that further research is needed to better understand the contribution of food taxation as part of a multifaceted prevention strategy (Thow et al. 2014).
- On the basis of 78 studies, Niebylski et al. (2015) also emphasize consistent evidence that food taxes and subsidies have the potential to influence dietary behaviours although the impact on health remains unclear. They suggest that "to maximize success and impact, [...] food taxes and subsidies should be a minimum of 10–15 % and preferably used in tandem". Finally, the review concludes on the importance to address substitution and equity concerns (see Chap. 8) (Niebylski et al. 2015, p. 3).
- Finally, on the basis of 55 papers (including experimental, observational and simulation studies), Maniadakis et al. (2013) also conclude that it is likely that taxes reduce the intake of the targeted foods and beverages (including SSBs), but the evidence seems globally equivocal as regards their potential effectiveness to reduce significantly energy intake and obesity prevalence. Substitution concerns were raised in many of the studies considered. When studies predicted positive effects on body weight, the authors highlight that these were generally modest. The complex aetiology of obesity and related diseases was also highlighted as a reason why an SSB tax alone may not have a significant influence. Equity concerns were also mentioned, since consumers may be differently affected by a tax depending on their socioeconomic status (see Chap. 8).

7.5 Concluding Comments

As a reminder, the evidence presented in Chap. 6 suggests that it is reasonable to expect SSB consumption to decrease as a result of a substantial tax (i.e. if accompanied by a 10 to 20 % price hike). This result is in itself encouraging as SSB overconsumption has been associated with an increased risk of type-2 diabetes and dental caries, independently of body weight gain (see Chap. 2). As regards energy balance, however, the evidence presented in this chapter is mixed.

On the one hand, simulations provide encouraging evidence to support decision-making, to design potential interventions and to plan their evaluation, but they present strong limitations in estimating how SSB taxation may influence energy balance and health in the "real world". On the other hand, empirical evidence on substitution effects from SSB taxation implemented across the world is limited and unclear.

Overall, the evidence presented suggests that substitution effect may partially compensate the SSB tax-induced reduction in energy intake from lower SSB consumption. The magnitude, the sustainability and the impact of this compensation on overall energy intakes and health remain complex to assess. The preliminary results of an evaluation study of the Mexican soda tax provide both the more robust and encouraging evidence. However, contextual factors may limit the generalisability of results from a particular country to others.

In sum, SSB taxation objectives on energy balance and health should be stated very carefully, mentioning at best a potential positive contribution to be tightly monitored in the long term. There is a clear need to track substitution effects precisely in countries where SSB taxes have been implemented.

Key messages

- A taxed-induced decrease in SSB consumption may be partially compensated by a higher consumption of other caloric foods and drinks. Therefore, SSB taxation objectives on energy balance and health should be stated very carefully.
- Studies simulating a 20 % price increase on SSBs and accounting for substitution effects generally predict a positive—although modest—impact on energy balance, but they are based on fragile assumptions. Therefore, substitution effects should be explored from soda taxes implemented across the world.
- Few "real-world" evaluations have been conducted, with the exception of Mexico, where evidence suggests that the tax has significantly reduced SSB purchases while favouring water consumption.
- If a tax is to be adopted in Canada, a detailed and robust evaluation plan should be designed ex-ante, and preliminary results should be carefully monitored in terms of price change, consumer demand and substitution effects before extending implementation in the longer term.

References

Berardi, N., Sevestre, P., Tepaut, M., & Vigneron, A. (2012). The impact of a "soda tax" on prices: Evidence from french micro data (Banque de France Working Paper No. 415) (p. 29). Banque de France Eurosystème. Direction Générale des Études et des Relations internationales. http://ssrn.com/abstract=2192470. Accessed February 24, 2016.

Boizot-Szantai, C., & Étilé, F. (2011). Taxer les boissons sucrées pour lutter contre l'obésité? Le point de vue de l'économie. *Obésité, 6*(4), 218–225. doi:10.1007/s11690-012-0306-4.

Bonnet, C., & Réquillart, V. (2012). Les effets de la réforme de la politique sucrière et des politiques de taxation sur le marché des boissons sucrées. *Cahiers de Nutrition et de Diététique, 47*(1), 35–41. doi:10.1016/j.cnd.2011.12.003.

Briggs, A. D. M., Mytton, O. T., Kehlbacher, A., Tiffin, R., Rayner, M., & Scarborough, P. (2013). Overall and income specific effect on prevalence of overweight and obesity of 20 % sugar sweetened drink tax in UK: Econometric and comparative risk assessment modelling study. *BMJ, 347*, f6189–f6189. doi:10.1136/bmj.f6189.

Cabrera Escobar, M. A., Veerman, J., Tollman, S. M., Bertram, M. Y., & Hofman, K. J. (2013). Evidence that a tax on sugar sweetened beverages reduces the obesity rate: A meta-analysis. *BMC Public Health, 13*(1), 1072. doi:10.1186/1471-2458-13-1072.

Colchero, M. A., Popkin, B. M., Rivera, J. A., & Ng, S. W. (2016). Beverage purchases from stores in Mexico under the excise tax on sugar sweetened beverages: Observational study. *BMJ*, h6704. doi:10.1136/bmj.h6704.

Colchero, M. A., Salgado, J. C., Unar-Munguía, M., Hernández-Ávila, M., & Rivera-Dommarco, J. A. (2015). Price elasticity of the demand for sugar sweetened beverages and soft drinks in Mexico. *Economics and Human Biology, 19*, 129–137. doi:10.1016/j.ehb.2015.08.007.

Dharmasena, S., & Capps, O. (2012). Intended and unintended consequences of a proposed national tax on sugar-sweetened beverages to combat the U.S. obesity problem. *Health Economics, 21*(6), 669–694. doi:10.1002/hec.1738.

Ecorys. (2014a). *Food taxes and their impact on competitiveness in the agri-food sector* (Final Report No. Ref. Ares(2014)2365745–16/07/2014). Rotterdam: DG Enterprise and Industry.

Ecorys. (2014b). *Food taxes and their impact on competitiveness in the agri-food sector*(Annexes to the Main Report No. Ref. Ares(2014)2604304–06/08/2014). Rotterdam: DG Enterprise and Industry.

Étilé, F. (2012). La taxation nutritionnelle comme outil de santé publique: justifications et effets attendus. *Cahiers de Nutrition et de Diététique, 47*(1), 25–34. doi:10.1016/j.cnd.2011.12.002.

Eyles, H., Ni Mhurchu, C., Nghiem, N., & Blakely, T. (2012). Food pricing strategies, population diets, and non-communicable disease: A systematic review of simulation studies. *PLoS Medicine, 9*(12), e1001353. doi:10.1371/journal.pmed.1001353.

Finkelstein, E. A., Zhen, C., Bilger, M., Nonnemaker, J., Farooqui, A. M., & Todd, J. E. (2013). Implications of a sugar-sweetened beverage (SSB) tax when substitutions to non-beverage items are considered. *Journal of Health Economics, 32*(1), 219–239. doi:10.1016/j.jhealeco.2012.10.005.

Fletcher, J. M., Frisvold, D. E., & Tefft, N. (2010). The effects of soft drink taxes on child and adolescent consumption and weight outcomes. *Journal of Public Economics, 94*(11–12), 967–974. doi:10.1016/j.jpubeco.2010.09.005.

Fletcher, J., Frisvold, D., & Tefft, N. (2013). Substitution patterns can limit the effects of sugar-sweetened beverage taxes on obesity. *Preventing Chronic Disease, 10*. doi:10.5888/pcd10.120195.

Hawkes, C. (2012). Food policies for healthy populations and healthy economies. *BMJ (Clinical research ed.), 344*, e2801.

Lin, B.-H., Smith, T. A., Lee, J.-Y., & Hall, K. D. (2011). Measuring weight outcomes for obesity intervention strategies: The case of a sugar-sweetened beverage tax. *Economics and Human Biology, 9*(4), 329–341. doi:10.1016/j.ehb.2011.08.007.

Maniadakis, N., Kapaki, V., Damianidi, L., & Kourlaba, G. (2013). A systematic review of the effectiveness of taxes on nonalcoholic beverages and high-in-fat foods as a means to prevent obesity trends. *ClinicoEconomics and Outcomes Research: CEOR, 5*, 519–543. doi:10.2147/CEOR.S49659.

Niebylski, M. L., Redburn, K. A., Duhaney, T., & Campbell, N. R. (2015). Healthy food subsidies and unhealthy food taxation: A systematic review of the evidence. *Nutrition, 31*(6), 787–795. doi:10.1016/j.nut.2014.12.010.

Powell, L. M., Chriqui, J., & Chaloupka, F. J. (2009). Associations between state-level soda taxes and adolescent body mass index. *The Journal of Adolescent Health: Official Publication of the Society for Adolescent Medicine, 45*(3 Suppl), S57–S63. doi:10.1016/j.jadohealth.2009.03.003.

Powell, L. M., Chriqui, J. F., Khan, T., Wada, R., & Chaloupka, F. J. (2013). Assessing the potential effectiveness of food and beverage taxes and subsidies for improving public health: a systematic review of prices, demand and body weight outcomes: Review of prices, demand and weight. *Obesity Reviews, 14*(2), 110–128. doi:10.1111/obr.12002.

Sharma, A., Hauck, K., Hollingsworth, B., & Siciliani, L. (2014). The effects of taxing sugar-sweetened beverages across different income groups: The impact of ssb taxes on different income groups. *Health Economics, 23*(9), 1159–1184. doi:10.1002/hec.3070.

Smith, T. A., Lin, B.-H., & Lee, J.-Y. (2010). Taxing caloric sweetened beverages: Potential effects on beverage consumption, calorie intake, and obesity. *ERR-100,* U.S. Department of Agriculture, Economic Research Service.

Sturm, R., Powell, L. M., Chriqui, J. F., & Chaloupka, F. J. (2010). Soda taxes, soft drink consumption, and children's body mass index. *Health Affairs, 29*(5), 1052–1058. doi:10.1377/hlthaff.2009.0061.

Thow, A. M., Downs, S., & Jan, S. (2014). A systematic review of the effectiveness of food taxes and subsidies to improve diets: Understanding the recent evidence. *Nutrition Reviews, 72*(9), 551–565. doi:10.1111/nure.12123.

Tiffin, R., Kehlbacher, A., & Salois, M. (2015). The effects of a soft drink tax in the UK. *Health Economics, 24*(5), 583–600. doi:10.1002/hec.3046.

Wang, Y. C., Coxson, P., Shen, Y.-M., Goldman, L., & Bibbins-Domingo, K. (2012). A penny-per-ounce tax on sugar-sweetened beverages would cut health and cost burdens of diabetes. *Health Affairs (Project Hope), 31*(1), 199–207. doi:10.1377/hlthaff.2011.0410.

Zhen, C., Finkelstein, E. A., Nonnemaker, J. M., Karns, S. A., & Todd, J. E. (2014). Predicting the effects of sugar-sweetened beverage taxes on food and beverage demand in a large demand system. *American Journal of Agricultural Economics, 96*(1), 1–25. doi:10.1093/ajae/aat049.

Zhen, C., Wohlgenant, M. K., Karns, S., & Kaufman, P. (2011). Habit Formation and Demand for Sugar-Sweetened Beverages. *American Journal of Agricultural Economics, 93*(1), 175–193. http://doi.org/10.1093/ajae/aaq155.

Chapter 8
Distribution of Sugar-Sweetened Beverage Taxation Effects in the Population

Abstract The evidence reviewed in Chaps. 5 and 6 suggests that a tax on soda/sugar-sweetened beverages (SSBs) causing a 10–20 % price increase could result in a lower-average SSB consumption in the population. But beyond average values, would the poorest consumers be disproportionally burdened by the tax? Would they benefit from larger health gains? Overall, the evidence suggests that the financial contribution to tax revenues would be relatively modest for most consumers and not much different across income groups. Low-income households consuming large SSB amounts and not prone to reduce their consumption may be more negatively impacted than other consumers. As regards SSB taxation health benefits, these are hardly predictable (see Chap. 7) and it remains unclear whether these could be greater in lower-income households and in the heaviest SSB consumers. Inequity concerns surrounding SSB taxation could be somewhat mitigated through the earmarking of tax proceeds for health promotion causes.

8.1 Introduction

This chapter questions how far a sugar-sweetened beverage (SSB) taxation policy is susceptible to generate differential financial costs, differential SSB intake variations and differential health outcomes across population groups. In particular, we analyse whether distributional effects across income strata are likely to raise significant equity concerns and whether the right population target is reached, i.e. those overconsuming SSBs (see Fig. 8.1).

The first concern emphasizes that a soda tax would be regressive, which means that it would unfairly impose a proportionally higher financial burden on lower-income households. This would hold to the fact that lower-income households tend to devote a higher proportion of their revenues to food budget and may also be higher SSB consumers (Drewnowski 2007; Étilé 2012). In Canada, the

Y. Le Bodo et al., *Taxing Soda for Public Health*,
DOI 10.1007/978-3-319-33648-0_8

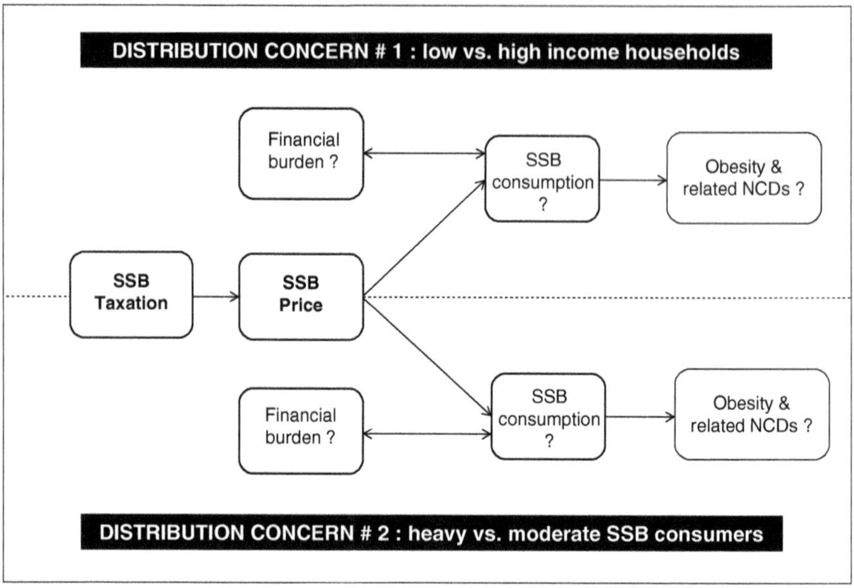

Fig. 8.1 Concerns related to the distribution of SSB taxation effects in the population

latter is not well documented. Nutrition data from the forthcoming CCHS 2015 should provide additional information. But there are some cues that a tax on SSBs will likely be somewhat regressive. For example, correlations have been observed between SSB intakes and socio-demographic characteristics in youth in the province of Quebec (but not in adults). Data collected in 2010–2011 from the Quebec survey on the health of more than 60,000 secondary school children (EQSJS) indicate that proportions of students consuming SSB at least once a day are significantly higher[1] in those who present higher social and material deprivation levels, whose self-assessment of financial situation and school performance is lower, and whose parents are unemployed and have lower education levels (Camirand et al. 2012). Another concern is to know whether SSB taxes actually reach the most-at-risk consumers, i.e. those over-consuming SSBs. These aspects will be discussed in the next 4 sections.

[1]Although the differences here mentioned are statistically significant ($p \leq 0.05$) in particular because of a very large sample size, they may be considered of modest amplitude. For example, 32.7 % of youth whose both parents are unemployed declare consuming SSBs at least once a day, versus 25.4 % in youth whose one parent is unemployed, and versus 23.1 % in youth whose both parents are employed.

8.2 Would a SSB Tax Impose a Larger Financial Burden on Lower-Income Households?

Would households' financial contribution to SSB taxation be substantial and would that make a significant difference between low-income and high-income households? This point is debatable.

In the USA, Finkelstein et al. (2010, p. 2032) consider that a tax-induced 40 % SSB price increase would have a modest impact on the annual food budget (i.e. approximately $28 per household).[2] They estimate that lower-income households purchase more SSBs than higher-income households. But considering that "they do so at lower average prices and because they are more price sensitive", households in the lowest income quartile would support approximately 20 % of total tax revenues versus 25 % in the 2 middle-income quartiles and 30 % in the highest income quartile. However, these estimations do not account for variations in spending power across quartiles. When such variations are taken into account, Lin et al. (2011) predict greater regressive effects. Still in the USA, they estimate that a 20 % tax on SSBs would, on average, lead to a slightly higher annual burden on lower-income individuals than on higher-income individuals ($19.97 vs. $18.84), which would represent 1 % of the annual food and beverage spending for the former and 0.6 % for the latter.[3] In the UK, a study modelling the impact of a 20 % SSB taxation also considers that the average financial burden per person per week would be low (around 8 pence), although greater increases in weekly expenditure on drinks are predicted as income decreases (Briggs et al. 2013a). Another recent simulation study from Australia, comparing the impacts of a 20 cents/litre excise tax or a 20 % valoric tax on SSBs, demonstrate that the average yearly per capita tax burden on low-income households would be $17.87 (0.21 % of income) compared with $15.17 for high-income households (0.07 % of income) in the case of the valoric tax, and $13.80 (0.15 %) and $10.10 (0.04 %) for the excise tax (Sharma et al. 2014). Using similar tax options and databases but methodological variations, Étilé and Sharma (2015) indicate that either case would represent a limited average tax burden per capita per year (i.e. ≈AU$38 in the case of a volumetric tax and ≈AU$26 in the case of a valoric tax).

Whatever the simulation, the burden per household seems low to moderate but may be underestimated either because only store-bought SSBs were considered[4] (Finkelstein et al. 2010; Sharma et al. 2014; Étilé and Sharma 2015) or because SSB intakes are based on self-reported consumption data, known to be underestimated (Lin et al. 2011; Briggs et al. 2013a). Indeed, others speculate that in a worst-case scenario based on US beverage market data, a penny-per-ounce SSB

---

[2]Based on after-tax consumption simulated from *Nielsen Homescan panel* (2006).

[3]Based on after-tax consumption simulated from US nationally representative SSB consumption data.

[4]In Canada, the "take-home" market accounts for an estimated two-thirds of sales (Agriculture and Agri-Food Canada 2015).

taxation could represent a $196 extra burden per year for an average family of four consumers, which may be much less in families with moderate consumers but much more in families counting heavy consumers (Butterworth 2012).

In sum, the available evidence tends to confirm that SSB taxation can be somewhat regressive, but robust estimations generally predict a modest financial contribution and a low differential burden between low-income and high-income households. From an ethical perspective, as argued by Barnhill and King (2013), the regressive nature of a soda tax does not necessarily rule it out: nearly all consumption taxes are regressive, be this in the form of sales taxes on various goods and services, tobacco taxes or alcohol taxes. The goal was not to devise a tax system in which all taxes are progressive, but for the overall tax system to be progressive. Generally, these regressive taxes are compensated by a progressive income tax, as it is the case in Canada (Barnhill and King 2013).

8.3 Would a SSB Tax Bring Larger Health Benefits to Lower-income Households?

Could the regressive effects of an SSB tax be somewhat compensated by a larger reduction in SSB intakes and, eventually, higher health gains in lower-income households? It has proved to be the case for tobacco taxation, which has shown to discourage consumption most particularly among low-income individuals, the young and new tobacco users (WHO 2012). As regards SSB taxation, the evidence remains scarce and unclear. Results from observational and simulation studies are summarized below.

On the one hand, a US observational study analysing the impact of existing state sales soda taxes (i.e. average rate of approx. 4 %) found that it did not have significant effect on SSB consumption when considering the population as a whole. However, it shows that subgroups of children already overweight or coming from low-income families "may be more sensitive than others to such taxes" (Sturm et al. 2010, p. 1052). The authors highlight that this is particularly true in the school context and hypothesize that it may be explained by the fact that tax-included prices are visible to consumers in school cafeterias and vending machines (whereas sales tax applies at the cash register in grocery stores) (Sturm et al. 2010). Stronger evidence recently came from an observational study on the Mexican SSB tax that took effect in January 2014 (Colchero et al. 2016). As mentioned before (see Sect. 6.4), the authors estimate that SSB taxation has been associated with per capita purchases of taxed beverages that are 6 % lower than what could have been expected on the basis of pre-existing trends. Additionally, their analysis (adjusted for seasonal trends and several macroeconomic factors) shows that these lower-than-expected purchases occurred across all socioeconomic groups, but were greater among households of low socioeconomic status (−9.1 %) than among households of middle and high socioeconomic status (−5.5 and −5.6 %, respectively) (Colchero et al. 2016).

On the other hand, results from simulation studies are mixed. A systematic review of food pricing simulations using price elasticity estimates and epidemiological models indicates that lower socioeconomic groups obtain nutritional benefits in 11 of the 14 studies considered. In half of the 14 studies, impacts were similar across socioeconomic groups. In the other half, four (three) studies showed greater (weaker) benefits for lower socioeconomic groups (Eyles et al. 2012). Considering SSBs only, Finkelstein et al. (2010) found slightly lower responsiveness to price change in lower-income households. Overall, they predict that a 20–40 % tax on carbonated SSBs would have similar effects across income categories, and that middle-income households would mainly drive the overall decrease in calories from beverages as a result of the tax. Using a similar approach, Lin et al. (2011) predict that a 20 % SSB tax would reduce the intake of beverage calories to a slightly greater extent in lower-income adults but also in children from high-income families. Finally, in the UK, a recent simulation study using price elasticity estimations from a large consumer panel does not predict different health gains across income groups in the hypothesis of a 20 % tax on SSBs (Briggs et al. 2013a). It should be noticed that these simulations are based on many assumptions (e.g. in terms of tax rate, tax base and price elasticity estimations). Therefore, their predictions should be cautiously considered.

All in all, depending on the context, it is unclear whether SSB taxation could bring a larger decrease in consumption and larger health benefits for lower-income households. Of particular interest is a pioneering US study on the association between household food insecurity and food prices, using the Early Childhood Longitudinal Study (n = 27900) and cost of living data from the Council for Community and Economic Research. The authors observe that increasing beverage prices (for soft drinks, orange juice and coffee) does not appear to foster food insecurity (independently of general food prices) and suggest that SSB taxes could even be protective in this regard. Further research is needed to confirm this hypothesis (Zhang et al. 2013).

8.4 Could Equity Concerns Be Mitigated?

Equity concerns may question the social and political acceptability of a soda tax. For instance, in the USA, an opinion survey using a probability-based sample of respondents from a large, nationally representative online panel (2011) considered that SSB taxation would be “harmful to the poor” (51 %) rather than “protecting the poor” (36 %) (Barry et al. 2013). In France, preliminary results from a large Internet-based survey conducted in 2012–2013 also indicate that half of the respondents considered SSB taxation to be unfair as the “poorest” will pay the same contribution as the “wealthy” (whereas a third disagreed with the statement). Additionally, more than 60 % considered that the tax would increase prices and reduce spending power (Julia et al., unpublished data). In Canada, the Montreal Economic Institute (MEI) is known for its position against SSB taxation. The organization stressed that a 70 cents increase on a 2-litre bottle of SSB would represent a greater financial burden on disadvantaged households (Elgrably-Lévy 2013). Although consumer groups have been discreet on the topic, a Quebec spokesperson from *Option Consommateurs* already told the press that there were better alternatives to promote health and that such a tax would reduce the spending power of low-income families (Bergeron 2010). These controversies suggest that equity concerns should rather be addressed before a soda tax is implemented. These concerns could somewhat be mitigated by truly earmarking[5] tax revenues for the health sector. For example, investing tax revenues to the funding of interventions aimed at promoting healthy lifestyles (especially towards youth) and at reducing health inequities could increase the social acceptability of the tax (Brownell et al. 2009, and see Chap. 13).

8.5 Would a SSB Tax Be Effective Towards Heavy SSB Consumers?

Another important concern is to know whether the right population target is reached, i.e. those overconsuming SSBs. Encouraging observational studies tend to indicate that young consumers, lower-income households and large SSB purchasers may be equally or even more responsive to price change but, overall, the evidence remains unclear (Gustavsen 2005; Sturm et al. 2010; Powell et al. 2013; Finkelstein et al. 2010, 2013; Briggs et al. 2013a, b; Tiffin et al. 2015; Colchero et al. 2015; Wada et al. 2015). Parallel evidence from tobacco and alcohol taxation suggests resistance phenomena to price hikes, in particular among the highest consumers (Peretti-Watel and Constance 2009; Ayyagari et al. 2009; Nghiem et al. 2013). Although SSB, tobacco and alcohol cannot be equally compared in terms of addiction, these observations suggest that greater attention should be devoted to SSB taxation effects in the heaviest consumers.

Interestingly, a recent study based on Australian consumer panel data collected in 2010 demonstrates that "although high consumers of SSBs have a less elastic demand for SSBs, their very high consumption levels imply that a tax would achieve higher reduction in consumption and higher health gains". The study also shows that a 20 cents/litre volumetric SSB taxation (i.e. indexed on fluid volume) would have a greater positive impact on heavy SSB consumers than a 20 % valoric tax (i.e. indexed on price), since the former would have more influence than the latter on the price of discounted multipacks items, particularly popular among these consumers (Étilé and Sharma 2015, p. 1). The choice of the tax instrument, either

[5]Earmarking consists in "assigning revenues from designated sources to finance designated expenditures" (WHO 2012, p. 8).

volumetric or valoric, may therefore deserve particular attention. Out of the 21 places where an SSB tax has been reported across the world, there were 12 cases of volumetric taxes and 9 cases of valoric taxes (see Table 1.1, column 5). Beyond effectiveness considerations, this choice also raises feasibility concerns that will be later discussed (see Chap. 12).

Key messages

- Evidence suggests that the financial contribution to tax revenues would be relatively modest for most consumers and not much different across income groups.
- Low-income households consuming large SSB amounts and not prone to reduce their consumption may be more negatively impacted than other consumers.
- As regards SSB taxation health benefits, these are hardly predictable (see Chap. 7) and it remains unclear whether these could be greater in lower-income households and the heaviest SSB consumers. Simulation studies are encouraging but more research and evaluations will be crucial in this area.
- Inequity concerns surrounding SSB taxation could be somewhat mitigated if the proceeds of a soda tax are truly earmarked for health promotion causes.

References

Agriculture and Agri-Food Canada. (2015). The Canadian Soft Drink Industry. http://www.agr.gc.ca/eng/industry-markets-and-trade/statistics-and-market-information/by-product-sector/processed-food-and-beverages/the-canadian-soft-drink-industry/?id=1172167862291. Accessed March 1, 2015.

Ayyagari, P., Deb, P., Fletcher, J., Gallo, W., & Sindelar, J. (2009). *Sin taxes: Do heterogeneous responses undercut their value?* (No. w15124). Cambridge, MA: National Bureau of Economic Research. http://www.nber.org/papers/w15124.pdf. Accessed February 25, 2016.

Barnhill, A., & King, K. F. (2013). Ethical agreement and disagreement about obesity prevention policy in the United States. *International Journal of Health Policy and Management, 1*(2), 117–120. doi:10.15171/ijhpm.2013.21.

Barry, C. L., Niederdeppe, J., & Gollust, S. E. (2013). Taxes on sugar-sweetened beverages: Results from a 2011 national public opinion survey. *American Journal of Preventive Medicine, 44*(2), 158–163. doi:10.1016/j.amepre.2012.09.065.

Bergeron, L. (2010). Alimentation: Boissons gazeuses: Coupables ? *Protégez-vous*, 31.

Briggs, A. D. M., Mytton, O. T., Kehlbacher, A., Tiffin, R., Rayner, M., & Scarborough, P. (2013a). Overall and income specific effect on prevalence of overweight and obesity of 20 % sugar sweetened drink tax in UK: Econometric and comparative risk assessment modelling study. *BMJ, 347*, f6189–f6189. doi:10.1136/bmj.f6189.

Briggs, A. D., Mytton, O. T., Madden, D., O'Shea, D., Rayner, M., & Scarborough, P. (2013b). The potential impact on obesity of a 10 % tax on sugar-sweetened beverages in Ireland, an effect assessment modelling study. *BMC Public Health, 13*(1), 860. doi:10.1186/1471-2458-13-860.

Brownell, K. D., Farley, T., Willett, W. C., Popkin, B. M., Chaloupka, F. J., Thompson, J. W., & Ludwig, D. S. (2009). The public health and economic benefits of taxing sugar-sweetened beverages. *New England Journal of Medicine, 361*(16), 1599–1605. doi:10.1056/NEJMhpr0905723.

Butterworth, T. (2012, January 12). Will a soda tax really save 26,000 lives, billions of dollars? *Forbes.* http://www.forbes.com/sites/trevorbutterworth/2012/01/12/will-a-soda-tax-really-save-26000-lives-billions-of-dollars/#40549e4f5bb1. Accessed February 24, 2016.

Camirand, H., Blanchet, C., & Pica, L. A. (2012). Habitudes alimentaires. In *L'enquête québécoise sur la santé des jeunes du secondaire, 2010–2011. Vol. Tome 1. Le visage des jeunes d'aujourd'hui : leur santé physique et leurs habitudes de vie* (pp. 71–96). Québec: Institut de la statistique du Québec.

Colchero, M. A., Popkin, B. M., Rivera, J. A., & Ng, S. W. (2016). Beverage purchases from stores in Mexico under the excise tax on sugar sweetened beverages: observational study. *BMJ*, h6704. doi:10.1136/bmj.h6704.

Colchero, M. A., Salgado, J. C., Unar-Munguía, M., Hernández-Ávila, M., & Rivera-Dommarco, J. A. (2015). Price elasticity of the demand for sugar sweetened beverages and soft drinks in Mexico. *Economics and Human Biology, 19*, 129–137. doi:10.1016/j.ehb.2015.08.007.

Drewnowski, A. (2007). The real contribution of added sugars and fats to obesity. *Epidemiologic Reviews, 29*(1), 160–171. doi:10.1093/epirev/mxm011.

Elgrably-Lévy, N. (2013, March 14). Une taxe imbuvable. *Le Journal de Montréal*, 25.

Étilé, F. (2012). La taxation nutritionnelle comme outil de santé publique: Justifications et effets attendus. *Cahiers de Nutrition et de Diététique, 47*(1), 25–34. doi:10.1016/j.cnd.2011.12.002.

Étilé, F., & Sharma, A. (2015). Do high consumers of sugar-sweetened beverages respond differently to price changes? A finite mixture IV-Tobit approach: Heterogeneity in response to soft drink tax. *Health Economics, 24*(9), 1147–1163. doi:10.1002/hec.3157.

Eyles, H., Ni Mhurchu, C., Nghiem, N., & Blakely, T. (2012). Food pricing strategies, population diets, and non-communicable disease: A systematic review of simulation studies. *PLoS Medicine, 9*(12), e1001353. doi:10.1371/journal.pmed.1001353.

Finkelstein, E. A., Zhen, C., Bilger, M., Nonnemaker, J., Farooqui, A. M., & Todd, J. E. (2013). Implications of a sugar-sweetened beverage (SSB) tax when substitutions to non-beverage items are considered. *Journal of Health Economics, 32*(1), 219–239. doi:10.1016/j.jhealeco.2012.10.005.

Finkelstein, E. A., Zhen, C., Nonnemaker, J., & Todd, J. E. (2010). Impact of targeted beverage taxes on higher- and lower-income households. *Archives of Internal Medicine, 170*(22), 2028–2034. doi:10.1001/archinternmed.2010.449.

Gustavsen, G. (2005). *Public policies and the demand for carbonated soft drinks: A censored quantile regression approach.* Presented at the European Association of Agricultural Economists-2005 International Congress, Copenhagen (Denmark). http://ageconsearch.umn.edu/bitstream/24737/1/cp05gu01.pdf. Accessed February 25, 2016.

Lin, B.-H., Smith, T. A., Lee, J.-Y., & Hall, K. D. (2011). Measuring weight outcomes for obesity intervention strategies: the case of a sugar-sweetened beverage tax. *Economics and Human Biology, 9*(4), 329–341. doi:10.1016/j.ehb.2011.08.007.

Nghiem, N., Wilson, N., Genç, M., & Blakely, T. (2013). Understanding price elasticities to inform public health research and intervention studies: Key issues. *American Journal of Public Health, 103*(11), 1954–1961. doi:10.2105/AJPH.2013.301337.

Peretti-Watel, P., & Constance, J. (2009). "It's All We Got Left". Why poor smokers are less sensitive to cigarette price increases. *International Journal of Environmental Research and Public Health, 6*(2), 608–621. doi:10.3390/ijerph6020608.

Powell, L. M., Chriqui, J. F., Khan, T., Wada, R., & Chaloupka, F. J. (2013). Assessing the potential effectiveness of food and beverage taxes and subsidies for improving public health: a systematic review of prices, demand and body weight outcomes: Review of prices, demand and weight. *Obesity Reviews, 14*(2), 110–128. doi:10.1111/obr.12002.

Sharma, A., Hauck, K., Hollingsworth, B., & Siciliani, L. (2014). The effects of taxing sugar-sweetened beverages across different income groups: The impact of SSB taxes on different income groups. *Health Economics, 23*(9), 1159–1184. doi:10.1002/hec.3070.

Sturm, R., Powell, L. M., Chriqui, J. F., & Chaloupka, F. J. (2010). Soda taxes, soft drink consumption, and children's body mass index. *Health Affairs, 29*(5), 1052–1058. doi:10.1377/hlthaff.2009.0061.

Tiffin, R., Kehlbacher, A., & Salois, M. (2015). The effects of a soft drink tax in the UK. *Health Economics, 24*(5), 583–600. doi:10.1002/hec.3046.

Wada, R., Han, E., & Powell, L. M. (2015). Associations between soda prices and intake: Evidence from 24-h dietary recall data. *Food Policy, 55*, 54–60. doi:10.1016/j.foodpol.2015.05.009.

World Health Organization (WHO). (2012). *Tobacco taxation and innovative health-care financing*. New Delhi, India: World Health Organization. http://www.searo.who.int/tobacco/documents/2012-pub1.pdf. Accessed January 22, 2016.

Zhang, Q., Jones, S., Ruhm, C. J., & Andrews, M. (2013). Higher food prices may threaten food security status among American low-income households with children. *The Journal of Nutrition, 143*(10), 1659–1665. doi:10.3945/jn.112.170506.

Chapter 9
Potential Undesirable Effects Related to Sugar-Sweetened Beverage Taxation

Abstract In spite of encouraging tax effects on prices and demand for soda/sugar-sweetened beverages (SSBs) (see Chaps. 5 and 6), undesirable effects should also be considered. It possibly includes increased administration costs (for the private sector and the government), adverse effects on employment, investment and competitiveness, as well as cross-border shopping. Overall, such effects have been weakly documented but observation studies, simulations and anecdotal evidence suggest that effects of a soda tax on these variables may be relatively limited and would not justify inaction.

Beyond equity concerns,[1] other undesirable effects may be associated with sugar-sweetened beverage (SSB) taxes. Studying the potential impact of food taxes on competitiveness in the agri-food sector in Europe, Ecorys identifies various areas where such effects may occur, including administrative costs, private sector's profitability, employment, investment and competitiveness. Their analysis confirms that such taxes may alter the profit margin for the taxed product, but this may not necessarily translate into net profit losses for a given company or the beverage sector as a whole (Ecorys 2014). Indeed, many factors may influence a tax's impact on business, such as the scope of items concerned, the tax rate and potential gains on SSB substitutes within the same company and in other companies. Other contextual factors may play an important role (e.g. price of raw materials, size of the company and width of its portfolio). For all these reasons, there is limited and sometimes conflicting evidence about the specific tax effects on business and administration (Ecorys 2014). Some key pieces of evidence are summarized here below.

[1]Please note that equity concerns could also be considered as "undesirable effects" and be discussed here. However, to avoid redundancy, we refer the reader to Chap. 8 that has been specifically devoted to these issues.

Y. Le Bodo et al., *Taxing Soda for Public Health*,
DOI 10.1007/978-3-319-33648-0_9

9.1 Tax-Related Administration Costs

Unfortunately, little is known about tax-related administration costs for the government and private actors. It may vary widely depending on the complexity of the tax base, the tax design and the pre-existence of administrative mechanisms for data monitoring, payment and confirmation of compliance (Preece 2013; Ecorys 2014). In the literature, the Danish "fat tax" implemented in 2011 on all foods containing more than 2.3 % saturated fat has been described as a "bureaucratic nightmare" (Stafford 2012, §2). Data acquisition on the percentage of saturated fat in foods as well as on the amount of saturated fat used in the process of a given food (including waste) proved to be particularly burdensome, especially for importers (Smed 2012; Ecorys 2014). Danish companies have claimed undesirable impact on competitiveness (Bødker et al. 2015). According to an ex post assessment by the Danish Commerce and Companies Agency's Department for Better Regulation (CKR) (quoted by Ecorys 2014, p. 58), the tax would have cost approx. €27 million to the retail and wholesale sector.

We can reasonably assume that such administrative burden may not equally apply to a tax restricted to SSB, especially if the tax rate is not based on the sugar content of the beverage, and if the beverage categorization is already used by the companies and the public administration (Bahl et al. 2003; Hespel and Berthod-Wurmser 2008; Ecorys 2014). Globally, the administrative burden also depends on the number of companies concerned by data reporting and tax collection: global costs may be smaller if the tax is applied on large companies and those positioned upstream in the supply chain (e.g. manufacturers and bottlers). At the company level, since administrative costs are usually considered fixed costs, they may represent a higher burden for small- and medium-sized companies than for large companies (Ecorys 2014). All these assumptions remain to be publicly documented in practice.

9.2 Tax Impact on Profitability

Companies' profitability on SSBs depends on many factors such as the price of raw materials, production costs, manufacturing/retail prices, sales volume and promotion strategies (Ecorys 2014). Therefore, it is complex to estimate the specific impact of SSB taxation. No comprehensive evaluation has been published on the topic, but empiric indications tend to refute the idea that SSB taxation may affect significantly the margins and profitability of economic actors. Additionally, it can reasonably be hypothesized that a lower profitability on some taxed items would be offset by a higher profitability on others.

In Denmark, for example, Ecorys data analysis indicates that excise taxes applied on colas and juices were generally accompanied by increased margins of retailers (data not available for manufacturers). Similar observations were made in

Finland, where the overshifting of excise taxes on soft drinks was accompanied by increased margins of retailers and manufacturers between 2011 and 2013. In Hungary also, the introduction of the tax on colas and fruit drinks in 2010 was followed by increased margins of manufacturers and retailers. Strategic pricing seemed to occur, since the margin on untaxed juices increased more than on taxed juices (Ecorys 2014, pp. 25–26; 30). Overall, the Ecorys data analysis concludes that the profitability in the SSB sector "slightly increased" in Hungary in 2011 (Ecorys 2014, p. 59).

In France, a refined data analysis shows that the tax enacted in 2012 has been progressively and fully shifted to the price of soda[2] in the following six months. The pass-through to prices of fruit drinks and flavoured waters was partial ($\approx$85 %, with variations depending on the time, brands and retailers), possibly due to the competition with untaxed substitutes (Berardi et al. 2012). Here again, these observations suggest strategic pricing decisions by companies among their portfolio of products in order to minimize the negative consequences of the tax. Ecorys data analysis also suggests pricing strategies in the case of colas: following the introduction of the tax in 2012, retailers' margins increased for low-calorie colas and decreased for regular colas although both were equally taxed (Ecorys 2014, p. 26). Overall, one year after the tax enactment, market data indicate that soft drink volume sales had decreased by 1.5 %, but monetary sales had kept increasing by 5 % (Rayon Boissons 2013). According to Ecorys (2014, p. 66), these observations suggest that the "strong brand loyalty among consumers" in the SSB sector may have allowed producers and retailers to pass the tax onto prices of premium products without too much risk in terms of competitiveness and profitability. Conversely, the tax may have placed private labels at greater risks (La Tribune 2011).

9.3 Tax Impact on Employment

Food taxes' impact on employment is difficult to assess, since employment is influenced by many contextual factors such as the overall economy, labour costs, profitability, innovations and consumption trends (Ecorys 2014). An analogy with the case of tobacco taxation may help in this regard. It suggests that, since the manufactured tobacco industry is highly capital-intensive, tobacco taxes have limited effects on employment in this sector but generally have, on net, a positive effect on overall employment (as smokers substitute their expenditures to other goods or services whose production is more labour-intensive). Would the same likely hold in the case of food and SSB taxes?

According to Ecorys (2014), some have claimed that food taxes may affect profitability and, consequently, may directly affect employment at the

[2]Including cola, energy, tonic and other soft drinks.

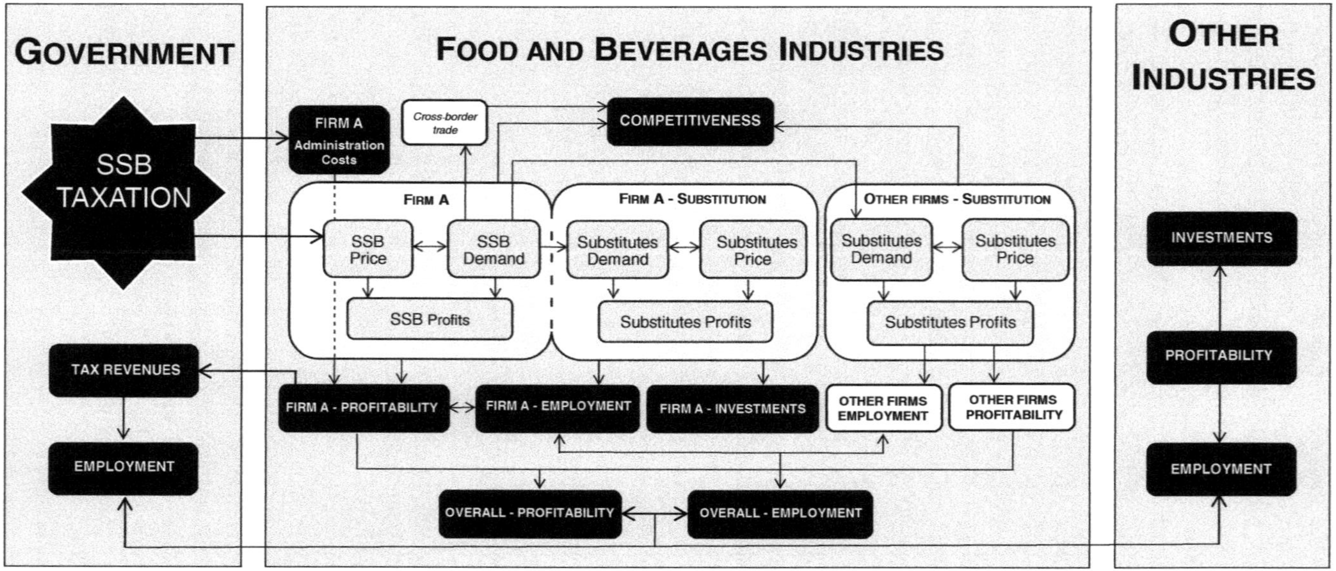

Fig. 9.1 Potential influences on administrative costs, profitability, investments, employment and competitiveness resulting from a SSB taxation policy (adapted from Ecorys 2014)

manufacturing level and indirectly affect employment along the supply chain (e.g. for bottlers and retailers). For example, Danish companies have claimed undesirable impact on employment following the world first tax on all foods containing more than 2.3 % saturated fat in 2011 (Vallgårda et al. 2015; Bødker et al. 2015). According to the Danish Agriculture and Food Council, if the tax had not been abolished, it could have caused 1300 job losses (Stafford 2012). Similar arguments have been used by the industry in the case of the Hungarian "Public Health Product Tax" (Bíró 2015). Conversely, others argue that employment losses related to taxed items would be compensated by gains in untaxed product lines of the same companies and others (see Fig. 9.1). Overall, an Ecorys case study concludes that the impact of the Danish fat tax on employment was "limited" due to its short time of implementation (15 months) (Ecorys 2014, p. 62).

Little is known on the specific case of SSB taxation impact on employment. In the USA, in 2012, a macroeconomic simulation model was used to predict the employment impact of a 20 % SSB tax in two states (Illinois and California) (Powell et al. 2014). First, the authors estimate the gross decline in the beverage industry employment accounting only for the estimated decrease in SSB demand (what the industry usually estimate). In this case, their model predicts an overall loss up to nearly 7000 jobs in Illinois and 15,000 jobs in California, which represents 0.09 and 0.07 % of all jobs, respectively. Among these jobs, nearly 6450 and 13,700 would be lost in the private sector in Illinois and California, respectively. However, when their model accounts for changes in SSB demand, substitution effects to other goods and services, consumers' income effects and government spending of tax revenues, the authors predict that the decline would be offset by new employment in other industries and government sectors leading to an overall net increase of about 4400 jobs in Illinois and 6650 jobs in California. In this case, the net decline in the private sector would be considerably lower and would only represent nearly 900 and 250 job losses in Illinois and California, respectively. It is true, however, that this last model shows larger job losses in beverage manufacturers: 1350 in Illinois and 2300 in California. But finally, when the authors account for specific SSB substitutions to other beverages (based on various cross-price elasticity hypotheses), job losses in the beverage industry are somewhat lower: between 1000 and 1300 in Illinois and between 1450 and 1850 in California, depending on the hypothesis (Powell et al. 2014). These findings suggest that job losses in the SSB industry resulting from a 20 % SSB tax would likely be more than offset by the accumulation of new jobs in the government sector (partially related to the spending of tax revenues), new jobs linked to SSB substitutions to other goods and services and new jobs linked to SSB substitutions to untaxed beverages (produced by the same beverage company and/or other ones). These flows are illustrated in Fig. 9.1.

In reality, empiric indications are mixed but also suggest that the influence of current SSB taxes on employment in comparison with other factors may be limited. In Finland, the Ecorys case study indicates that the beer and soft drink industry suffered a 12 % reduction in employment between 2010 and 2013. Some industry stakeholders claimed that SSB taxation was partly responsible for job losses, but the specific impact of the tax remained unclear. In France and Hungary, it seems that employment trends in the SSB industry did not change substantially over the period concerned by the tax, but Ecorys indicates that the aggregated nature of their data and the complexity to isolate specific tax effects preclude drawing firm conclusions (Ecorys 2014, p. 63).

9.4 Tax Impact on Investments

SSB taxation has sometimes been presented as a way to catalyze product reformulation. But industry stakeholders from several countries have stressed that many improvements had already started prior to the introduction of the tax and that food taxes could actually be counterproductive (see also Sect. 11.2). Actually, available information on the impact of food taxes on sectorial investments is scarce (Ecorys 2014).

In France, the threats perceived by SSB companies were obvious during the legislative debates on the soda tax in 2011. The largest bottler even stated that it would reconsider investing 17 million € in a bottling plant to demonstrate its opposition to the tax announcement made by the French government. After suffering a surge of indignation from the political sphere, they finally retracted by invoking a communication mistake (Challenges 2011). After all, although the French non-alcoholic beverage industry acknowledged modest growth in 2012–2013 partly because of the soda tax, they anticipated a rebound in 2014 thanks to a clement weather, the soccer world cup and a successful diversification in no-calorie drinks, natural flavours and lower-calorie options based on natural alternatives to artificial sweeteners (e.g. stevia) (FNB 2014). In Finland, the upward investment trend in the soft drink sector did not appear to have been impacted by the introduction of the tax in 2011 (Ecorys 2014, p. 60). Finally, in Hungary, a survey of the National Institute for Health Development (NIHD) quoted by Ecorys (2014, p. 61) indicates that 23 % of manufacturers concerned by food taxes had claimed to have cancelled planned investments. However, this may not specifically apply to SSB. According to the aggregated data analysis by Ecorys, it seems that the investment activity in the SSB sector increased in the year of enactment of the tax, whereas investments had decreased over the previous years (Ecorys 2014, p. 60).

9.5 Tax Impact on Competitiveness

It has been suggested that food taxes (depending on their scope and design) may unfairly distort competitiveness within and across sectors because of substitution effects. This concern may not strongly apply, for example, when manufacturers produce both taxed and untaxed beverages, both premium brands and private labels, or both beverages and other goods. But this is not necessarily the case in the SSB sector.

For example, in 2011 in France, the 2nd bigger SSB manufacturer, whose 90 % of products contained added sugars (versus 60 % for the leader, who invests more in NCSBs), successfully lobbied for an extension of the tax scope to NCSBs (La Tribune 2011). This “internal squabbling” between the largest manufacturers did not seem to strongly affect profitability of the sector at large, but it is estimated that small- and medium-sized manufacturers, who generally have less diversity in their portfolio of products, may be more impacted by the tax (Bonnet and Réquillart 2012; Ecorys 2014). In the USA, with the support of the American Beverage Association, small-business owners have claimed that they would be strongly burdened by a tax on SSBs (Harrington 2014). To mitigate such effects, SSB taxation in Berkeley includes a “small-business exemption” for some retailers (city of Berkeley 2014). In Finland, the SSB taxation also exempts producers with an annual production volume of fewer than 50,000 l (Ecorys 2014).

Premium and non-premium products may also have been differently affected. In France, data analyses by Berardi et al. (2012) suggest that, in most cases, retailers increased more the price of their own private labels than they did for premium brands, possibly because these beverages were already cheaper and offered less flexibility in terms of pricing. The authors also observe that the largest retailing groups, who may have more bargaining power over producers than the smallest ones, may have fewer difficulties to negotiate prices and preserve their profits. In sum, as Berardi et al. (2012, p. 13) put it, “the diversity of competitive situations is then likely to induce the same diversity regarding the way retailers (and producers) of beverages have shifted the SSB tax to their prices”.

9.6 Tax Impact on Cross-border Trade and Smuggling

It has been suggested that food taxes may favour cross-border trade, depending on price variations between regions and countries and depending on the cost (in time and travel) of shopping abroad. Overall, evidence gathered by Ecorys on food taxes’ impact on competitiveness in the agri-food sector suggests that additional cross-border shopping resulting from new taxes is actually limited (Ecorys 2014).

An often-quoted example is the case of the Danish tax on saturated fat. According to industry stakeholders and the Danish Chamber of Commerce (quoted by Ecorys 2014), this tax would have significantly fostered cross-border shopping in Sweden and Germany. This has been tempered by the Danish Ministry of Taxation, who indicated that the tax may have caused cross-border shopping up to approx. 13 million € from 2010 to 2012 (Ecorys 2014). Other causes may be concurrently responsible for cross-border shopping, such as other tax variations on sweets, confectioneries, alcohol and tobacco (Ecorys 2014, p. 68). Some argue that the cheap nature of SSB may not justify cross-border shopping on its own, but may occur concurrently with the purchase of other products abroad. Little is known about soft drink cross-border trade related to taxation, with the exception of Bergman and Hansen's analysis (2013) of successive tax increases (1998 and 2001)[3] and a tax cut (2003)[4] on soft drinks in Denmark. Since soft drinks were known to be cheaper in Germany, observing a lower tax pass-through in Danish outlets located near the German border would have been a clue of cross-border shopping. But their analysis did not confirm this hypothesis (Bergman and Hansen 2013).

In Finland, an Ecorys case study (2014, p. 69) indicates that breweries and the soft drink industry claimed the tax had stimulated cross-border trade in Estonia. The Finnish beverage industry also suggests that the soda tax may have favoured cross-border shopping and illegal soda importation into Finland (Weston 2014). Well-documented evidence on a case-by-case basis would be required to confirm such statements.

In Canada, cross-border shopping risks may only marginally concern population-dense border areas, where a tax could be enacted in the jurisdiction located on one side of the border and not on the other. For instance, this could be the case of the Ottawa–Gatineau region, positioned at the border of Ontario and Quebec provinces. However, in such situations, manufacturers and retailers may develop adaptation strategies and not fully pass the tax onto soda prices to dissuade customers from doing their grocery shopping out of the tax perimeter. This possibility has been mentioned in the case of the municipality of Berkeley, California, where SSB retail prices rose by less than half of expected price increase following the implementation of the soda tax in March 2015 (Cawley and Frisvold 2015). Finally, it should be noticed that the Ottawa-Gatineau case is an exception and not the rule in Canada. The great majority of Canadians live quite far from provincial or US borders.

Considering parallel evidence from tobacco taxation also brings interesting lessons as regards smuggling risks, particularly in Canada (IARC 2011; Galbraith and Kaiserman 1997). In February 1994, the federal government and some provincial governments undertook to counteract smuggling by reducing tobacco taxation, which caused a sudden and significant increase in the sales of taxed cigarettes and a significant decrease in the sales of untaxed cigarettes[5] (Galbraith and Kaiserman 1997). However, the Canadian experience also showed that a reduction in smuggling might not necessarily lead to greater health benefits for the population. A tobacco consumption survey showed that, between January 1994 and February 1995, the decline in smoking prevalence had been slower in provinces where the federal tobacco tax cut had been supplemented by a provincial tobacco tax cut (i.e. in Ontario, Quebec, New Brunswick, Nova Scotia and Prince Edward

---

[3]From 0.8 DKK (eq. 0.17 CAD)/l to 1 DKK (0.21 CAD) in 1998 and to 1.65 DKK (eq. 0.34 CAD) in 2001.

[4]Back to 1.15 DKK (eq. 0.24 CAD)/l in 2003.

[5]These changes reflect the fact that consumers who had previously been purchasing contraband cigarettes actually "returned" to the "legal market" (Galbraith and Kaiserman 1997, p. 293).

Island). The authors conclude that the 1994 tobacco tax rollbacks may have actually "slowed the reduction in smoking prevalence in Canada" (Hamilton et al. 1997, p. 191). As the International Agency for Research on Cancer concludes in its review of evidence on tobacco control (2011, p. 356), "tax avoidance and tax evasion reduce, but do not eliminate, the public health and revenue impact of tobacco tax increases". In other terms, although this risk should be carefully considered, it does not necessarily justify tax repeal or inaction. Furthermore, in the case of a soda tax, smuggling practices would likely be more limited owing to the relatively low soda retail prices and tax rates considered (see Table 1.1).

Key messages

- Decision-makers have to weigh all SSB taxation potential benefits against potential risks. Among other things, this includes taking into account undesirable effects in terms of administration costs, employment, investment, competitiveness and cross-border shopping.
- Overall, such effects have been weakly documented, but observational studies, simulations and anecdotal evidence suggest that effects of a soda tax on these variables may be relatively limited.
- Provided other uncertainties are addressed (see other Chapters) and an evaluation of preliminary effects is well designed in advance, these risks may not justify inaction.

References

Bahl, R., Bird, R., & Walker, M. B. (2003). The uneasy case against discriminatory excise taxation: Soft drink taxes in Ireland. *Public Finance Review, 31*(5), 510–533. doi:10.1177/1091142103253753.

Berardi, N., Sevestre, P., Tepaut, M., & Vigneron, A. (2012). *The impact of a "soda tax" on prices: Evidence from French micro data* (Banque de France Working Paper No. 415) (p. 29). Banque de France Eurosystème. Direction Générale des Études et des Relations internationales. http://ssrn.com/abstract=2192470. Accessed February 24, 2016.

Bergman, U. M., & Hansen, N. L. (2013). *Are excise taxes on beverages fully passed through to prices? The Danish Evidence* (Working paper) (p. 33). http://web.econ.ku.dk/okombe/MBNLH.pdf. Accessed January 22, 2016.

Bíró, A. (2015). Did the junk food tax make the Hungarians eat healthier? *Food Policy, 54*, 107–115. doi:10.1016/j.foodpol.2015.05.003.

Bødker, M., Pisinger, C., Toft, U., & Jørgensen, T. (2015). The rise and fall of the world's first fat tax. *Health Policy (Amsterdam, Netherlands), 119*(6), 737–742. doi:10.1016/j.healthpol.2015.03.003

Bonnet, C., & Réquillart, V. (2012). Les effets de la réforme de la politique sucrière et des politiques de taxation sur le marché des boissons sucrées. *Cahiers de Nutrition et de Diététique, 47*(1), 35–41. doi:10.1016/j.cnd.2011.12.003.

Cawley, J., & Frisvold, D. (2015). *The incidence of taxes on sugar-sweetened beverages: The case of Berkeley, California* (No. w21465). Cambridge, MA: National Bureau of Economic Research. http://www.nber.org/papers/w21465.pdf. Accessed February 25, 2016.

Challenges. (2011, September 15). Le patron Europe désavoue le patron France de Coca-Cola. *Challenges*, (0268). http://www.challenges.fr/magazine/20110915.CHA4213/le-patron-europe-desavoue-le-patron-france-de-coca-cola-le-duo-de-choc.html. Accessed April 1, 2013.

City of Berkeley. (2014). Imposing a general tax on the distribution of sugar-sweetened beverage products. Pub. L. No. Measure D § Ballot question.

Ecorys. (2014). *Food taxes and their impact on competitiveness in the agri-food sector. Final report* (No. Ref. Ares(2014)2365745—16/07/2014). Rotterdam: Client: DG Enterprise and Industry.

Fédération Nationale des Boissons (FNB). (2014). Les BRSA et la période estivale, une belle histoire ! *E-Mag. L'actualité des entrepositaires grossistes en boissons*, *26*(Édition estivale). http://www.fnb-info.fr/fr/Entrez-dans-la-tendance-620.html?mbID=2668. Accessed February 25, 2016.

Galbraith, J. W., & Kaiserman, M. (1997). Taxation, smuggling and demand for cigarettes in Canada: Evidence from time-series data. *Journal of Health Economics, 16*(3), 287–301.

Hamilton, V. H., Levinton, C., St-Pierre, Y., & Grimard, F. (1997). The effect of tobacco tax cuts on cigarette smoking in Canada. *CMAJ: Canadian Medical Association journal = journal de l'Association medicale canadienne, 156*(2), 187–191.

Harrington, E. (2014, September 9). Billboards feature small business owners to fight San Francisco soda tax. *abc7news.com*. San Francisco. http://abc7news.com/politics/billboards-feature-small-business-owners-to-fight-sf-soda-tax/302056/. Accessed February 25, 2016.

Hespel, V., & Berthod-Wurmser, M. (2008). *Rapport sur la pertinence et la faisabilité d'une taxation nutritionnelle*. Inspection générale des Finances—Numéro 2008-M-002-01. Inspection générale des Affaires sociales—Numéro RM-2008-073. http://www.igf.finances.gouv.fr/webdav/site/igf/shared/Nos_Rapports/documents/2008/Taxe_nutritionnelle.pdf. Accessed February 25, 2016.

International Agency for Research on Cancer (IARC) (Ed.). (2011). *IARC handbooks of cancer prevention, tobacco control, volume 14, Effectiveness of tax and price policies for tobacco control*. Lyon: IARC.

La Tribune. (2011, October 13). La « taxe soda» attise la guerre entre industriels des boissons sucrées. *La Tribune (France)*.

Powell, L. M., Wada, R., Persky, J. J., & Chaloupka, F. J. (2014). Employment impact of sugar-sweetened beverage taxes. *American Journal of Public Health, 104*(4), 672–677. doi:10.2105/AJPH.2013.301630.

Preece, R. (2013). The effective contribution of excise taxation on non-alcoholic beverages to government revenues and social objectives: A review of the literature. *World Customs Journal, 7*(1), 21–38.

Rayon Boissons. (2013). Boissons rafraîchissantes sans alcool : les soft-drinks entraînent la catégorie à la baisse. *Rayon Boissons. Liquides BRSA & Eaux (France), 216*, 52.

Smed, S. (2012). Financial penalties on foods: The fat tax in Denmark: The fat tax in Denmark. *Nutrition Bulletin, 37*(2), 142–147. doi:10.1111/j.1467-3010.2012.01962.x.

Stafford, N. (2012). Denmark cancels "fat tax" and shelves "sugar tax" because of threat of job losses. *BMJ, 345*, e7889–e7889. doi:10.1136/bmj.e7889

Vallgårda, S., Holm, L., & Jensen, J. D. (2015). The Danish tax on saturated fat: why it did not survive. *European Journal of Clinical Nutrition, 69*(2), 223–226. doi:10.1038/ejcn.2014.224.

Weston, S. (2014, June 2). Tax increases affect drinks sales in Finland. *FoodBev.com*. http://www.foodbev.com/news/tax-increases-affect-drinks-sales-in-fin/#.VTAntWTBwXB. Accessed March 1, 2015.

Chapter 10
Raising Revenues from Sugar-Sweetened Beverage Taxation

Abstract Another potential logic of soda/sugar-sweetened beverage (SSB) taxation is to raise fiscal revenues, which may be earmarked for health care, health promotion or for the reduction of health inequities. In this chapter, we argue that a specific tax on SSBs is fiscally defensible because overconsumption of these beverages represents a significant public health issue and because targeting a limited scope of products is generally more acceptable and feasible. We also underline that soda taxes tend to live up to expectations in terms of revenues generated. We also refute the apparent antagonism between fiscal and behavioural logics and suggest that these may be concurrently pursued. Finally, building on international lessons from tobacco and soda taxation, we argue that earmarking soda tax revenues for health-related causes appears to be feasible and would likely increase public health benefits as well as the acceptability of the tax in the population.

10.1 Introduction

As previously described in Chap. 4 of this book, three sugar-sweetened beverage (SSB) taxation intervention logics generally emerge in the literature and could be concurrently pursued: a price-induced behavioural logic; a fiscal logic; and a signal-induced denormalization logic. Whereas Chaps. 5–9 have considered the price-induced behavioural logic, the present chapter will discuss the fiscal logic. It postulates that SSB taxation could be a way to raise fiscal revenues, which may be earmarked for health care, health promotion or for the reduction of health inequities. Several uncertainties are represented on Fig. 10.1.

A first issue to consider is to clarify whether targeting specifically SSBs (vs. other foods and beverages) is fiscally defensible. Secondly, it is essential to consider if tax collectors (e.g. SSB manufacturers or retailers) would actually collect and transfer the revenues generated to the public administration. Thirdly, since SSB taxation can reduce SSB demand (see Chap. 6), it is important to consider whether this may significantly alter the amount of revenues generated. Fourthly, SSB tax revenues shall be weighed against other potential tax-induced economic losses.

Y. Le Bodo et al., *Taxing Soda for Public Health*,
DOI 10.1007/978-3-319-33648-0_10

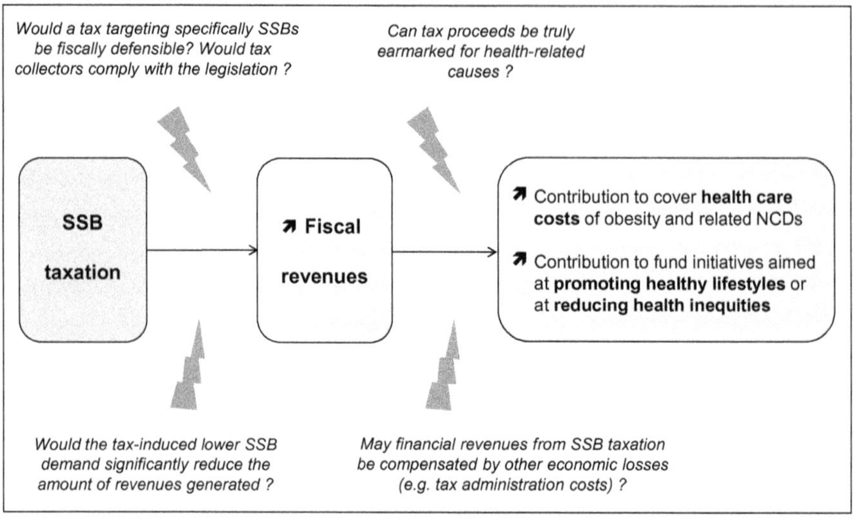

Fig. 10.1 Uncertainties about the propensity of SSB taxation to generate significant tax revenues and for these revenues to be earmarked for health-related causes

Finally, it is essential to question what are the prospects for tax revenues to be truly earmarked for health-related causes. All these aspects will be briefly discussed in the next sections.

10.2 Justifying the Fiscal Focus on SSBs

From a fiscal perspective, if the main objective of a soda tax is to raise revenue (and not necessarily to change behaviours), the reason for taxing SSBs only (vs. other foods and beverages) has to be argued (Preece 2013). In other terms, such a tax would be deemed discriminatory unless negative externalities specifically generated by SSB overconsumption are demonstrated. Chapter 2 of this book underlines that SSB overconsumption is not the only dietary risk factor to focus on. Yet, we also emphasize that the health risks related to their overconsumption, the high intakes often observed in youth, and the aggressive marketing practices of manufacturers justify specific prevention efforts. Although this rationale remains debatable (see Chap. 13), we consider that it is sound.

Preece (2013) highlights two other criteria to be taken into account for an SSB excise tax to be considered appropriate: the clarity of the tax scope (i.e. compatibility with the fiscal nomenclature) and the simplicity of tax administration (i.e. as regards data monitoring, compliance checks, revenue collection, etc.). On this particular point, large nutrition tax systems may raise acceptability and feasibility concerns (Hespel and Berthod-Wurmser 2008). Some taxes targeting a broad range of products proved to be viable at a large scale (e.g. in Hungary, Mexico), but others not. This has been illustrated by the rollback of the Danish "fat tax" only a few months after its implementation. The difficulty to report the saturated fat content in processed foods proved to be an important issue, and industries, politicians and the public opinion expressed significant concerns about this tax (Bødker et al. 2015). Therefore, on a fiscal ground, taxes targeting a limited scope of products tend to be more acceptable and feasible, as illustrated by several cases of taxes targeting specifically SSBs across the world (see Table 1.1). At least, such taxes seem to be relevant starting points (Hespel and Berthod-Wurmser 2008).

10.3 Revenues Potentially Raised from SSB Taxation

The fiscal objective of a soda tax (i.e. raising revenues) is often advocated on the basis of the substantial revenues raised from tobacco taxes. For example, in California, due to regular tobacco tax increases since 1989, tax revenues kept increasing despite decreasing smoking prevalence (Chaloupka et al. 2012). Evidence about revenues generated from soda taxes in French Polynesia, Samoa Islands, Hungary, Finland, Denmark, France, Mexico and the city of Berkeley (California) suggest that these taxes lived up to fiscal expectations (Thow et al. 2011; Landon and Graff 2012; Bíró 2015, WHO 2015a; Rosenberg 2015; Raguso 2016). For instance, in France, the soda tax raised 280 million € in 2012 and 288 million € in 2013, close to what was initially planned (Chauliac 2014). In the province of Quebec alone, for example, our macroscopic estimations based on apparent consumption data from Statistics Canada (2010) indicate that a tax equivalent to the French soda tax (i.e. 0.1 CAD/l) could rise between 75 and 79 million CAD in the year following implementation. These amounts could reach between 139 and 153 million CAD in the case of a tax twice as much elevated (i.e. 0.2 CAD/l) (see Box 10.1).

Box 10.1: Macro-estimation of the revenues that could be raised from a soda tax in the province of Quebec

In 2009, the Canadian domestic market for carbonated soft drinks (including SSBs and NCSBs) was estimated at 3.5 billion litres sold annually, using per capita consumption data and adjusting for trade (Agriculture and Agri-Food Canada 2015). Assuming that sales volumes are equally distributed across the Canadian territory proportionally to the population, since the province of Quebec accounts for 23 % of the Canadian population (Institut de la statistique du Quebec 2013), the Quebec carbonated soft drinks market can be estimated at 805 million litres sold annually. As an example, on the basis of price data collected in 2011 in a Quebec supermarket by the Quebec Weight Coalition (2012b, p. 21), the price of carbonated soft drinks varies from 1.17 to 3.36 CAD/l depending on the format (excluding special offers). Assuming a full pass-through of a volumetric tax of 0.1 CAD/l onto prices, the price increase would vary between 3 and 8.5 %, depending on the drink format. Assuming a conservative own-price elasticity of carbonated soft drinks demand at −0.79 (Andreyeva et al. 2010), the decrease in carbonated soft drinks demand would vary between 2.4 and 6.8 %. As a result, all things being equal, the Quebec carbonated soft drinks market would lower between 750 and 786 million litres the following year. According to these macro-estimations, a tax of 0.1 CAD/l on carbonated soft drinks could rise between 75 and 78.6 million CAD in the year following implementation in the province of Quebec. Replicating the same calculations and hypotheses in the case of a volumetric tax twice as much elevated (0.2 CAD/l), soda prices could increase between 6 and 17.1 %. As a result, the decrease in carbonated soft drinks demand could vary between 4.7 and 13.5 %, and the tax could rise between 139 and 153 million CAD in the year following implementation.

Therefore, Sects. 10.2 and 10.3 tend to indicate that the fiscal SSB taxation logic can stand alone. However, the evidence presented in Chaps. 5 and 6 also indicates that it could reasonably be accompanied with a behavioural objective to reduce SSB consumption in the population. The next section argues that fiscal and behavioural logics can be concurrently pursued.

10.4 Refuting Apparent Antagonism Between Fiscal and Behavioural Logics

If changing behaviours and raising revenues are concomitant objectives, the two logics may be perceived to enter somewhat in contradiction. Indeed, one may be tempted to argue that the more SSB consumption will decrease as a result of taxation, the less fiscal revenues could be raised in the long term (Bahl et al. 2003; Preece 2013). Such concerns should be addressed. First, it may be useful to hierarchize clearly the two priorities. For example, it could be explicitly acknowledged that the priority is to achieve a decrease in SSB overconsumption, whereas raising revenues is a collateral benefit but not the main goal of the intervention. Secondly, it seems important to remind that even if SSB demand decreases significantly overtime, a new tax will at worst generate fewer revenues, but it will always be more than if the tax was absent. Indeed, SSB demand is considered to be relatively sensitive to price (see Chap. 6), but it seems very unlikely that the new SSB taxation rate practised (e.g. 0.2 CAD/l) would fully drain the SSB market in the short to medium term. Additionally, periodical increases of the tax rate could compensate for the decline in SSB sales and maintain sizeable tax revenues in the long term. This approach already proved successful, e.g. in California, where, since

1989, regular tobacco tax increases made it possible to increase tax revenues despite decreasing smoking prevalence. This mechanism has permitted to extend the funding of tobacco control programmes (Chaloupka et al. 2012).

10.5 Weighing Fiscal Revenues Against Other Potential Economic Losses

Preece (2013) warns that the revenues raised from SSB taxation should be weighed against potential taxed-induced economic losses. First, these losses may concern manufacturers and retailers through new administration costs and lower profitability due to lower SSB sales. These aspects were briefly covered in Chap. 9. Although little documented, preliminary evidence indicates that the nuisance would likely be limited for the industry at large. Secondly, introducing a new excise tax may create additional administration costs for the government (e.g. for revenue collection, monitoring) and generate fewer revenues from other taxes (e.g. goods and services tax) as SSB sales decrease (Preece 2013). For example, in Ireland, Bahl et al. (2003) have demonstrated that tax revenues lost because of the abolishment of the excise tax on soft drinks in 1992 had been partly offset by greater sales tax and income tax revenues and by savings in administration costs. Making a very conservative extrapolation from this analysis, we could assume that a new SSB excise tax might be, at worst, revenue-neutral from a governmental perspective.

However, beyond direct fiscal revenues from SSB taxation, two other potential economic benefits related to a new SSB excise tax should be highlighted: (1) the creation of new jobs in governmental organizations (see Chap. 9); and (2) health-care cost savings resulting from reduced SSB consumption and from the ear-marking of SSB tax revenues for health promotion initiatives (see Sect. 10.6). Considering the latter, cost-effectiveness evaluations indicate that taxes could be among the most cost-saving and health-improving obesity interventions because of their relative low cost of implementation, the critical mass of the population they may reach, the revenues they generate, and their sustainability (Cecchini et al. 2010; Lehnert et al. 2012; Ananthapavan et al. 2014). According to modelling works by the Organization for Economic Cooperation and Development and the World Health Organization (Sassi 2010, p. 196), fiscal policies could even be the interventions most likely to "pay for [themselves], i.e. the only one[s] which generate larger savings in health expenditure than costs of delivery".

To better weigh health and economic benefits of SSB taxation against its costs, further economic evaluations will be useful and will gain robustness as new data from large-scale SSB taxes implemented across the world become available (see Chap. 14). In particular, cost–benefit analyses could help evaluate consumer welfare gains or losses from SSB taxation.

10.6 Earmarking SSB Tax Revenues for Health Promotion

Earmarking consists in "assigning revenues from designated sources to finance designated expenditures" (WHO 2012, p. 8). As previously mentioned (see Sect. 8. 4), earmarking soda tax revenues to interventions aimed to promote health and to reduce health inequities could somewhat mitigate equity concerns and increase the social acceptability of a tax (Caraher and Cowburn 2005; CQPP 2012a; Allday 2013) (see also Chap. 13). For example, some have suggested that tax revenues be earmarked for implementing fruit and vegetables subsidies towards disadvantaged households or for setting up water fountains in public spaces (Cash and Lacanilao 2007; Thow et al. 2011; Faulkner et al. 2011; Capacci et al. 2012).

Precedents exist in other domains, such as tobacco control. In its 2015 Report on the global tobacco epidemic, the WHO inventories 29 countries across the world earmarking parts of their excise tax revenues for health purposes. To name a few, this includes medical care and emergency activities, health facility enhancement programmes, cancer prevention and treatment, tobacco control, social programmes, health promotion and sports programmes, AIDS-related care activities, development of drugs for rare diseases (WHO 2015b). In North America, in 2005, 26 out of the 50 US states where excise taxes were applied on tobacco products earmarked some of these revenues for particular causes (Pomeranz 2012). In Canada, the Quebec tobacco tax revenues had been legally earmarked for the funding of the Olympic stadium built in Montreal in 1976 (Quebec Government 1976). Today, these revenues still contribute to funding sports, physical activity and healthy lifestyle promotion programmes. Beyond tobacco, revenues from other taxes are used for societal causes. For instance, in the USA, many of the excise taxes already applied by the federal government on fuel or vaccines are dedicated to health-, education- or environment-related programmes (Pomeranz 2012). In the Canadian province of Quebec, a fee paid by alcohol manufacturers (proportionally to beverage sales) finances the Éduc'alcool non-for-profit organization, whose aim is to raise awareness and educate youth and adults to make responsible decisions about drinking (Éduc'alcool 2016).

What about revenues from existing health-related soda taxes across the world? In about 2/3 of the 21 countries where such a tax was enacted (see Table 1.1, column 6), we have not found explicit mentions of earmarking of tax revenues for health-related causes. Conversely, eight jurisdictions provide interesting examples where such earmarking has been announced (although we did not research if these commitments had actually been fulfilled). In some cases, the earmarking was presented in a healthcare perspective. In Hungary, in 2011, revenues from taxes applied on SSBs and other foods and beverages have purportedly been earmarked for health care, via bonus payment granted to health workers (Hungarian Ministry of human resources 2011). In France, since 2013, all the revenues generated from the soda tax are earmarked for the National public health insurance system (Chauliac 2014). In Barbados as well, the proceeds of the tax on sugary drinks adopted in 2015 is to go to the health sector (WCRF 2015). In other cases, the earmarking was rather presented in a preventive or health promotion perspective. In French Polynesia, the revenues collected from a soft drink tax between 2002 and 2005 were used to support a variety of preventive health and "citizenship" projects (Thow et al. 2011). In the Navajo Nation (USA), revenues from the tax on SSBs and other energy-dense foods adopted in 2014 were announced to go for a diversity of health-promoting projects at community level (e.g. creation of sports and recreation infrastructures, vegetable gardens, healthy cooking classes, etc.) (Navajo Nation Council 2014). In Dominica, the proceeds of the SSB tax adopted in 2015 are to be earmarked for the funding of the national "Get Healthy" campaign (Skerrit 2015). In Berkeley, the City Council recently announced the funding of large public health and nutrition programmes, in particular to tackle SSB overconsumption[1] (Raguso 2016). Finally, in both a preventive and healthcare perspective, revenues from the tax on carbonated beverages adopted in Algeria in 2012 were to be earmarked for initiatives aimed to treat, prevent, detect and raise awareness on cancers (Official Journal of the Algerian Republic 2012).

Generally, little information is available on the portion of revenues truly earmarked for the mentioned causes and on the manner the money has been used. Therefore, the risk that tax revenues are diverted from health matters to fill general budget deficits has been highlighted in the case of tobacco and soda taxes (Fletcher et al. 2011; Marlow and Shiers 2010; Thow et al. 2011; WHO 2012). In the Pacific, for example, revenues from soda taxes initially earmarked for public health causes have sometimes been redirected over time as governmental priorities have changed (Thow et al. 2011). Such well-known issues jeopardize the acceptability of the tax in the public opinion (Barry et al. 2013). To avoid such diversion, planning explicitly in advance the earmarking of tax revenues for healthcare or health promotion activities has been increasingly recommended (IOM 2012; WHO 2012). Mechanisms that would be legally feasible and politically acceptable on a Canadian ground remain to be discussed (see Sect. 12.5).

Key messages

- From a public health perspective, taxing SSBs as a way to raise fiscal revenues could be acceptable, realistic and rapidly achievable.
- This fiscal logic could reasonably be accompanied with a behavioural objective to reduce SSB consumption in the population.
- If SSB demand decreases significantly overtime, a new tax will at worst generate fewer revenues, but it will always be more than if the tax was absent.

[1]The ballot of November 2014 mentioned that the revenues generated from the soda tax would be allocated to "[…] the general governmental needs" of the Berkeley population (City of Berkeley 2014, p. 2). However, a "Sugar Sweetened Beverage Products Panel of Experts" was then appointed to make recommendations about how the proceeds of the tax may be used from the general budget (Raguso 2016).

– Earmarking tax revenues for healthcare or health promotion initiatives could bring additional public health benefits, somewhat mitigate inequity concerns, and increase the acceptability of the tax. It is desirable to ensure that tax revenues will be truly earmarked for such initiatives.

References

Agriculture and Agri-Food Canada. (2015). The Canadian Soft Drink Industry. http://www.agr.gc.ca/eng/industry-markets-and-trade/statistics-and-market-information/by-product-sector/processed-food-and-beverages/the-canadian-soft-drink-industry/?id=1172167862291. Accessed March 1, 2015.

Allday, E. (2013, February 13). More support soda tax if for kids' health. *SFGate.com*. http://www.sfgate.com/health/article/More-support-soda-tax-if-for-kids-health-4277195.php. Accessed March 1, 2013.

Ananthapavan, J., Sacks, G., Moodie, M., & Carter, R. (2014). Economics of obesity—Learning from the past to contribute to a better future. *International Journal of Environmental Research and Public Health, 11*(4), 4007–4025. doi:10.3390/ijerph110404007.

Andreyeva, T., Long, M. W., & Brownell, K. D. (2010). The impact of food prices on consumption: A systematic review of research on the price elasticity of demand for food. *American Journal of Public Health, 100*(2), 216–222. doi:10.2105/AJPH.2008.151415.

Bahl, R., Bird, R., & Walker, M. B. (2003). The uneasy case against discriminatory excise taxation: Soft drink taxes in Ireland. *Public Finance Review, 31*(5), 510–533. doi:10.1177/1091142103253753.

Barry, C. L., Niederdeppe, J., & Gollust, S. E. (2013). Taxes on sugar-sweetened beverages: Results from a 2011 national public opinion survey. *American Journal of Preventive Medicine, 44*(2), 158–163. doi:10.1016/j.amepre.2012.09.065.

Bíró, A. (2015). Did the junk food tax make the Hungarians eat healthier? *Food Policy, 54*, 107–115. doi:10.1016/j.foodpol.2015.05.003.

Bødker, M., Pisinger, C., Toft, U., & Jørgensen, T. (2015). The rise and fall of the world's first fat tax. *Health Policy (Amsterdam, Netherlands), 119*(6), 737–742. doi:10.1016/j.healthpol.2015.03.003

Capacci, S., Mazzocchi, M., Shankar, B., Brambila Macias, J., Verbeke, W., Pérez-Cueto, F. J., et al. (2012). Policies to promote healthy eating in Europe: A structured review of policies and their effectiveness. *Nutrition Reviews, 70*(3), 188–200. doi:10.1111/j.1753-4887.2011.00442.x.

Caraher, M., & Cowburn, G. (2005). Taxing food: Implications for public health nutrition. *Public Health Nutrition*, *8*(08). doi:10.1079/PHN2005755

Cash, S., & Lacanilao, R. (2007). Taxing food to improve health: Economic evidence and arguments. *Agricultural and Resource Economics Review, 36*(2), 174.

Cecchini, M., Sassi, F., Lauer, J. A., Lee, Y. Y., Guajardo-Barron, V., & Chisholm, D. (2010). Tackling of unhealthy diets, physical inactivity, and obesity: Health effects and cost-effectiveness. *The Lancet, 376*(9754), 1775–1784. doi:10.1016/S0140-6736(10)61514-0.

Chaloupka, F. J., Yurekli, A., & Fong, G. T. (2012). Tobacco taxes as a tobacco control strategy. *Tobacco Control, 21*(2), 172–180. doi:10.1136/tobaccocontrol-2011-050417.

Chauliac, M. (2014, December 17). *Faut-il taxer les boissons sucrées pour contribuer à prévenir le surpoids et l'obésité: l'expérience française*. Quebec Heart and Lung Institute, Quebec city (Quebec), Canada.

City of Berkeley. (2014). Imposing a general tax on the distribution of sugar-sweetened beverage products. Pub. L. No. Measure D § Ballot question.

Éduc'alcool. (2016). About us. Governance. http://educalcool.qc.ca/en/about-us/governance/#.VqJbshgrK-w. Accessed January 22, 2016.

Faulkner, G. E., Grootendorst, P., Nguyen, V., Andreyeva, T., Arbour-Nicitopoulos, K., Auld, M. C., et al. (2011). Economic instruments for obesity prevention: Results of a scoping review and modified delphi survey. *International Journal of Behavioral Nutrition and Physical Activity, 8* (1), 109. doi:10.1186/1479-5868-8-109.

Fletcher, J. M., Frisvold, D. E., & Tefft, N. (2011). Are soft drink taxes an effective mechanism for reducing obesity? *Journal of Policy Analysis and Management: [the Journal of the Association for Public Policy Analysis and Management], 30*(3), 655–662.

Hespel, V., & Berthod-Wurmser, M. (2008). *Rapport sur la pertinence et la faisabilité d'une taxation nutritionnelle.* Inspection générale des Finances – Numéro 2008-M-002-01, Inspection générale des Affaires sociales – Numéro RM-2008-073. http://www.igf.finances.gouv.fr/webdav/site/igf/shared/Nos_Rapports/documents/2008/Taxe_nutritionnelle.pdf. Accessed February 25, 2016.

Hungarian Ministry of human resources. (2011). Hungary's public health tax on food products a hit: 42,000 health care workers to receive bonuses this year. http://www.kormany.hu/en/ministry-of-human-resources/news/hungary-s-public-health-tax-on-food-products-a-hit-42-000-healthcare-workers-to-receive-bonuses-this-year. Accessed March 1, 2013.

Institut de la Statistique du Quebec (ISQ). (2013). Population et structure par âge et sexe. http://www.stat.gouv.qc.ca/statistiques/population-demographie/structure/index.html. Accessed November 1, 2013.

Institute of Medicine (U.S.), Glickman, D., Committee on Accelerating Progress in Obesity Prevention, & Food and Nutrition Board (Eds.). (2012). *Accelerating progress in obesity prevention: Solving the weight of the nation.* Washington, DC: National Academies Press.

Landon, J., & Graff, H. (2012). What is the role of health-related food duties? In *A report of a National Heart Forum meeting*. London: National Heart Forum. http://www.worldobesity.org/site_media/uploads/UKHF_duties.pdf. Accessed February 28, 2016.

Lehnert, T., Sonntag, D., Konnopka, A., Riedel-Heller, S., & König, H.-H. (2012). The long-term cost-effectiveness of obesity prevention interventions: Systematic literature review. *Obesity Reviews: An Official Journal of the International Association for the Study of Obesity, 13*(6), 537–553. doi:10.1111/j.1467-789X.2011.00980.x.

Marlow, M. L., & Shiers, A. (2010). Would soda taxes really yield health benefits? *Regulation. Health and Medicine, Fall, 2010*, 34–38. doi:10.2139/ssrn.1691104.

Navajo Nation Council. (2014). President Shelly signs Healthy Dine' Nation Act of 2014 into law. Resolution of the Navajo Nation Council. 22nd Navajo Nation Council, Fourth Year, 2014. http://www.navajo-nsn.gov/News%20Releases/OPVP/2014/nov/Healthy%20Dine%20Nation%20Act%20of%202014.pdf. Accessed December 17, 2015.

Official Journal of the Algerian Republic. (2012). Conventions et accords internationaux – loi et décrets, arrêtés, décisions, avis, communications et annonces. Parution Numéro 52, 51ème année, 19 Septembre 2012. http://www.joradp.dz/FTP/jo-francais/2012/F2012052.pdf. Accessed March 1, 2013.

Pomeranz, J. L. (2012). Advanced policy options to regulate sugar-sweetened beverages to support public health. *Journal of Public Health Policy, 33*(1), 75–88. doi:10.1057/jphp.2011.46.

Preece, R. (2013). The effective contribution of excise taxation on non-alcoholic beverages to government revenues and social objectives: A review of the literature. *World Customs Journal, 7*(1), 21–38.

Quebec Government. (1976). Loi modifiant la Loi de l'Impôt sur le tabac. Chapitre 21. Sanctionnée le 30 Juin 1976.: http://www.bibliotheque.assnat.qc.ca/DepotNumerique_v2/AffichageFichier.aspx?idf=99424. Accessed February 25, 2016.

Quebec Weight Coalition (CQPP). (2012a). Sondage omnibus web effectué par Ipsos, pour le compte de la Coalition québécoise sur la problématique du poids, auprès d'un échantillon représentatif de 1001 répondants francophones du Quebec, âgés de 18 à 74 ans (du 12 au 20 novembre 2012). http://www.cqpp.qc.ca/fr/dossiers/boissons-sucrees/appuis. Accessed March 1, 2015.

Quebec Weight Coalition (CQPP). (2012b). *Sugar-sweetened beverage marketing unveiled. Volume 2: Price, an argument that pays off.* http://www.cqpp.qc.ca/en/sugar-sweetened-beverages/marketing. Accessed March 1, 2013.

Raguso, É. (2016, January 20). Council approves $1.5M to fight soda consumption. *Berkeleyside*. http://www.berkeleyside.com/2016/01/20/berkeley-council-approves-1-5m-to-fight-soda-consumption/. Accessed January 22, 2016.

Rosenberg, T. (2015, November 3). How one of the most obese countries on earth took on the soda giants. *TheGuardian.com*. http://www.theguardian.com/news/2015/nov/03/obese-soda-sugar-tax-mexico. Accessed December 16, 2015.

Sassi, F. (2010). *Obesity and the economics of prevention: Fit not fat* (p. 265). Organisation for Economic Co-operation and Development (OECD).

Skerrit, R. (2015). *Budget address 2015–2016—"Keeping it Real" by Honourable Roosevelt Skerrit*. http://finance.gov.dm/index.php/budget/budget-addresses/file/22-budget-address-2015-2016-keeping-it-real-by-honourable-roosevelt-skerrit-july-24-2015. Accessed December 17, 2015.

Statistics Canada. (2010). *Food Statistics 2009* (No. Catalogue no. 21-020-X). http://www.statcan.gc.ca/pub/21-020-x/21-020-x2009001-eng.pdf. Accessed March 1, 2012.

Thow, A. M., Quested, C., Juventin, L., Kun, R., Khan, A. N., & Swinburn, B. (2011). Taxing soft drinks in the Pacific: Implementation lessons for improving health. *Health Promotion International, 26*(1), 55–64. doi:10.1093/heapro/daq057.

World Cancer Research Fund (WCRF) (Ed.). (2015). WCRF international nourishing framework. Use economic tools to address food affordability and purchase incentives. Table last updated November 24, 2015. http://www.wcrf.org/sites/default/files/3_Economic%20Tools_Final.pdf. Accessed December 17, 2015.

World Health Organization (WHO). (2012). *Tobacco taxation and innovative health-care financing*. New Delhi, India: World Health Organization. http://www.searo.who.int/tobacco/documents/2012-pub1.pdf. Accessed January 22, 2016.

World Health Organization (WHO). (2015a). *Using price policies to promote healthier diets.* Copenhagen (Denmark): World Health Organization (WHO).

World Health Organization (WHO). (2015b). *WHO report on the global tobacco epidemic, 2015. Raising taxes on tobacco. Fresh and alive. MPOWER. Annexes.* http://www.who.int/tobacco/global_report/2015/appendix2.pdf?ua=1. Accessed December 20, 2015.

Chapter 11
Potential "Signal" Effects from Sugar-Sweetened Beverage Taxation

Abstract Many characteristics of the market, school and family environments contribute to making or maintaining frequent soda consumption a social norm, particularly among youth. In such circumstances, soda/sugar-sweetened beverage (SSB) taxation could be used as "alarm signal" facilitating the emergence of a "moderate consumption" norm, provided the price hike is sufficiently salient, the population aware of the tax, and these two conditions sustained by a communication campaign. Unfortunately, such effects remain weakly documented and counterproductive effects cannot be ruled out. SSB taxation could also be used as a coercive signal pushing manufacturers to move beyond voluntary commitments, be this in terms of product reformulation and/or in terms of marketing to children. On this particular point, the evidence remains mixed and anecdotal. A tax linearly indexed on the sugar content of the beverages or based on a threshold of sugar content above which beverages are taxed may provide additional benefits. Here again, it remains to be demonstrated.

Beyond the price-induced behavioural logic and the fiscal logic associated with sugar-sweetened beverage (SSB) taxation (see Chaps. 5–10), a third logic is sometimes discussed in the literature. It consists in taxing SSBs to send a salient signal to the population and to the manufacturers in order to encourage, in a longer term, social norms changes and product reformulation (see Fig. 11.1).

However, the propensity of consumers and manufacturers to actually perceive the signal, to be influenced by the signal and to react in the desired direction is uncertain. This chapter provides an overview of these issues.

11.1 An "Alarm Signal" to the Population

Reviews on SSB consumption in youth show that it has been significantly associated with many factors at the family level (e.g. SSB availability at home, parental SSB consumption, parental rules, TV watching), at the school level (e.g.

Y. Le Bodo et al., *Taxing Soda for Public Health*,
DOI 10.1007/978-3-319-33648-0_11

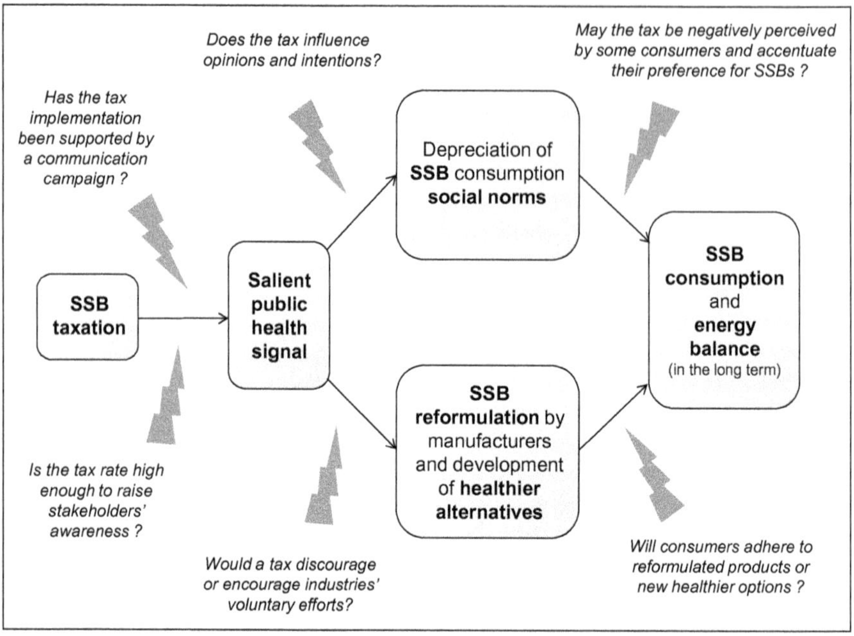

Fig. 11.1 Uncertainties about the propensity of SSB taxation to generate a salient public health signal positively affecting SSB consumption's social norms and fostering product reformulation

availability) and at the market level (e.g. low SSB price, living near a fast-food or a convenience store) (Hector et al. 2009; Mazarello Paes et al. 2015; Tak et al. 2011; Verloigne et al. 2012). For example, in Australia, a study based on a large-scale cross-sectional survey conducted in 2010 showed that children aged 4–10 years were almost three times as likely to be high SSB consumers (≥ 5 cups/week) if they purchased SSBs from their school canteen. For youth aged 9–16, the ratio rose to 5 times if SSBs were "usually available at their home" and to 10 times if SSBs were "usually consumed with meals at home" (Hebden et al. 2013). In sum, many characteristics of the market, school and family environments may contribute to making or maintaining SSB consumption a social norm[1] in some groups of the population. This social norm seems to be particularly salient among adolescents. For example, in the USA and the UK, two studies highlighted by Baril et al. (2014) have shown that adolescents tend to overestimate SSB consumption of their peers and to overestimate how positive their peers feel about it (Perkins et al. 2010; Lally et al. 2011). This social norm may contribute to stimulating SSB overconsumption.

In such circumstances, among other instruments, SSB taxation could be used as "alarm signal" counteracting SSB consumption social norms, i.e. making it less

[1]Baril et al. (2012, p. 6) define social norms as "socially shared rules and behavioural models, based on common values and implying a pressure in favour of the adoption of a given behaviour, at the risk of disapproval from the society or the reference group" (free translation).

acceptable and desirable in the population (Fédération québécoise du sport étudiant [FQSE] 2010). In other words, taxes might alleviate the current "high consumption" norm and reinforce instead the emergence of a "moderate consumption" norm, as advocated by public health proponents and experts. Some prerequisites may be necessary to produce such a signal: (1) the price hike would have to be sufficiently salient; (2) the population would have to be sufficiently aware of the tax; and (3) the two precedent conditions may have to be sustained by a large communication campaign focused on the tax. For example, in Hungary, a study conducted by the National Institute for Health Development (NIHD) (quoted by Ecorys 2014, p. 219) indicates that among the consumers having reduced their consumption of pre-packaged sweets and salty snacks following the introduction of taxes in 2011, 80 % reported having done so primarily because of price changes, but 20 % reported having done so because "they became aware of the fact that the consumption of such products is unhealthy". The NIHD concluded that the tax may have been more effective if greater public communication efforts had been committed. This area deserves further attention. For example, it would be worth analysing if (and how far) the controversies and strong advocacy campaigns that have surrounded the soda tax bills in Mexico (in 2013) and Berkeley (in 2014) have fostered lower SSB purchases (beyond the influence of price hikes once the tax was implemented).

However, using soda taxes as a signal may also have counterproductive effects: research has shown that banning the consumption of some foods (including SSBs) tends to accentuate children's preference for it (Baril et al. 2014). Consequently, the preference for SSBs in children may get reinforced if the tax is explicitly presented as a way to restrict consumption (Farley et al. 2012). Actually, data from the *Enquête Québécoise sur la Malbouffe* conducted in 2010–2011 in 9941 Quebec secondary school children[2] show that, among the students willing to change their "junk food habits" and having carbonated beverages at least once or twice a week, nearly 4 out of 10 students (38 %) indicate their intention to reduce their SSB intake. For those students, may SSB taxation work as a facilitator? This is uncertain as, overall, in this survey, soft drink taxation is the most unpopular option to reduce "junk food" consumption (28 % of agreement) in comparison with others such as improvements in nutrition labelling (62 %) or improvements in the food offered in the places students often visit (corner shops, movie theatres, etc.) (60 %) (Hovington and RESQ 2012). This is part of the reasons why Just and Wansink consider that "voluntary approaches (e.g. making soft drinks less visible and less convenient) are much more likely than regulations to create long-term behavioural habits and much less likely to create a class of soft-drink freedom fighters" (Farley et al. 2012, p. 1466).

---

[2]This study aimed to collect a minimum number of questionnaires by school (30) and by region (100) in order to strengthen the statistical validity of the results. In fact, 36 out of the 87 schools and 2 out of the 14 regions surveyed did not reach these minimum numbers. Design weights were not consequently adjusted (Hovington and RESQ 2012). Therefore, even if the survey is based on a very large sample of students, it should be noted that it is not a probability-based survey.

From another angle, the "tax signal" may have greater impact on adults' SSB consumption norms, which may indirectly impact consumption in youth. Indeed, research indicates that children whose parents regularly consume SSBs are more likely to consume SSBs themselves (Baril et al. 2014). Furthermore, data indicate that SSB taxation may be more acceptable for adults than for youth. Further research should be conducted to better estimate SSB taxation awareness and acceptance in the population as well as related effects on intentions and behaviours (see also Chap. 13).

For example, a cross-sectional study analysed Belgian citizens' ($n = 565$) awareness and acceptance of fiscal measures implemented in neighbouring countries (i.e. "fat tax" in Denmark, "soda tax" in France and "public health nutrition tax" in Hungary) (Mustafa 2013). Among cola consumers ($n = 405$), 30 % reported that a tax similar to the French soda tax would make them reduce their cola consumption, whereas 60 % reported that they would likely not change their habit and 10 % would actually increase their consumption.[3] However, overall, the results indicate that the awareness of these food taxes was relatively low[4] in the sample of Belgian citizens under study and was significantly correlated neither with their level of acceptance of these taxes nor with the purchase intention of taxed products (Mustafa 2013). It suggests that the "signal" conveyed by food taxes in a population operates in a complex environment and may contribute to diverging opinions, intentions and behaviours within and across consumers. For example, in this study, high-income individuals showed greater awareness (not acceptance) of fiscal measures, but their purchase intention of taxed products did not significantly differ from other consumers. Similarly, those with higher nutrition knowledge showed greater level of acceptance (but not awareness) of fiscal measures, but their purchase intention of taxed products did not significantly differ from other consumers. The author suggests to explore these variables during the evaluation of food taxes actually implemented at a large scale (Mustafa 2013). The use of multivariate models and larger samples would likely be helpful to disentangle the respective contributions of these different factors (see Sect. 11.4).

11.2 A "Coercive Signal" to Manufacturers

On the other hand, taxing SSBs could send a coercive signal to engage manufacturers into moving beyond voluntary commitments (Boizot-Szantai and Étilé 2011), be this in terms of product reformulation and/or in terms of marketing to children.

---

[3]The propensity to reduce consumption was higher for "special foods" (e.g. energy drinks) than for common foods (e.g. butter) (Mustafa 2013).

[4]For example, 7.4 % reported being very "well aware" of the French soda tax (implemented since January 2012), whereas 32.2 % reported "having heard" about it without being "sure", and 60.4 % said they had "never heard" about it (Mustafa 2013).

Firstly, SSB taxation could be used as a signal to catalyse product reformulation across all SSB brands and categories. The recent downward trend in SSB demand in many countries seems to have already favoured this innovation process, e.g. through the replacement of added sugars by low-calorie intense sweeteners (e.g. stevia) (Euromonitor 2014). In parallel, some of the biggest SSB companies have already launched large-scale corporate programmes to demonstrate their willingness and efforts in developing and promoting a healthier portfolio of beverages (Arthur 2015). A new tax or increases in the rate of existing taxes may further pressure manufacturers into reducing the sugar content of their products and could contribute to accelerate the development and promotion of healthier alternatives. In Hungary, for example, a case study conducted by Ecorys indicates that the "Public Health Product Tax" (PHPT) explicitly raised attention of the industry on public health issues and may have favoured a number of pro-health initiatives in terms of reformulation, health impact assessment and public–private negotiations (Ecorys 2014, p. 222). At the opposite, according to Ecorys, industry stakeholders from several countries have claimed that many improvements had already started prior to the introduction of health-related food taxes and that these new taxes may actually have slowed this progress for two reasons: (1) taxes may be perceived as a signal discouraging voluntary efforts, e.g. those aimed to reduce (but not eliminate) SSBs' sugar content[5], and (2) in some cases, additional taxes may reduce the capacity of investment in product innovation (Ecorys 2014, p. 208, 242).

From this perspective, a tax linearly indexed on the sugar content of the beverages (i.e. an "input tax") may prove much more virtuous than a tax hitting indifferently all types of calorically sweetened beverages. In the case of alcohol, a specific tax amount based on the quantity of alcohol (rather than the volume of liquid) has proved to encourage the production of lower alcohol beverages and to foster healthier consumption patterns (e.g. case study on beers in South Africa) (Blecher 2015). However, administratively speaking, such a tax is more likely to raise feasibility concerns (see Sect. 12.4). Among the cases of SSB taxation listed in Table 1.1, Mauritius (excise tax of 3 cents per gram of sugar content in the targeted beverage) is the only case (WCRF 2015). An alternative likely easier to administer would consist in fixing a threshold of sugar content above which beverages are taxed. For example, such an approach has been used for SSB taxation in Denmark (≥ 0.5 g of sugar/100 mL), in Finland (>0.5 g of sugar/100 g or mL), in Hungary (>8 g of added sugars/100 mL), in St Helena (≥ 15 g of sugar/L) and in Chile (>6.25 g of sugar/100 mL) (see Table 1.1, column 4). Such proposals are attractive in that they may encourage manufacturers to reformulate existing products (provided technological and taste requirements are met) or to develop lower-calorie options. Such effects remain to be demonstrated.

Secondly, SSB taxation could also be used as a signal contributing to counteract and "denormalize" SSB marketing to youth (FQSE 2010). Indeed, the marketing of

[5]Reducing SSBs' sugar content is recognized to be a technological and marketing challenge, since taste is a critical choice parameter for SSB consumers.

SSBs is characterized by a very diversified offer, a high availability in multiple settings, very competitive prices and a strong promotional pressure (especially towards youth) (see Sect. 2.3). According to the above-mentioned survey *Enquête Québécoise sur la Malbouffe,* youth seem well aware of the marketing influence on dietary habits: a majority of respondents (79 %) think that "junk food" advertising influences students to eat more of it, and 84–95 % acknowledge that "SSB manufacturers only go after profits and not youth's health and well-being". However, they appear to be weakly aware of the type of marketing strategies used by this industry (Hovington and RESQ 2012). Consequently, an SSB tax could be part of a multicomponent strategy denouncing SSB marketing to youth instead of laying too much emphasis on SSB consumers.

Although direct evidence supporting tax-induced signal effects on manufacturers is lacking, large-scale natural experiments have demonstrated that regulatory policies aimed at reducing specific nutrients in food (e.g. trans fat, salt) can effectively influence industrial practices and improve the population's dietary habits. Congruent informational policies (e.g. front-of-pack icons and nutrient labels) may favour industrial compliance to such regulations (Mozzafarian et al. 2012).

11.3 Arguments for a Comprehensive "denormalization" Strategy

In sum, there is a limited scientific evidence about how SSB taxation could be used as a signal contributing to "denormalize" SSB overconsumption. This is not surprising: changing social norms is a complex process that should be based on a thorough behavioural and contextual analysis and the intense involvement of multiple stakeholders. This generally requires a combination of various strategies such as mass media campaigns, social network strategies, legitimization by opinion leaders and social marketing actions over a long period of time (Baril et al. 2014). To reduce SSB overconsumption, Baril et al. (2014, p. 15) suggest adopting a multisectoral strategy that may include a policy limiting access to SSBs and increasing access to healthier options (especially in schools and sports settings); the mobilization of local advocacy groups; the denormalization of SSB industry marketing practices; the correction of potential overestimation by the public of SSBs consumed by their peers; the normalization of the consumption of healthy beverages via social marketing techniques (e.g. water, milk); a social debate on SSBs (working with coalitions); a regulation of SSB marketing to children; a taxation of SSBs and price decreases on healthier options; the withdrawal of SSB vending machines; and addition of water fountains in public places. Thus, rather than an isolated "signal", SSB taxation would be better positioned within a comprehensive "denormalization" strategy aimed at influencing knowledge, opinions, attitudes and intentions, as well as manufacturers' practices (see also Sect. 3.2).

11.4 The Complex Evaluation of Signal Effects

Characterizing SSB taxation signal effects and their impact on social norms is a challenge, especially as the strength of the signal does not remain constant over time: it may reach a peak at the time of adoption, but its resonance in the long term may be less detectable, especially if there are no "boosters" (e.g. regular tax increases). Using qualitative or mixed research designs would be desirable on these issues.

Among other indicators, measuring the acceptability of SSB taxation over time in the population can be used as an informative proxy. Encouraging results from the NutriNet-Santé cohort study conducted in France between November 2012 and April 2013 show that 60 % of respondents are aware of the SSB tax and most of them recall that SSBs are the primary target of the tax (Julia et al., unpublished data). While some polls (sponsored by the industry) rather showed low public support of SSB taxation prior to its implementation (Harris Interactive 2011), half of the NutriNet-Santé survey respondents "support imposing a tax on sweetened beverages" and 58 % consider that "having a tax on sweetened beverages would help improve population health" (Julia et al. 2015, p. 4). The authors note that younger respondents and those with lower educational levels were less likely to support the tax. Moreover, participants reporting moderate and high consumption of sweetened beverages were less likely to feel that the tax would contribute to improving the population's health (Julia et al. 2015). It suggests that on average, those consumers may be less receptive to a tax-induced "alarm signal". Finally, the authors highlight that many contextual factors may have influenced survey results. In particular, many actions (including mass communication campaigns) implemented as part as of the National public health nutrition programme since 2000 may have contributed to raise the public's awareness on SSB-related health risks (Julia et al. 2015). Replicating this survey would certainly help monitor the population's perception of the tax over time as well as the propensity of this perception to encourage actual behaviour change.

As regards manufacturers' practices, observations from countries having recently adopted SSB taxes at a large scale will be critical to appreciate how far SSB taxation could be a "coercive signal". For example, in France, although causal inferences cannot be demonstrated, SSB taxation has been concomitant with the development of healthier innovations on the soft drink market, e.g. beverages blending sugar with low-calorie sweeteners (e.g. stevia), fruit-based beverages with fewer added sugars and new flavours (FNB 2014).

Key messages

- Social norms are a significant driver of SSB consumption behaviours, especially in adolescents.
- Parallel data (e.g. from tobacco control) indicate that SSB taxation may work as a signal altering “SSB consumption” norms, but evidence is limited and counterproductive effects cannot be ruled out.
- Rather than an isolated “signal”, SSB taxation would be better positioned within a comprehensive “denormalization” strategy aimed at influencing knowledge, opinions, attitudes and intentions.
- SSB taxation may also work as a “coercive signal” pushing manufacturers to reformulate existing products or to develop healthier options. Such effects should be further documented. A tax linearly indexed on the sugar content of the beverages or based on a threshold of sugar content above which beverages are taxed may be particularly inciting.

References

Arthur, R. (2015, March 5). All Coke is equal: Coca-Cola GB shifts Life, Zero, and Diet to centre stage in new “one brand” strategy. *Beveragedaily.com.* http://www.beveragedaily.com/Manufacturers/All-Coke-is-equal-Coca-Cola-GB-shifts-Life-Zero-and-Diet-to-centre-stage-in-new-one-brand-strategy. Accessed 5 March 2015

Baril, G., Laguë, J., Paquette, M.-C., Institut national de santé publique du Québec, & Direction du développement des individus et des communautés. (2012). *Les normes sociales et l’alimentation analyse des écrits scientifiques*. [Montréal, Qué.]: Direction du développement des individus et des communautés, Institut national de santé publique Québec. http://site.ebrary.com/id/10617088. Accessed 25 February 2016

Baril, G., Trudel, C., Le May, D., Mongeau, L., & Noreau, C. (2014). *Les saines habitudes de vie, c’est bien normal!* (p. 46). [Montréal, Qué.]: Rapport sous la responsabilité de la Coordination conjointe (Québec en Forme-Directions régionales de santé publique-Ministère de la Santé et des Services sociaux du Québec). http://veilleaction.org/documents/Les-saines-habitudes-de-vie-cest-bien-normal.pdf

Blecher, E. (2015). Taxes on tobacco, alcohol and sugar sweetened beverages: Linkages and lessons learned. *Social Science and Medicine, 136–137*, 175–179. doi:10.1016/j.socscimed.2015.05.022.

Boizot-Szantai, C., & Étilé, F. (2011). Taxer les boissons sucrées pour lutter contre l’obésité? Le point de vue de l’économie. *Obésité, 6*(4), 218–225. doi:10.1007/s11690-012-0306-4.

Ecorys. (2014). *Food taxes and their impact on competitiveness in the agri-food sector. Annexes to the Main report* (No. Ref. Ares(2014)2604304—06/08/2014). Rotterdam: Client: DG Enterprise and Industry.

Euromonitor International. (2014). *The Sugar Backlash and its Effects on Global Consumer Markets*. http://www.euromonitor.com/the-sugar-backlash-and-its-effects-on-global-consumer-markets/report. Accessed 24 February 2016

Farley, T., Just, D. R., & Wansink, B. (2012). Clinical decisions. Regulation of sugar-sweetened beverages. *The New England Journal of Medicine, 367*(15), 1464–1466. doi:10.1056/NEJMclde1210278.

Fédération Nationale des Boissons (FNB). (2014). Les BRSA et la période estivale, une belle histoire ! *E-Mag. L'actualité des entrepositaires grossistes en boissons, 26*(Édition estivale). http://www.fnb-info.fr/fr/Entrez-dans-la-tendance-620.html?mbID=2668. Accessed 25 February 2016

Fédération québécoise du sport étudiant (FQSE). (2010). *La dé-normalisation de la malbouffe auprès des jeunes – Constats issus de la littérature. Un projet du Sport Étudiant.*

Harris Interactive. (2011, September 5). La réaction des Français suite aux annonces de François Fillon du 24 août 2011. Sondage Harris Interactive pour le Syndicat National des Boissons Rafraîchissantes. Note détaillée. http://harris-interactive.fr/wp-content/uploads/sites/6/2015/09/results_HIFR_economie_07092011.pdf. Accessed 25 February 2016

Hebden, L., Hector, D., Hardy, L. L., & King, L. (2013). A fizzy environment: availability and consumption of sugar-sweetened beverages among school students. *Preventive Medicine, 56* (6), 416–418. doi:10.1016/j.ypmed.2013.02.017.

Hector, D., Rangan, A., Louie, J., Flood, V., & Gill, T. (2009). *Soft drinks, weight status and health: a review.* Sydney: Cluster of Public Health Nutrition, the University of Sydney. http://ro.uow.edu.au/cgi/viewcontent.cgi?article=1317&context=hbspapers. Accessed 25 February 2016

Hovington, J., & Réseau du Sport Étudiant du Québec (RESQ). (2012). *Projet de dénormalisation de la malbouffe auprès des jeunes: enquête québécoise sur le marketing de la malbouffe : 10,000 jeunes se prononcent!* http://ll.rseq.ca/download/attachments/15958040/Rapport%20d'enquete-FRA-1-page.pdf?version=1. Accessed 28 February 2016

Julia, C., Méjean, C., Vicari, F., Péneau, S., & Hercberg, S. (2015). Public perception and characteristics related to acceptance of the sugar-sweetened beverage taxation launched in France in 2012. *Public Health Nutrition, 18*(14), 2679–2688. doi:10.1017/S1368980014003231.

Lally, P., Bartle, N., & Wardle, J. (2011). Social norms and diet in adolescents. *Appetite, 57*(3), 623–627. doi:10.1016/j.appet.2011.07.015.

Mazarello Paes, V., Hesketh, K., O'Malley, C., Moore, H., Summerbell, C., Griffin, S., et al. (2015). Determinants of sugar-sweetened beverage consumption in young children: A systematic review. *Obesity Reviews: An Official Journal of the International Association for the Study of Obesity, 16*(11), 903–913. doi:10.1111/obr.12310.

Mozaffarian, D., Afshin, A., Benowitz, N. L., Bittner, V., Daniels, S. R., Franch, H. A., et al. (2012). Population approaches to improve diet, physical activity, and smoking habits: A scientific statement from the American Heart Association. *Circulation, 126*(12), 1514–1563. doi:10.1161/CIR.0b013e318260a20b.

Mustafa, T. S. (2013). *Food and nutrition policy interventions for healthier consumer food choice* (Master's dissertation submitted in partial fulfillment of the requirements for the degree of Master of Science in Nutrition and Rural Development: Human Nutrition. Major: Public Health Nutrition). Universiteit Gent. Faculteit Bio-ingenieurswetenschappen., Ghent, Belgium. Retrieved from http://lib.ugent.be/catalog/rug01:002063678

Perkins, J. M., Perkins, H. W., & Craig, D. W. (2010). Misperceptions of peer norms as a risk factor for sugar-sweetened beverage consumption among secondary school students. *Journal of the American Dietetic Association, 110*(12), 1916–1921. doi:10.1016/j.jada.2010.09.008.

Tak, N. I., Te Velde, S. J., Oenema, A., Van der Horst, K., Timperio, A., Crawford, D., & Brug, J. (2011). The association between home environmental variables and soft drink consumption among adolescents. Exploration of mediation by individual cognitions and habit strength. *Appetite, 56*(2), 503–510. doi:10.1016/j.appet.2011.01.013.

Verloigne, M., Van Lippevelde, W., Maes, L., Brug, J., & De Bourdeaudhuij, I. (2012). Family- and school-based correlates of energy balance-related behaviours in 10-12-year-old children: a systematic review within the ENERGY (European Energy balance Research to prevent excessive weight Gain among Youth) project. *Public Health Nutrition, 15*(8), 1380–1395. doi:10.1017/S1368980011003168.

World Cancer Research Fund (WCRF) (Ed.). (2015). WCRF international nourishing framework. Use economic tools to address food affordability and purchase incentives. Table last updated 24 November 2015. http://www.wcrf.org/sites/default/files/3_Economic%20Tools_Final.pdf. Accessed 17 December 2015

Part III
How: Applicability of Sugar-Sweetened Beverage Taxation

Chapter 12
Feasibility of Sugar-Sweetened Beverage Taxation in Canada

Abstract Exploring the feasibility of soda/sugar-sweetened beverage (SSB) taxation in Canada requires considering federal and provincial powers and prerogatives in terms of public health and taxation. In this chapter, the applicability of various types of soda taxes is considered for both jurisdictions. Overall, excise duties generally are the more documented and recommended option to tax soda. Introducing such a tax in Canada is likely feasible, but at the federal level only. At provincial level, the introduction of a special tax on SSBs sold at retail according to the tobacco and alcohol special tax models is an avenue deserving particular attention. Defining the tax scope would likely raise some issues since generally, in the Canadian fiscal nomenclature, all carbonated beverages are considered regardless of their calorie content and composition. As regards taxation rate, Chaps. 6 and 7 of this book suggest that it is reasonable to expect SSB consumption to decrease as a result of a significant tax (i.e. provided it generates a 10–20 % price hike). Finally, a robust mechanism facilitating the administration and the earmarking of tax revenues for health promotion initiatives and social programmes would likely be desirable and feasible in Canada.

12.1 Introduction

Evidence indicates that the enactment and the implementation of a sugar-sweetened beverage (SSB) tax are facilitated when it relies on existing legislative mechanisms (Thow et al. 2011). In this chapter, we will consider the mechanisms that have been more commonly described, recommended or used in the context of SSB taxation. We will also discuss the feasibility of such mechanisms to be used on a Canadian ground. For a purpose of clarity and interpretation, Table 12.1 gives a definition of several terms that will be used along this chapter. Building on Chriqui et al. (2013)'s typology of beverage taxation and other references, the table has been subdivided into several themes: tax targeting; tax base; tax type; tax visibility; tax scope and rate; and earmarking of tax revenues (Sharma et al. 2014a; WHO 2012; von Tigerstrom 2012; Pomeranz 2014; Thow et al. 2011; Preece 2013).

Y. Le Bodo et al., *Taxing Soda for Public Health*,
DOI 10.1007/978-3-319-33648-0_12

Table 12.1 Definition of terms frequently used to describe SSB tax instruments

Tax target	
Direct tax[a]	Tax "imposed on the person who is intended to bear the burden of the tax" (i.e. not supposed to be passed on to another person)
Indirect tax[a]	Tax imposed on a person who is not intended to bear the burden of the tax, i.e. a person "who is expected to pass it on to the person who ultimately will bear it"
Tax base	
Volumetric or specific tax[b,c]	Tax or levy based on the volume or unit of a product, e.g. 10 cents per litre or 2 cents per gram of added sugar in a beverage
Valoric or *ad valorem* tax[b,c]	Tax applied as a percentage of the value of a product, e.g. 20 % of the pre-tax price
Tax type	
Excise tax or duty[a,c,d]	Tax "levied on the manufacture, sale, use or distribution" of a good, either as a **volumetric** or **valoric** tax. Their primary goal generally is to raise revenues and, eventually, to discourage the consumption of some goods, e.g. for health reasons (excise taxes on tobacco, alcohol, SSB). Since these taxes are generally passed to the final consumer but are applied earlier in the commercial chain, they are generally considered to be **indirect**
Canadian sales taxes[a,c,d]	In Canada, the goods and services tax (GST), the harmonized sales tax (HST), the provincial sales tax (PST) and the Quebec sales tax (QST) are sales taxes working as value added or cascading taxes, meaning that they apply to the "production and distribution of commercial goods [and are] charged …at each stage in the production/distribution chain… as a percentage of price". Their primary goal is to raise revenue. They are ultimately paid by the final consumer at the point of purchase. These taxes generally apply "as a single rate on a broad range of goods", although reduced rates or exemptions may be used (e.g. on basic groceries)
Special tax[a,e]	Tax generally applied on the consumption of specific goods, e.g. for health or environmental reasons (e.g. tobacco, alcohol and fuel). It can be **valoric** or **volumetric**. The mechanism can be very similar to an excise tax or a sales tax, even if not named so. Instead, it may be called a "special consumption tax" or a "specific tax". When the tax directly applies to the final consumer (at the point of retail), it is considered to be **direct**
Import duty[d,e,f]	Tax "on a selected good imported in a country and destined for domestic consumption". These duties are generally "collected from the importer at the point of entry into the country". Their mechanism can be very similar to an excise tax, even if not named so. Their primary goal is to "protect domestic production and raise revenue"
Minimum pricing[g]	"Addition of a minimum percentage markup to the base price of a good by the retailer to increase its retail price to consumers" (e.g. minimum price laws on alcohol or tobacco)
Tax visibility	
Tax included in the shelf price[c]	A tax is said to be included in the shelf price when the price displayed on the shelf where the consumer selects the product (e.g. store or vending machine) already includes the amount of the tax

(continued)

Table 12.1 (continued)

Tax included at the point of purchase[c]	A tax is said to be included at the point of purchase when it is applied to the price when the consumer actually pays the product (e.g. at the cash register or checkout). This means that the tax is generally not visible on the shelf and only appears on the receipt (i.e. once the consumer has already made his decision)
Tax scope and rate	
Tax scope[c]	The scope of the tax determines what types of beverages are impacted or not by the tax. For a tax scope to be legally applicable, it has to be compatible with the fiscal nomenclature currently used in the jurisdiction considered
Tax rate[c]	In the case of a **volumetric** tax, the rate corresponds to a fixed amount per volume or unit of a good (e.g. 10 cents per litre). In the case of a **valoric** tax, the rate is expressed as a percentage of the value of a good (e.g. 20 % of the pre-tax price)
Earmarking of tax revenues	
Earmarking[d]	Earmarking consists in "assigning revenues from designated sources to finance designated expenditures"
Substantive earmarking[d]	Earmarking is qualified as substantive when the earmarked revenues "flow into a special fund and are the sole, or at least at the margin the incremental, source of funding for a particular expenditure item"
Symbolic earmarking[d]	Earmarking is qualified as "symbolic" when "taxes are designated to help pay for particular government services, but the revenues from these taxes flow into a general (or consolidated revenue) fund and finances only a portion of the government service in question"

Sources Parts or whole of the definitions are extracted or adapted from: [a]von Tigerstrom (2012, p. 57); [b]Sharma et al. (2014a, p. 1160); [c]Chriqui et al. (2013, p. 407); [d]WHO (2012, pp. 5–9); [e]Preece (2013, p. 22); [f]Thow et al. (2011); [g]Pomeranz (2014, p. e13)

As a starting point, these definitions raise important implications when discussing the feasibility and applicability of a soda tax. First, depending on the way the person supposed to bear the burden of the tax is reached, two taxing approaches are distinguished: direct or indirect. This has important implications in Canada since federal and provincial jurisdictions have distinct prerogatives in this regard (von Tigerstrom 2012) (see Sect. 12.2).

Secondly, the table of definitions presents various types of taxes. Out of the 21 places where an SSB tax has been reported across the world, excise taxes are the more common (see Table 1.1, column 5). We will discuss the prospect of these different options on a Canadian ground (see Sect. 12.3). According to the type of tax considered, a particular attention will be given to the tax base, which can be valoric (e.g. in the case of sales taxes), or either valoric or volumetric (in the case of excise taxes, import duties, special taxes). Depending on the tax type considered, we will also devote particular attention to the visibility of the tax to the consumers, as it may have important implications on purchasing decisions (Chriqui et al. 2013).

Thirdly, we will discuss feasibility concerns related to the tax scope, rate and base. In parts I and II of this book, we have summarized evidence suggesting that a 10–20 % tax rate applied on beverages containing added sugars could be effective in reducing SBB consumption while encouraging substitution towards low- or no-calorie options. However, among the 21 places where an SSB tax has been reported across the world, the tax rates and scopes are highly variable (see Table 1.1, columns 4 and 5). It suggests that contextual influences on a legal, administrative and political ground play an important role (see Sect. 12.4).

Finally, we will discuss the feasibility of earmarking SSB tax revenues for health- and social-related causes on a Canadian ground. Lessons from the tobacco case will be briefly described (see Sect. 12.5).

12.2 Public Health and Taxation Prerogatives in Canada's Political System

The Canadian political system includes three levels of governments: federal, provincial/territorial[1] and municipal.[2] This has important legal and political implications in terms of health-related and taxing powers (see Table 12.2).

Table 12.2 shows that the planning and delivery of healthcare and social services clearly is a responsibility of provincial/territorial governments, who receive administrative and financial support from the federal government. In public health, although jurisdictional uncertainties confuse the distribution of federal and provincial powers, provincial responsibilities generally predominate and cover areas such as health promotion and chronic disease prevention (PHAC 2004). This may explain why, so far, soda tax proposals have been more frequently considered at provincial level (see Sect. 13.4). Notwithstanding, the federal government keeps responsibilities for some public health and health protection programmes (Parliament of Canada 2015a, b, c). Among those, Health Canada is in charge of important nutrition policies such as the Canada's food guide (Health Canada 2011), food safety regulations (Health Canada 2012) and nutrition labelling (Health Canada 2015a). Theoretically, the enactment and the implementation of a soda tax could be connected to such competencies. Finally, in Canada, soda taxes would most probably be out of municipal jurisdiction.[3]

---

[1]Canada is a federal parliamentary democracy counting ten provinces (Alberta, British Columbia, Manitoba, New Brunswick, Newfoundland and Labrador, Nova Scotia, Ontario, Prince Edward Island, Quebec, and Saskatchewan) and three territories (Northwest Territories, Nunavut, and Yukon).

[2]Elected "band councils" governing First Nations communities are distinct from municipal governments even though they have similar responsibilities and make decisions affecting their local communities (Government of Canada 2010).

[3]This situation differs significantly from the US situation where some municipalities have powers to tax soda (Pomeranz 2012).

Table 12.2 Division of main powers between the three levels of Canadian governments

	Level of government		
	Federal	***Provincial/territorial***	***Municipal*** (usually based in a city, town or district)
Shared powers (Parliament of Canada 2015a, b, c)	Agriculture, Immigration, certain aspects of natural resources		
Distinctive powers (Government of Canada 2015a; Parliament of Canada 2015a, b, c)	• Responsible for areas of national concern such as follows: national defence, foreign affairs, employment insurance banking, **direct and indirect taxes**, the post office, fisheries, shipping, railways, telephones and pipelines, Aboriginal lands and rights and criminal law • Additionally, responsible for "equalization payments (extra money) to provinces to **make sure that the standards of health**, education and welfare **are the same for every Canadian**"	• Provinces are responsible for areas such as follows: **direct taxes**, education, **health care** (including hospitals), property and civil rights (including labour), administration of justice, fines and penalties for breaking provincial laws, some natural resources, road regulations and prisons • Territories "have their own governments, with responsibilities that are given to them by the federal government"	• Receive authority from the provincial government • Responsible for areas such as libraries, parks, community water systems, local police, roadways and parking
Specific role in health care (Government of Canada 2015b)	• Setting and administration of national principles for the healthcare system through the **Canada Health Act** • Assistance in the **financing** of provincial/territorial **healthcare services through fiscal transfers** • **Delivery** of healthcare services to **specific groups** (e.g. First Nations and Inuit and veterans) • Responsibilities for other health-related	• Administration and delivery of **healthcare services** (including hospitals) under the guidance of the Canada Health Act and with funding contributions from the federal government (fiscal transfers) • Responsibility for providing some groups with "**supplementary health benefits**" not covered by the act such as prescription drugs	

(continued)

Table 12.2 (continued)

	Level of government		
	Federal	***Provincial/territorial***	***Municipal*** (usually based in a city, town or district)
	functions such as some **public health** and **health protection** programmes (see below) and **health research**		
Specific role in public health (PHAC 2004)	• There are jurisdictional uncertainties about the distribution of federal and provincial powers in public health (not explicitly covered by the Canada Health Act). However, owing to their responsibilities in property/civil rights and for matters of a local or private nature, "**Provincial/territorial presence predominates in public health** with most of the delivery of services occurring locally or regionally". These responsibilities vary and may cover **prevention, community development, health promotion, workplace health, chronic diseases, and infectious diseases**. Some of these responsibilities and others (e.g. surveillance) may be shared at various degrees with the federal government • According to the Department of health Act, the federal government keeps some responsibilities in areas such as "the **promotion of the physical, mental and social well-being**, the **protection against risks to health** and the **spreading of diseases**, the **investigation and research into public health**, including the **monitoring of diseases"**. Health Canada also covers domains such as **emergency preparedness** and **response, healthy human development, health products** and **food safety, food inspection, foodborne zoonoses, healthy environments** and **consumer safety**		• Municipal and local governments **sometimes** have **public health responsibilities** under provincial public health legislation or as part of their general municipal powers

Considering this, would both provincial and federal governments have the legal capacity to adopt soda taxes? Actually, in Canada, taxing powers are shared between federal and provincial governments. However, under the Constitution Act (1867), federal powers include "the raising of Money by any Mode or System of Taxation", whereas provincial powers are limited to "Direct Taxation within the Province in order to the raising of a Revenue for Provincial Purposes" (Parliament of Canada 2015a, b, c). This difference significantly conditions the fiscal mechanisms federal and provincial governments may use if they wish to enact a soda tax.

Whereas the constitution leaves much room for manoeuvre at federal level, it seems to be much less the case in provinces. At this level, the constitution seems to preclude the enactment of an excise tax or a special tax levied on the manufacturer, since these two types of taxes are usually deemed indirect. Nonetheless, a special tax levied directly on the consumer may be legally acceptable (von Tigerstrom 2012). Implications will be discussed in the next sections.

12.3 Tax Types

This section presents the main types of SSB taxes described in the literature or used across the world as well as their prospect for enactment on a Canadian ground.

12.3.1 Modulation of Sales Taxes

In Canada, sales taxes apply to the sale of all kinds of goods and services along the food chain. The amount of the tax is calculated as a percentage of the price (valoric taxes) and is ultimately paid by the consumer (see Table 12.1). In Canada, sales taxes apply at both the federal level and the provincial level. This includes the goods and services tax (GST, federal level), the provincial sales taxes (PST, or QST in Quebec) and the harmonized sales tax (HST, when GST and PST are combined in some provinces) (Canada Revenue Agency 2015).

In the context of a health-related SSB tax, the key question is as follows: Could sales tax rates be modulated to impose a higher burden on SSBs in comparison with healthier beverage options? Out of the 21 places where such a tax has been reported across the world (see Table 1.1, column 5), there may be one case of such mechanism.[4] In Chile, before 2015, a 13 % valoric tax was applied on all non-alcoholic beverages. In 2015, this tax rate was lowered at 10 % except for sugar-sweetened beverages, for which the tax rate was raised at 18 % (Ministerio de Hacienda 2015). Could a similar mechanism be used in Canada?

To answer this question, it is necessary to consider how sales taxes currently apply in Canada and how SSBs are impacted. It is important to note that, as in many other countries, the GST/HST/PST or QST nomenclature does not distinguish non-alcoholic beverages according to nutrition criteria (e.g. according to the content in added sugars). From a legal perspective, categories generally include spring water; milk; milk-based beverages; fruit juices (containing ≥ 25 % of natural fruit

[4]It should be noted that this special tax (*Impuesto a las Bebidas Alcohólicas, Analcohólicas y Productos Similares*) is distinct from the general sales tax (*Impuesto a las Ventas y Servicios*) but both work the same way, i.e. as value added taxes: they are paid and passed on at each step of the commercialization process and ultimately affect the final consumer (Servicio de impuestos internos 2016).

juice); non-carbonated fruit-flavoured and fruit juice-based beverages (containing <25 % of natural fruit juice); carbonated beverages (including colas, ginger ale, tonic, aerated mineral water and fruit-based carbonated drinks containing <25 % of natural fruit juice); ice teas; tea and coffee (Canada Revenue Agency 2007).

Let us focus on fruit juices, fruit-flavoured and fruit juice-based beverages, and carbonated beverages, since these are the three main categories that include SSBs (as defined in this book (see Sect. 2.1)). Table 12.3 summarizes how sales taxes apply on these beverages in Canadian provinces and territories.

Table 12.3 shows that in many cases, unlike basic groceries, SSBs are often taxed at a full rate in Canada (Ries and von Tigerstrom 2011). But this is not always

Table 12.3 Application of sales taxes on fruit juices, non-carbonated fruit-flavoured and fruit juice-based beverages, and carbonated beverages in Canada

	Sales taxes		
Provinces and territories	**Fruit juices** (containing ≥ 25 % of natural fruit juice)	**Non-carbonated fruit-flavoured and fruit juice-based beverages** (containing < 25 % of natural fruit juice)	**Carbonated beverages** (including aerated mineral water and fruit juice-based carbonated drinks)
Alberta[a,c]	GST exempted (except single servings, taxed at 5 %)	GST at 5 %	GST at 5 %
	No provincial sales tax (PST)		
British Columbia[b,c]	GST exempted (except single servings, taxed at 5 %)	GST at 5 %	GST at 5 %
	Carbonated drinks and non-carbonated drinks **exempt** from the 7 % PST		
Manitoba[c,d]	GST exempted (except single servings taxed at 5 %)	GST at 5 %	GST at 5 %
	PST exempted (except single servings taxed at 8 %)	PST at 8 %	PST at 8 %
New Brunswick[c]	HST exempted (except single servings taxed at 13 %)	HST at 13 %	HST at 13 %
Newfoundland and Labrador[c,e]	HST exempted (except single servings taxed at 13 %)	HST at 13 %	HST at 13 %

(continued)

Table 12.3 (continued)

	Sales taxes		
Northwest Territories[a]	GST exempted (except single servings taxed at 5 %)	GST at 5 %	GST at 5 %
	No territorial sales tax		
Nova Scotia[c]	HST exempted (except single servings taxed at 15 %)	HST at 15 %	HST at 15 %
Nunavut[a]	GST exempted (except single servings taxed at 5 %)	GST at 5 %	GST at 5 %
	No territorial sales tax		
Ontario[c,f]	HST exempted (except single servings taxed at 13 %, or 5 % if ready for immediate consumption and price $\leq$4$)	HST at 13 % (5 % if ready for immediate consumption and price $\leq$4$)	HST at 13 % (5 % if sold with a food or beverage qualifying for the 8 % rebate of the Ontario part of the HST)
Prince Edward Island[c,g]	HST exempted (except single servings taxed at 14 %)	HST at 14 %	HST at 14 %
Quebec[c]	GST exempted (except single servings taxed at 5 %)	GST at 5 %	GST at 5 %
	QST exempted (except single servings taxed at 9.975 %)	QST at 9.975 %	QST at 9.975 %
Saskatchewan[c,h]	GST exempted (except single servings taxed at 5 %)	GST at 5 %	GST at 5 %
	Carbonated, hot and cold beverages **exempt** from the 5 % PST		
Yukon[a]	GST exempted (except single servings taxed at 5 %)	GST at 5 %	GST at 5 %
	No territorial sales tax		

Sources Parts or whole of the tax scopes described are extracted or adapted from: [a]Alberta Treasury Board and Finance (2015); [b]BC Ministry of Finance (2013); [c]Revenu Quebec and Canada Revenue Agency (2013); [d]Manitoba Finance (2014); [e]Newfoundland and Labrador Department of Finance (2015); [f]Canada Revenue Agency (2010); [g]PEI Department of Finance (2013); [h]Saskatchewan Ministry of Finance (2015) the case. First, in three provinces (Saskatchewan, British Columbia, and Ontario to some extent), carbonated drinks and/or fruit-flavoured drinks are exempt from provincial sales taxes. Cancelling this exemption may be feasible and beneficial.[5] Secondly, in Alberta, the Northwest Territories, Nunavut and Yukon, there is no PST or HST tax scheme: this leaves little room for improvement. Thirdly, fruit juices (containing ≥ 25 % of natural fruit juice) are most of the time exempt from GST/PST/HST, unless they are sold as single servings. From a nutritional perspective, changing the 25 % threshold into a 100 % threshold may be a step to reinforce taxation of beverages with added sugars instead of 100 % fruit juices. Finally, in the carbonated beverage category, the fiscal nomenclature does not distinguish SSBs, NCSBs and sparkling water, meaning that all these beverages are equally taxed. Taxing SSBs at a full rate and exempting sparkling water and NCSBs may be a way to favour SSB substitution towards healthier choices.

Another concern, when discussing how sales taxes on SSBs may be reinforced, is to consider also how taxes on healthier alternatives (e.g. water, milk) could be reduced. As a case in point, in the province of Quebec, basic groceries[6] are tax-free, i.e. neither concerned by the federal goods and services tax (GST) nor by the Quebec sales tax (QST).[7] However, as summarized in Table 12.3, it is not the case for carbonated beverages and fruit juice-based beverages, subjected to GST and QST (Revenu Quebec 2010). As regards the other beverage categories (see Table 12.4), we observe that individual servings of spring water and milk are subject to GST and QST. An exemption on these two subcategories and formats would add consistency from a nutritional standpoint. Conversely, the exemption of large ice tea servings may be cancelled.

In sum, where sales taxes apply on SSBs in Canada, it is not explicitly based on nutritional criteria. This is also the case in many other countries (Bahl et al. 2003; Hespel and Berthod-Wurmser 2008). We have briefly identified some areas from improvement. Overall, as von Tigerstrom puts it (2012, p. 51), "to target SSB more specifically, it would be necessary to redefine the categories that are zero-rated or exempt, to apply a higher rate of sales tax to SSB, or both". Although theoretically feasible, no doubt that such a "sales tax" reform would require significant discussions and seems unlikely to happen on a short term either at the provincial level or at the federal level. The strong emergence of newer products (e.g. flavoured waters) may add to the complexity of the task (Food in Canada 2009).

[5]Ries and von Tigerstrom (2011, p. 381, 411) note that such move was already proposed in the past in Ontario and British Columbia, but faced too much opposition to be passed.

[6]Including essentially meat (beef, poultry, pork, lamb, processed meats, sausages, etc.), cereals, fruits, vegetables, eggs, bread, fish and milk products (unflavoured milk, cheese, butter, cream, yogurt, etc.) (Revenu Quebec 2012b).

[7]GST and QST are collected upon most goods and services in the province of Quebec. GST is applied at a rate of 5 % of the retail price, and QST is added at a rate of 9.5 % of the retail price including GST (Revenu Quebec 2012a).

Table 12.4 Sales taxes applied to various types of beverages in the province of Quebec (Revenu Quebec 2010)

Beverage type	Format	Distribution	Subject to GST and QST	Untaxed
Spring water	>600 ml	All types		*
	Individual serving	All types	*	
Milk	>600 ml	All types		*
	Individual serving	Regular		*
	Individual serving	Prepared and served at the point of purchase, including via vending machine	*	
Milk-based beverages	>600 ml	All types		*
	Individual serving	All types	*	
Fruit juices (containing $\geq$ 25 % of natural fruit juice)	>600 ml	All types		*
	Individual serving	All types	*	
Non-carbonated fruit juice-based beverages (containing <25 % of natural fruit juice)	All types	All types	*	
Carbonated beverages including aerated mineral water and fruit-based carbonated drinks (containing <25 % of natural fruit juice)	>600 ml	All types	*	
Mixes for ice tea	>600 ml	All types		*
	Individual serving	All types	*	
Tea, coffee	Individual serving	Prepared and served at the point of purchase, including via vending machine	*	

Furthermore, from a public health perspective, using sales taxes as a way to "overtax" SSBs has several drawbacks. First, in Canada and the USA, sales taxes are not included in the shelf price but only appear to consumers on the cash register receipt.[8] In experimental conditions (computer-based trial), the fact that an

[8]This is not the case in all countries. In Europe, for instance, the value added tax imposed on goods and services is a general sales tax that is generally included in the shelf price.

"unhealthy food tax" be "exclusive" (i.e. neither included in prices shown on a lunch menu nor explicitly attached to specific items on this menu) has proved to have less impact on the nutritional quality of food and beverages chosen by the consumer than if the tax was "inclusive" (included in the price of the items displayed on the menu) (Chen et al. 2015). This deficit of perception may somewhat be "mitigated through labelling or public education strategies" making sales taxes more salient when consumers make decisions, but this raises feasibility concerns (von Tigerstrom et al. 2012, p. 62). Secondly, as emphasized by Chriqui et al. (2013) and the Yale Rudd Center (2012), since general sales taxes in Canada are valoric, SSB "overtaxation" may encourage some consumers: (1) either to buy larger containers (since multipacks offers make the price per volume lower, it makes the tax per volume of beverages lower as well) or (2) to buy cheaper products (e.g. private labels), on which tax impact would be proportionally lower. Yet, an advantage of valoric taxes, indexed on price, is that their effect on price rises automatically with inflation. However, it may affect the predictability of tax revenues depending of the economic growth and the pricing strategies implemented by private actors (Yale Rudd Center 2012; Chriqui et al. 2013).

All these reasons likely contribute to explaining why, beyond sales taxes, other types of SSB taxes have been proposed and enacted across the world (See Table 1.1). The next subsection will discuss the feasibility of the most commonly discussed and adopted SSB tax option: excise taxes.

12.3.2 Increase or Introduction of an Excise Tax

Excise taxes or duties apply to specific types of goods and are generally levied on the manufacturer, as a fixed amount per unit of product (see Table 12.1). Out of the 21 "real-world" cases of SSB taxation that we have inventoried (see Table 1.1), 15 are excise taxes or duties. In some cases, these were pre-existing and have been increased for health and fiscal reasons (e.g. in Denmark in 2009 or in Finland in 2011). In other cases, these taxes have been newly introduced, often inspired by existing schemes imposed on other goods such as alcohol or tobacco (e.g. in Nauru in 2007, in India in 2014). They have all been implemented at country level, with the exception of the soda tax enacted in the city of Berkeley (California) in 2015. Most of these taxes (10) are volumetric and generally use the liquid volume as a unit of calculation. For example, the soda tax adopted in France in 2011 was implemented by extending the excise tax rate from 0.0054 € to 0.072 €/litre (French Ministry of Budget 2012). Mauritius makes an exception: the SSB tax is based on the sugar content in the drink (3 cents per gram of sugar) (Mauritius Revenue Authority 2012). The other cases (5) are valoric taxes. For example, a 10 % excise tax has recently been adopted in Dominica and Barbados (Sinckler 2015; Skerrit 2015). Preece (2013) indicates that volumetric excise taxes may be easier to administer than valoric taxes, since the latter is more subject to interpretation by the taxpayer as regards the excisable value of the good used to calculate the amount of the tax.

From a public health perspective, excise taxes are generally preferred to sales taxes for several practical and strategic reasons, particularly identified by Chriqui et al. (2013) and the Yale Rudd Center (2012): (1) depending on the fiscal nomenclature, a new excise tax can generally focus on specific categories of non-alcoholic beverages, whereas modulating sales taxes may require considerations for a wider range of products (see Sect. 12.3.1); (2) excise taxes are generally levied on manufacturers or retailers, which makes revenues easier to collect; (3) most of the time, excise taxes are volumetric, which means that they are indexed on the volume of products manufactured: this is likely to influence all taxed beverages equally, whatever their price is; (4) since an excise tax is generally levied on the manufacturer or the retailer, it is more likely to be included in shelf prices and, consequently, to impact consumer choices. Indeed, as mentioned in the previous section, a recent US laboratory experiment confirms that "inclusive" taxes (i.e. included in prices when consumers make their choice) may better encourage consumers to make healthier choices than "exclusive" taxes (i.e. whose precise amounts only appear at the checkout) (Chen et al. 2015). However, a drawback of excise taxes is that their fixed amount is not indexed on inflation (Yale Rudd Center 2012; Chriqui et al. 2013). As a result, planning regular adjustment to keep pace with the consumer price index has been strongly recommended, e.g. in the case of alcohol excise taxes in Canada (Stockwell et al. 2006; Thomas and Canadian Centre on Substance Abuse 2012). It appears to be even more important that sales taxes (i.e. GST/HST/PST/QST) and other special taxes are calculated on the basis of the price of alcoholic beverages including excise taxation (Stockwell et al. 2006). For example, in France, where the excise tax rate on soda has been substantially increased in 2011, an annual update of the tax is mandatory to keep pace with inflation (French Ministry of Budget 2012).

If excise taxation appears to be the more recommended and implemented option, what is its feasibility on a Canadian ground? Legally speaking, the Canadian federal government's jurisdiction could apply excise taxation on SSBs. Actually, under the Excise Tax Act, excise duties are already imposed on several goods such as spirits,[9] wine[10] and tobacco products[11] (Canadian Minister of Justice 2015; Canada Revenue Agency 2012, 2014). Conversely, the prospect for an excise tax at provincial level appears to be weak. According to the seminal and comprehensive legal analysis conducted by von Tigerstrom (2012), provinces only have the power to apply direct taxes, i.e. taxes generally imposed on the last purchaser of an article (i.e. not supposed to be passed onto any other person) (see Sect. 12.2). Yet, from a

[9]For example, spirits containing more than 7 % of absolute ethyl alcohol by volume are imposed an excise duty of $11.696 per litre of absolute ethyl alcohol.

[10]For example, wine containing more than 7 % of absolute ethyl alcohol by volume is imposed an excise duty of $0.62 per litre.

[11]For example, cigarettes are imposed an excise duty of $0.52575 per five cigarettes.

public health perspective, most proposals of SSB excise tax levied on the manufacturer anticipate that it will be passed onto retail prices and, consequently, impact consumers. The burden of evidence presented in Chap. 5 of this book clearly indicates that excise taxes are likely to be partially or fully shifted to retail prices. As a result, such an excise tax is generally considered to be indirect and would likely fall outside of current provincial and municipal jurisdictions (von Tigerstrom 2012). From a legal standpoint, this is somewhat unfortunate since, so far, most SSB taxation proposals were made at the provincial level: this was the case in Quebec (Canada Newswire [CNW] 9 May 2013), in Ontario (Toronto Star 23 October 2012), in British Columbia (Korstrom 2013) and in Alberta (Marketwire 28 April 2011).

12.3.3 Creation of a Special Tax

An alternative to traditional sales taxes and excise duties is the creation of a special tax. Among SSB taxes implemented across the world, two cases have raised particular attention, since they encompass not only SSBs but also a broader range of foods and beverages of low nutritive value (see Table 1.1). The first was the Hungarian "Public Health Product Tax" (PHPT) adopted in 2011. This tax is imposed on pre-packaged foods with high salt and sugar content, including crisps, chocolates, sweets, biscuits, ice creams and alcoholic and non-alcoholic beverages. For SSBs, the scheme imposes a volumetric tax of 7 Hungarian florint (Eq. 0.03 CAD) per litre (WHO 2015a). The second interesting case is the Mexican "Special tax on production and services" (IEPS) adopted in 2014. The scheme includes a tax of 8 % on energy-dense foods (except basic groceries) as well as a special tax of 1 peso (Eq. 0.08 CAD) per litre on SSBs (WCRF 2015). These two special taxes have in common to be levied on the manufacturer or the importer. Since they are both explicitly aimed at promoting healthier dietary patterns, it seems reasonable to consider that the legislator expected that these taxes would be passed onto the consumer. In these circumstances, this fiscal scheme would likely be considered indirect. As such, it may only be applicable at federal level in Canada (see Sect. 12.2). In all cases, such a broad scheme would likely require well-founded public health and budgetary rationales, a strong political will, social acceptability and administrative readiness.

The list of SSB taxes implemented across the world (see Table 1.1) reveals another case of special tax that would more likely be considered direct taxation. Indeed, the "Healthy Diné' Nation Act" adopted by the Navajo Nation (US) in 2014 introduces a 2 % tax levied on the retailers from sales to the consumer. The tax is imposed on SSBs as well as on other foods of low nutritive value (Navajo Nation

Council 2014). In such a case, the revenues are collected at the point of retail sales, which is a common point with the special taxes imposed on tobacco and alcohol in Canadian provinces, also considered to be direct taxes (see Box 12.1).

Box 12.1: Snapshot of tobacco and alcohol taxation in Canada

In Canada, as mentioned in the previous section, a federal excise duty is imposed on tobacco products in all provinces and territories. Additionally, the federal goods and services tax (GST) or the harmonized sales tax (HST) applies in all jurisdictions. Where the HST is not used, Saskatchewan and Manitoba apply a provincial sales tax (PST), but tobacco products are exempt of such a tax in Alberta, British Columbia and Quebec (QST). Finally, a special tobacco tax is applied in each province or territory (Physicians for a Smoke-Free Canada 2015, BC Ministry of Finance 2013).

As an example of this special tobacco tax, in Quebec, it represents $0.149 per cigarette sold at retail (Revenu Quebec 2014). This amount is included in the retail price, which is not noticeable for the consumers since cigarettes' display in convenience stores is banned in all provinces (Li et al. 2013). To facilitate the collection of revenues, retailers generally pay the amount of the tax to their wholesale vendor (who works as a collection officer), who generally remits the sum to the government or pays it to the manufacturer, who remits the sum himself to the government on a monthly basis. The retailer reimburses himself when recouping the tax from the final consumer (Revenu Quebec 2015b).

As regards alcohol taxation policies, they are heterogeneous across Canadian provinces and territories. As mentioned before (Sect. 12.3.2), a federal excise tax applies on all alcoholic beverages. The GST or HST also applies in all jurisdictions. Additionally, where the HST is not used, the general provincial sales tax (PST or QST in Quebec) is imposed in some cases. Most provinces and territories also apply an additional sales tax on alcohol sold at retail (as a % of price) except in the province of Quebec, where this tax is volumetric, i.e. expressed as a fixed amount per volume of beverage sold (as for the special tobacco tax). Above all, a diversity of special levies and markups (e.g. for health- and environment-related reasons) are used and often represent a significant proportion of the total price (Stockwell et al. 2006). Finally, minimum pricing policies are applied to various extents (depending on the jurisdiction and the beverage), which is facilitated by the fact that alcoholic beverages are mainly sold in government-owned liquor stores (Thomas and Canadian Centre on Substance Abuse 2012; National Alcohol Strategy Advisory Committee 2015). In some cases, all taxes are included in the shelf price (e.g. in Quebec), whereas, in others, sales taxes are added at the checkout (e.g. in British Columbia) (SAQ 2016; BC Liquor Stores 2016).

When considering how some special levies on alcohol may inspire SSB taxation schemes, the Quebec case is interesting for two reasons. First, it is the only Canadian jurisdiction where the sale of beer and wine is allowed in grocery or corner stores at a large scale[12] (Thomas and Canadian Centre on Substance Abuse 2012). In this case, the special alcohol tax is included in the shelf price, but the GST and QST are added at the checkout. Retailers play the role of tax collection officer for the Quebec government (Revenu Québec 2015a). Another interesting feature in Quebec is the existence of the special levy paid by alcohol manufacturers (proportionally to beverages sales) to finance the Éduc'alcool non-for-profit organization, whose aim is to raise awareness and educate youth and adults to make responsible decisions about drinking (Éduc'alcool 2016).

Therefore, in Canadian provinces, the feasibility to adopt the special tax scheme currently used for tobacco and alcohol to the case of SSB could be explored. The special tax imposed on wine and beer in Quebec grocery and convenience stores is particularly interesting in this regard for three reasons suggesting transferability to SSB: (1) the tax is applied on beverages sold at retail, which is considered to be direct taxation; (2) special taxes are included in the shelf price, which is more likely to influence consumers' decisions than a tax added at the cash register; (3) SSBs are largely distributed via grocery and convenience stores. Nonetheless, the technical and administrative burden that a specific tax on SSBs sold at retail may represent for retailers should be carefully considered. The jurisprudence indicates that, although imposed at the point of retail, revenues may be collected and remitted at the wholesale or distributor level (von Tigerstrom 2012). This appears to be an interesting option to consider.

12.3.4 Other Options

Beyond sales taxes, excise duties and special taxes, some other options have been discussed in the literature or have been actually implemented.

First, this includes specific import or customs duties. For example, such taxes have been imposed on SSBs in French Polynesia (since 2002), in Fiji (since 2006), in Nauru (since 2007) and in the Cook islands (since 2012). Rationales for such taxes may be linked to a low domestic production of the targeted items, a low cost

[12]In other Canadian provinces, the sale of alcoholic beverages is essentially authorized in specific government-owned stores or, in some cases (depending on the beverage and the jurisdiction), in privately owned liquor stores. New Brunswick and Newfoundland and Labrador allow the sale of beer and/or wine in a limited number of convenience/grocery stores. Since December 2015, the sale of beer in Ontario has been allowed in a large number of grocery stores (The Canadian Press 2015; Government of Ontario 2015).

of implementation (in comparison to other tax schemes) and the importance to rely on such taxes from a budgetary standpoint (Thow et al. 2011). These conditions do not equally apply to the Canadian situation.

Secondly, the feasibility to levy a direct tax on manufacturers has also been discussed. For example, in Algeria, in 2012, a tax of 0.5 % has been imposed on the revenue of manufacturers and importers of carbonated drinks (Official Journal of the Algerian Republic 2012). In Canada, von Tigerstorm (2012, p. 61, 63) suggests that a tax on SSB manufacturers charged as “a flat-fee or lump sum, rather than a volumetric or per unit amount” may be considered as a direct tax since it is “more likely to have a diffuse impact as part of the cost of doing business rather than being passed on (…) [SSB] units that are sold”. Consequently, such a tax may fall within provincial jurisdictions provided any attempt to have the tax passed on SSB prices and impact consumers is not pursued or, minimally, not “traceable”. This is questionable, since impacting consumers generally is at the core of most SSB taxation proposals made in Canadian provinces (von Tigerstorm 2012, p. 63). Therefore, the prospect for such a tax may be higher at federal level.

Finally, another option mentioned in the literature as a complement to taxation is “minimum pricing”, i.e. the “addition of a minimum percentage markup to the base price of a good by the retailer to increase its retail price to consumers” (Pomeranz 2014, p. e13). Such a policy seems promising from a public health perspective, especially in heavy drinkers of alcoholic beverages and heavy smokers relying on promotional offers (Thomas and Canadian Centre on Substance Abuse 2012; Pomeranz 2014; Sharma et al. 2014b). In Canada, the National Alcohol Strategy Advisory Committee (2015) has recently proposed to strengthen such policies in provinces where it is not or little implemented. It is particularly recommended to apply minimum prices on all types of alcoholic beverages (more intensively on beverages with a high alcohol content) and to update minimum prices regularly to keep pace with inflation. As a complement, the ban of various in-store discounting strategies that may help circumvent minimum prices is also recommended (National Alcohol Strategy Advisory Committee 2015). Since health-related soda taxes have only been enacted recently (in comparison with tobacco and alcohol taxes), the discussions on the possibility to use minimum pricing are quite new but very relevant (Pomeranz 2014). Indeed, as discussed in Chap. 5 of this book, when an excise SSB tax is levied on the manufacturer or the retailer, it still has the possibility to undershift or overshift the tax to SSB prices. For example, he may decide to absorb the cost of the tax or spread it across his portfolio of products. Consequently, Pomeranz (2014) recommends the application of minimum prices on SSBs as a complementary strategy helping to secure the pass-through of a tax onto prices. To prevent the frequent use of promotional offers by retailers (see Sect. 5.3), Pomeranz also suggests exploring the feasibility to implement prohibitions on coupons and discounting, as already done as part of tobacco control strategies (IARC 2011; Pomeranz 2012, 2014). These propositions could be further explored as evaluations of SSB taxes actually bring additional evidence.

12.3.5 Summary

Table 12.5 reports the conclusions of the previous four sections. In summary, excise duties generally are the more documented and recommended option to tax soda. Introducing such a tax in Canada is likely feasible, but at the federal level only. At provincial level, the introduction of a special tax on SSBs sold at retail according to the tobacco and alcohol special tax models is an avenue deserving particular attention. Taxation at both federal and provincial levels may be complementary, as shown in other public health domains such as tobacco control (Ries and von Tigerstrom 2011).

Table 12.5 Feasibility of commonly suggested options to reinforce soda taxation or to increase soda price in Canada

	Modulation of sales taxes	Excise duty	Special tax	Other options
Federal level	**Large reform unlikely feasible** *GST already applies at a maximum rate on carbonated beverages and fruit-based beverages* (containing *<25 % of a natural fruit juice)*	**Likely feasible** *This mechanism is already used on tobacco products and alcoholic beverages*	**Possibly feasible,** *although a new mechanism would require well-founded public health and budgetary rationales, a strong political will, social acceptability and administrative readiness*	**To be explored** *For example, the feasibility to impose a flat-fee on manufacturers could be explored*
Provincial (P) or territorial (T) level	**Improvements likely feasible in some P/T** *PST/QST or HST applies at a maximum rate on carbonated beverages and fruit-based beverages (containing <25 % of a natural fruit juice) in some P/T, but not in others*[a]	**Unlikely feasible** *The tax would likely be considered "indirect" and, consequently, beyond provincial/territorial powers*	**Likely feasible,** *provided the tax is validated as "direct" (already used on tobacco products and alcoholic beverages in most provinces, be this as a valoric tax (% of price) or as a volumetric tax (per unit or per ml))*	**To be explored** *For example, minimum pricing and regulation of price promotions could be explored*

[a]Saskatchewan, British Columbia and Ontario (in some circumstances) list carbonated drinks and/or fruit-flavoured drinks among non-taxable items. Therefore, in these 3 provinces, taxing SSBs at higher rate via the PST or HST may be feasible

Sources von Tigerstrom (2012), Thomas and Canadian Centre on Substance Abuse (CCSA) (2012), CCSA (2012), National Alcohol Strategy Advisory Committee (2015), Stockwell et al. (2006), Physicians for a smoke-free Canada (2015), Chriqui et al. (2013)

12.4 Considerations on SSB Taxation Scope, Rate and Base

12.4.1 SSB Taxation Scope

From a nutritional perspective, we previously argued that taxing all non-alcoholic beverages containing added sugars (exempting 100 % fruit juice and NCSBs) would probably be the most relevant option (see Sect. 2.5). Using such a tax base may also favour healthy substitution effects (see Sect. 7.3). However, currently, the Canadian fiscal nomenclature considers all carbonated beverages jointly regardless of their calorie content and composition (von Tigerstrom 2012). Furthermore, the nomenclature distinguishes two main categories of fruit-based beverages: those containing fewer than 25 % of natural fruit juice and those containing at least 25 % of natural fruit juice (whatever their content in added sugars is). Therefore, ideally, taxing only sugar-sweetened beverages would require adjustments in this nomenclature.

If we consider the diversity of tax scopes applied in countries where a soda tax has been enacted, similar administrative and legal constraints have likely been encountered (see Table 1.1, column 4). Some tax scopes seem to be close to the SSB definition provided earlier in this book (see Sect. 2.1), which suggests a deliberate focus on added sugars (e.g. in Hungary, the Cook Islands, Mauritius, Mexico, the municipality of Berkeley, Chile and Barbados). In a few cases, 100 % fruit juices have been included (e.g. in Denmark, Finland). In others, flavoured milk is explicitly targeted (e.g. in France, Nauru). Cases of tax scopes including NCSBs have been particularly controversial (e.g. in France, Belgium). Finally, some tax scopes are little discriminatory (e.g. in India, where bottled waters were seemingly included).

Interestingly, some tax scopes have used a threshold of sugar content, such as in Denmark ($\geq$0.5 g of sugar/100 mL), in Finland (>0.5 g of sugar/100 g or mL), in Hungary (>8 g of added sugars/100 mL), in St Helena ($\geq$ 15 g of sugar/L) and in Chile (>6.25 g of sugar/100 mL) (see Table 1.1, column 4). Such proposals are attractive in that they may encourage consumers to switch to low-calorie drinks and could push manufacturers to reformulate or develop lower-calorie options (provided technological and taste requirements are met). However, from an administrative and industrial vantage, such models would likely be more complex to handle (von Tigerstorm 2012).

12.4.2 SSB Taxation Rate

The evidence presented in Chaps. 6 and 7 of this book suggests that it is reasonable to expect SSB consumption to decrease as a result of a significant tax (i.e. accompanied by a 10–20 % price hike), although substitution effects may partially compensate for the reduction of energy intake from SSBs. This is consistent with the advocacy efforts of SSB taxation proponents (especially in the USA), who generally claim that high rates, up to 15–20 % of the average purchase price, are necessary to expect significant reduction in SSB intakes. However, such an increase may cause acceptability concerns (see Chap. 13) and, so far, has only been passed in a minority of cases (see Table 1.1, column 5).

Comparing the tax rates across countries is complex since more than half of the cases inventoried are volumetric taxes (i.e. fixed monetary amount per volume of beverage). Therefore, translating the volumetric tax rate into a valoric one (i.e. as a % of pre-tax price) requires estimating the country-specific average SSB price per volume. In Table 1.1, if we consider valoric rates (per se) and volumetric rates converted into valoric values by other authors, we observe that the rates were often set between 5 and 10 %. This is the case in Fiji (5 % import tax), in Samoa (excise tax eq. to a 5–8 % tax), in France (excise tax eq. to a 6 % tax), in Mexico (special tax eq. to a 9 % tax), in India (5 % excise tax), in Chile (8 % special tax), in Barbados (10 % excise tax) and in Dominica (10 % excise tax) (Thow et al. 2011; Berardi et al. 2012; Grogger 2015). The tax adopted by the Navajo Nation is lower (2 % special tax), whereas import duties enacted in the Cook Islands (15 %) and in Nauru (30 %) are well above. Finally, the excise tax adopted in the city of Berkeley (California) is closer to the recommendation, as it was deemed to represent an average price increase of 11, 31 and 25 % for a 20-oz.bottle, a 2-litre bottle and a 12-pack of 12-oz.cans, respectively (Cawley and Frisvold 2015).

It will be essential to conduct evaluation of these different situations to detect how variations in the tax type and tax rate contribute to explaining different effects on SSB price and demand. For example, on the one hand, data from the USA indicate that moderate soft drink taxation (at 4–5 %) has led to a "somewhat modest" reduction in soft drink consumption by children and adolescents (Fletcher et al. 2010). On the other hand, preliminary evidence from France confirms that the soda tax implemented since 1 January 2012 may have contributed to a 2.2 % downward trend in non-alcoholic soft drinks sales volume in 2013. Comparing such results is complex and will require further evaluation efforts (see Chap. 14). Some parameters are likely to add complexity, such as differential tax rates between imported and locally manufactured products (e.g. in French Polynesia) or variations in the tax rate and/or scope overtime (e.g. in Finland, Hungary and Mauritius). As demonstrated in the case of alcohol and tobacco, the importance to increase regularly the tax rate in the case of SSB taxation deserves further attention (Blecher 2015).

12.4.3 SSB Taxation Base

All kinds of tax bases (% of price or fixed amount per liquid volume or sugar content) may not equally impact manufacturers and consumers (Blecher 2015). In the case of alcohol, a specific tax amount based on the quantity of alcohol (rather than the volume of liquid) has been shown to encourage the production of lower alcohol beverages and to foster healthier consumption patterns (e.g. case study on beers in South Africa) (Blecher 2015). Consequently, the advantage of a specific soda tax linearly indexed on the amount of sugar in the beverage has been suggested. In comparison with a fixed threshold (see Sect. 12.4.1), this progressive approach may create greater incentive for manufacturers to reduce sugar content in their products and for consumers to purchase lower-calorie options (Blecher 2015). In Table 1.1, the Mauritius excise tax of 3 cents per gram of sugar content falls into this category. Applying an excise duty on sweetener inputs at the production level (e.g. on sugar or corn syrup) rather than a sales tax on the consumption of a final product (e.g. on SSB) may also decrease energy intake at lower costs for consumers (Miao et al. 2012). However, taxing specific ingredients or nutrients rather than food categories is generally considered less feasible and administratively more burdensome (Thow et al. 2011; Ecorys 2014). Indeed, taxes on the production or final consumption of food and beverage categories (including SSB) already exist in most countries, which make them easier and less expensive to implement (Hespel and Berthod-Wurmser 2008). Conversely, nutrient- or ingredient-based taxes have been less common and may be difficult to handle. For example, in Canada, the current nutrition labelling nomenclature does not distinguish between naturally occurring and added sugars, which would likely raise some technical and administrative barriers as well as evaluability concerns if the amount of added sugars were to be used as a tax base (Canadian Food Inspection Agency 2014). This situation may change as Health Canada recently proposed to single out added sugars in the nutrition facts table of processed foods (Health Canada 2015b).

12.5 Feasibility to Earmark SSB Tax Revenues for Health-Related Causes

12.5.1 General Considerations

As we previously argued (see Sect. 10.6), earmarking tax revenues for healthcare or health promotion initiatives seems desirable: it could bring additional public health benefits, could somewhat mitigate inequity concerns and could increase the acceptability of the tax in the public opinion. In principle, it seems feasible, as demonstrated in several countries where a soda tax has been enacted (See Table 1.1 and Sect. 10.6). However, little information is generally available on the amount of revenues truly earmarked for the mentioned causes in the long term and on the manner the money has been used (WHO 2012). Mandatory and "substantive" earmarking (e.g. via the creation of a dedicated fund or a foundation) may offer greater guarantees for a significant, regular, predictable and sustainable funding than a "symbolic" earmarking, where the linkage between the amount of revenues generated and the volume of specific health services delivered is much weaker

(WHO 2012). Indeed, the risk that tax revenues be diverted from health matters to fill general budget deficits has often been highlighted in the case of tobacco and soda taxes (see Sect. 10.6). Therefore, planning explicitly in advance the earmarking of tax revenues for healthcare or health promotion activities has been increasingly recommended (IOM 2012; WHO 2012). It requires a strong political will on a public health ground, since substantive earmarking may add budgetary and administrative constraints and, consequently, may create political resistance commonly expressed by Ministries of Finance (Doetinchem 2010; McIntyre 2015). Therefore, when exploring and designing potential earmarking mechanisms, it seems essential to involve the Ministry of Finance to anticipate and prevent these issues. Other important questions to address in advance are the term of earmarking (fixed time period or indefinite), the possibility (or not) to adjust the rate of the earmarked tax over time (e.g. to keep pace with inflation) and the period during which the earmarked tax revenues can be used on purpose (WHO 2012). At the implementation stage, a strong intersectoral coordination appears to be a critical factor for a successful earmarking (WHO 2012).

Table 1.1 suggests some examples of soda tax revenues' "substantive earmarking" such as in French Polynesia (creation of a preventive health fund), in Algeria (creation of a fund against cancer) or in Barbados (funding of a national campaign). It would be worth learning from these experiences. In the meantime, the tobacco case brings interesting lessons.

12.5.2 Learning from the Tobacco Case

Interesting cases of tobacco taxation revenues' earmarking should be mentioned. In Australia, in 1987, VicHealth is known to have been the world's first health promotion foundation. It has been established as part of the Tobacco Act of 1987, which permitted to earmark revenues of the 5 % tobacco excise tax for tobacco control efforts as well as for sport and arts initiatives that were formerly sponsored by the tobacco industry. Today, VicHealth more largely contributes to health promotion research, practice and projects in a diversity of areas (e.g. physical activity, mental health, health at the workplace, violence issues, and NCD prevention) (VicHealth 2015). Since 1987, this best practice spread across the world. In its 2015 Report on the global tobacco epidemic, the WHO inventories 29 countries across the world earmarking part of their tobacco excise tax revenues for health purposes. Among those, the consolidation of specific health funds or foundations is mentioned in several cases (see Table 12.6). Such foundations generally provide a security for "long-term funding, relatively independent governing boards, and acceptance by a wide range of political and other stakeholders" (WHO 2012, p. 23). Such mechanisms also apply to other taxes. In the USA, for example, revenues from the excise tax on coal are earmarked for the Black Lung Disability Benefits Trust Fund devoted to miners' health (Pomeranz 2012).

Table 12.6 A few cases of specific health funds or foundations consolidated from tobacco taxation revenues

Country	Tax revenues targeted	Earmarking description
Algeria	6 dinars per pack of cigarettes	Emergency fund and medical care activities
Côte d'Ivoire	Extra tax of 5 % on tobacco	AIDS solidarity fund
India	Specific amount of revenues from excise taxation of most tobacco products	Health Cessation Fund
Jamaica	20 % of the special consumption tax (SCT) on tobacco and 5 % of the SCT on all products including tobacco	National Health Fund
Madagascar	6 ariary per pack of cigarettes	National Fund for the promotion and development of youth, sports and recreation
Mongolia	A proportion of tobacco (2 %) and alcohol (1 %) excise tax revenues	Health promotion foundation
Nepal	All the tobacco tax revenues	Health Tax Fund for the prevention and the treatment of NCDs
Republic of Korea	354 won per pack of cigarettes	Health Promotion Fund which finances health promotion research and projects
Switzerland	0.26 francs per pack of cigarettes	Tobacco Prevention Fund
Thailand	2 % surtax on tobacco products and alcoholic beverages (over and above the existing taxation)	Thai Health Foundation devoted to a variety of health promotion activities
USA	For example, excise tax revenues in California (2012) NB: Similar approaches have been adopted in other states, e.g. in Kentucky, Louisiana, Massachusetts and Oregon	57 % going to the Children and Families First Trust Fund; 29 % to the health education, hospital services, physician services and research; 2 % to the Breast Cancer Fund

Sources WHO (2012), (2015b)

12.5.3 The Canadian Situation

Interesting precedents of tobacco tax revenues' earmarking also exist in Canada. For example, since the 1970s, parts of the Quebec tobacco tax revenues have been legally earmarked for the funding of the Olympic stadium (Quebec Government 1976). Despite the fact that a significant part of these revenues has not been earmarked as initially planned (Phaneuf 2010), the reimbursement of the Olympic infrastructures has been achieved since 2006 (Régie des Installations Olympiques 2006). Currently, the Quebec tobacco law does not specify any earmarking conditions for the revenues generated (Quebec Government 2015c). According to the Quebec coalition for tobacco control *(Coalition québécoise pour le contrôle du tabac* (CQCT) 2014, p. 6), the proportion of these revenues contributing to financing tobacco control efforts would not exceed 5 % of all revenues raised from provincial tobacco taxes. Legally speaking, the laws instituting the Physical Activity and Sport Development Funds *(Fonds pour le développement du sport et de l'activité physique)* (Quebec Government 2015b) and the Healthy Lifestyles Promotion Funds (*Fonds pour la promotion des saines habitudes de vie*) (Quebec Government 2015a) also mention explicitly earmarking from tobacco tax revenues. These examples demonstrate the feasibility to earmark health-related tax revenues for health promotion causes in Canadian provinces.

What if a soda tax were to be implemented at federal level? As shown in Table 12.2, provincial responsibilities predominate in public health, but some responsibilities remain at federal level in areas such as food safety, food inspection and the prevention of chronic diseases. Indeed, as underlined by the Public Health Agency of Canada (2004) "the federal government can involve itself in public health by providing conditional funding for public health programs or by entering into legal contracts to develop public health initiatives". Consequently, soda tax revenues could theoretically be earmarked for these areas. Alternatively, the revenues from a federal soda tax might be earmarked for public health responsibilities at provincial level. However, it seems that Health Canada's programme of grants and contributions is generally "not directed to other levels of government, but to non-profit and non-governmental organizations" (PHAC 2004). Notwithstanding, precedents exist, e.g. in the immunization area. In 2007, the federal government has announced a 300 million CAD budget to support the provinces and territories in the implementation of a human papillomavirus (HPV) immunization programme aimed to decrease the morbidity and mortality of cervical cancer in Canadian women. The fund was transferred to a third-party trust, who made the money available to the provinces and territories in proportion to their population, provided the fund was used for such a programme. Within this framework, the distribution and the use of the budget were deemed flexible (Norris 2011).

If a soda tax is to be enacted in Canada, all the evidence presented in this section could inspire the set-up of a robust mechanism facilitating the administration and the earmarking of tax revenues for health promotion initiatives and social programmes. By default, if political and/or legal considerations preclude the set-up of an earmarking mechanism by regulation, stating officially that SSB tax revenues would be used for such purposes would be the least public authorities could do. In such a case, however, "watchdogs" would have a key role to play to make sure earmarking engagements are upheld.

Key messages

- Excise duties generally are the more documented and recommended option to tax soda. Introducing such a tax in Canada is likely feasible, but at the federal level only.
- At provincial level, the introduction of a special tax on SSBs sold at retail according to the tobacco and alcohol special tax models is an avenue deserving attention. In particular, the special tax imposed on wine and beer in Quebec grocery and convenience stores suggests transferability to SSB. Nonetheless, the burden it may represent for retailers should be carefully considered.
- Defining the taxation scope would likely raise some issues in Canada since, in the nomenclature, all carbonated beverages are considered regardless of their calorie content and composition.
- Finally, a robust mechanism facilitating the administration and the earmarking of tax revenues for health promotion initiatives and social programmes would likely be desirable and feasible in Canada.

References

Alberta Treasury Board and Finance. (2015). Tax and revenue administration links for information provincial sales tax and harmonized sales tax. http://www.finance.alberta.ca/publications/tax_rebates/provincial_sales_tax_links.html. Accessed 22 January 2016.

Bahl, R., Bird, R., & Walker, M. B. (2003). The uneasy case against discriminatory excise taxation: Soft drink taxes in Ireland. *Public Finance Review, 31*(5), 510–533. doi:10.1177/1091142103253753.

BC Liquor Stores. (2015, April 1). Shelf prices do not include taxes. http://www.bcliquorstores.com/promotion/shelf-prices-do-not-include-taxes. Accessed 22 January 2016.

Berardi, N., Sevestre, P., Tepaut, M., & Vigneron, A. (2012). *The impact of a "soda tax" on prices: Evidence from French micro data* (Banque de France Working Paper No. 415) (p. 29). Banque de France Eurosystème. Direction Générale des Études et des Relations internationales. http://ssrn.com/abstract=2192470. Accessed 24 February 2016.

Blecher, E. (2015). Taxes on tobacco, alcohol and sugar sweetened beverages: Linkages and lessons learned. *Social Science and Medicine, 136–137*, 175–179. doi:10.1016/j.socscimed.2015.05.022.

British Columbia Ministry of Finance. (2013). Grocery and drug stores. Provincial sales tax act. *Provincial Sales Tax (PST) Bulletin, October 2013*(Bulletin PST 206). http://www.sbr.gov.bc.ca/documents_library/bulletins/pst_206.pdf. Accessed 22 January 2016.

Canada Revenue Agency. (2007, January). GST/HST memoranda series. 4.3. Basic groceries. http://www.cra-arc.gc.ca/E/pub/gm/4-3/4-3-e.pdf. Accessed 22 January 2016.

Canada Revenue Agency. (2010, July). GST/HST info sheet. Harmonized sales tax for Ontario—point-of-sale rebate on prepared food and beverages. http://www.cra-arc.gc.ca/E/pub/gi/gi-064/gi-064-e.pdf. Accessed 22 January 2016.

Canada Revenue Agency. (2012, October). Excise duty memorandum. The excise duty program. http://www.cra-arc.gc.ca/E/pub/em/edm1-1-1/edm1-1-1-e.pdf. Accessed 22 January 2016.

Canada Revenue Agency. (2014, September). Excise duty memorandum. Rates of excise duty. http://www.cra-arc.gc.ca/E/pub/em/edm1-5-1/edm1-5-1-e.pdf. Accessed 22 January 2016.

Canada Revenue Agency. (2015). Learning the basics of the GST/HST. *Government of Canada*. http://www.cra-arc.gc.ca/tx/bsnss/tpcs/gst-tps/gnrl/menu-eng.html. Accessed 22 January 2016.

Canadian Centre on Substance Abuse (CCSA). (2012). *Alcohol price policy series: reducing harm to Canadians*. http://www.ccsa.ca/Resource%20Library/CCSA-Alcohol-Price-Policy-Brief-Canada-2012-en.pdf. Accessed 11 December 2015.

Canadian Food Inspection Agency. (2014). Information within the nutrition facts table. Mandatory information and serving size. *Government of Canada*. http://www.inspection.gc.ca/food/labelling/food-labelling-for-industry/nutrition-labelling/information-within-the-nutrition-facts-table/eng/1389198568400/1389198597278?chap=1. Accessed 1 March 2015.

Canadian Minister of Justice. (2015). Excise tax act. Consolidation. R.S.C., 1985, c. E-15. http://laws-lois.justice.gc.ca/PDF/E-15.pdf. Accessed 24 February 2016.

Cawley, J., & Frisvold, D. (2015). *The incidence of taxes on sugar-sweetened beverages: The Case of Berkeley, California* (No. w21465). Cambridge, MA: National Bureau of Economic Research. http://www.nber.org/papers/w21465.pdf. Accessed 25 February 2016.

Chen, X., Kaiser, H. M., & Rickard, B. J. (2015). The impacts of inclusive and exclusive taxes on healthy eating: An experimental study. *Food Policy, 56*, 13–24. doi:10.1016/j.foodpol.2015.07.006.

Chriqui, J. F., Chaloupka, F. J., Powell, L. M., & Eidson, S. S. (2013). A typology of beverage taxation: Multiple approaches for obesity prevention and obesity prevention-related revenue generation. *Journal of Public Health Policy, 34*(3), 403–423. doi:10.1057/jphp.2013.17.

CNW Telbec. (2013, May 9). Le ministre des Finances doit voir clair dans les tentatives d'influence déguisées de l'industrie des boissons sucrées. Montréal (Quebec). http://www.newswire.ca/fr/news-releases/le-ministre-des-finances-doit-voir-clair-dans-les-tentatives-dinfluence-deguisees-de-lindustrie-des-boissons-sucrees-512391561.html. Accessed 24 February 2016.

Coalition québécoise pour le contrôle du tabac (CQCT). (2014). Letter to Mr Carlos Leitão. Minister of Finance. 13 May 2014. http://www.cqct.qc.ca/Documents_docs/DOCU_2014/

Doetinchem, O. (2010). *Hypothecation of tax revenue for health* (Background Paper No. 51). Geneva, Switzerland: World Health Organization. http://www.who.int/healthsystems/topics/financing/healthreport/51Hypothecation.pdf. Accessed 20 December 2015.

Ecorys. (2014). *Food taxes and their impact on competitiveness in the agri-food sector. Final report* (No. Ref. Ares(2014)2365745 - 16/07/2014). Rotterdam: Client: DG Enterprise and Industry.

Éduc'alcool. (2016). About us. Governance. http://educalcool.qc.ca/en/about-us/governance/#.VqJbshgrK-w. Accessed 22 January 2016.

Fletcher, J. M., Frisvold, D. E., & Tefft, N. (2010). The effects of soft drink taxes on child and adolescent consumption and weight outcomes. *Journal of Public Economics, 94*(11–12), 967–974. doi:10.1016/j.jpubeco.2010.09.005.

Food in Canada. (2009, March 2). GST rules on beverages can drive retailers to drink. http://www.foodincanada.com/food-business/gst-rules-on-beverages-can-drive-retailers-to-drink-1575/. Accessed 22 January 2016.

French Ministry of Budget. (2012). Contributions sur les boissons et préparations liquides pour boissons sucrées et édulcorées. Contributions indirectes. Circulaire du 24 janvier 2012. NOR : BCRD 1202351C.

Government of Canada. (2010). Information for First Nation Administration. https://www.aadnc-aandc.gc.ca/eng/1100100021851/1100100021852. Accessed 7 December 2015.

Government of Canada. (2015a). Constitution Acts, 1867 to 1982. VI. DISTRIBUTION OF LEGISLATIVE POWERS. Powers of the Parliament. Sections 91 and 92. http://laws-lois.justice.gc.ca/eng/const/page-4.html. Accessed 7 December 2015.

Government of Canada. (2015b). Federal role in health/provincial/territorial role in health: http://healthycanadians.gc.ca/health-system-systeme-sante/cards-cartes/federal-role-eng.php. Accessed 22 September 2015.

Government of Ontario. (2015). Beer sales in grocery stores. https://www.ontario.ca/page/beer-sales-grocery-stores. Accessed 22 January 2016.

Grogger, J. (2015). *Soda taxes and the prices of sodas and other drinks: Evidence from Mexico* (No. w21197). Cambridge, MA: National Bureau of Economic Research. http://www.nber.org/papers/w21197.pdf. Accessed 28 February 2016.

Health Canada. (2011). Canada's food guide. http://www.hc-sc.gc.ca/fn-an/food-guide-aliment/index-eng.php. Accessed 23 September 2015.

Health Canada. (2012). Food and nutrition. Food safety. http://www.hc-sc.gc.ca/fn-an/securit/index-eng.php. Accessed 23 September 2015.

Health Canada. (2015a). Food and nutrition. Nutrition labelling. http://www.hc-sc.gc.ca/fn-an/label-etiquet/nutrition/index-eng.php. Accessed 23 September 2015.

Health Canada. (2015b). Health Canada's proposed changes to the core nutrients declared in the Canadian nutrition facts table. http://www.hc-sc.gc.ca/fn-an/consult/2014-core-nutrients-principaux-nutriments/document-consultation-eng.php#a71. Accessed 15 February 2016.

Hespel, V., & Berthod-Wurmser, M. (2008). *Rapport sur la pertinence et la faisabilité d'une taxation nutritionnelle*. Inspection générale des Finances – Numéro 2008-M-002-01, Inspection générale des Affaires sociales—Numéro RM-2008-073. http://www.igf.finances.gouv.fr/webdav/site/igf/shared/Nos_Rapports/documents/2008/Taxe_nutritionnelle.pdf. Accessed 25 February 2016.

Institute of Medicine (U.S.), Glickman, D., Committee on Accelerating Progress in Obesity Prevention, & Food and Nutrition Board (Eds.). (2012). *Accelerating progress in obesity prevention: solving the weight of the nation*. Washington, DC: National Academies Press.

International Agency for Research on Cancer (IARC) (Ed.). (2011). *IARC handbooks of cancer prevention, tobacco control, volume 14, Effectiveness of Tax and Price Policies for Tobacco Control*. Lyon: IARC.

Korstrom, G. (2013, July 23). Push for sugary drinks tax ignites heated debate. *Business in Vancouver*, p. 9. Vancouver.

Li, L., Borland, R., Fong, G. T., Thrasher, J. F., Hammond, D., & Cummings, K. M. (2013). Impact of point-of-sale tobacco display bans: Findings from the International Tobacco Control Four Country Survey. *Health Education Research, 28*(5), 898–910. doi:10.1093/her/cyt058.

Manitoba Finance. (2014). *The Retail Sales Tax Act. May 2000. Revised June 2014* (Information Bulletin No 29). https://www.gov.mb.ca/finance/taxation/pubs/bulletins/029.pdf. Accessed 22 January 2016.

Marketwire. (2011, April 28). Canadian obesity network-international consensus: Take action to prevent childhood obesity. *Marketwire*.

Mauritius Revenue Authority. (2012, December 11). Communiqué. Excise tax on sugar content of soft drink. http://mra.gov.mu/download/CommuniqueSoftDrinks2012.pdf. Accessed 25 February 2014.

McIntyre, D. (2015). Domestic financing for NCDs. In *Policy brief*. Presented at the WHO global coordination mechanism on the prevention and control of noncommunicable diseases. Working Group on how to realize governments' commitment to provide financing for NCDs, Geneva, Switzerland: World Health Organization (WHO).

Miao, Z., Beghin, J. C., & Jensen, H. H. (2012). Taxing sweets: Sweetener input tax or final consumption tax? *Contemporary Economic Policy, 30*(3), 344–361. doi:10.1111/j.1465-7287.2011.00278.x.

Ministerio de Hacienda (Chili). (2015). Reforma Tributaria. Principales modificaciones. http://reformatributaria.gob.cl/principales-modificaciones.html. Accessed 17 December 2015.

National Alcohol Strategy Advisory Committee. (2015). *Social reference prices for alcohol: a tool for Canadian governments to promote a culture of moderation*. Ottawa. http://www.deslibris.ca/ID/247663. Accessed 28 February 2016.

Navajo Nation Council. (2014). President Shelly signs Healthy Dine' Nation Act of 2014 into law. Resolution of the Navajo Nation Council. 22nd Navajo Nation Council, Fourth Year, 2014. http://www.navajo-nsn.gov/News%20Releases/OPVP/2014/nov/Healthy%20Dine%20Nation%20Act%20of%202014.pdf. Accessed 17 December 2015.

Newfoundland and Labrador Department of Finance. (2015). Tax information bulletin. Cancellation of the harmonized sales tax rate increase. http://www.fin.gov.nl.ca/fin/tax_programs_incentives/pdf/hst_cancellation.pdf. Accessed 22 February 2015.

Norris, S. (2011, November 9). The human papillomavirus Vaccine. Background paper No. 2011-114-E. *Parliament of Canada*. Library of Parliament of Canada. http://www.lop.parl.gc.ca/content/lop/ResearchPublications/2011-114-e.htm. Accessed 1 December 2015.

Official Journal of the Algerian Republic. (2012). Conventions et accords internationaux – loi et décrets, arrêtés, décisions, avis, communications et annonces. Parution Numéro 52, 51ème année, 19 Septembre 2012. http://www.joradp.dz/FTP/jo-francais/2012/F2012052.pdf. Accessed 1 March 2013.

Parliament of Canada. (2015a). Canada's system of Government. The division of powers. http://www.parl.gc.ca/About/Parliament/Education/OurCountryOurParliament/html_booklet/division-powers-e.html. Accessed 23 September 2015.

Parliament of Canada. (2015b). Canada's system of Government. The three levels of government. http://www.parl.gc.ca/About/Parliament/Education/OurCountryOurParliament/html_booklet/three-levels-government-e.html. Accessed 23 September 2015.

Parliament of Canada. (2015c). How Canadians govern themselves. Powers of the National and Provincial Governments. http://www.parl.gc.ca/About/Parliament/SenatorEugeneForsey/book/chapter_3-e.html. Accessed 23 September 2015.

Phaneuf, C. (2010, May). Les revenus du tabac, déficit ou profit ? la dette éternelle. *La Vraie Vérité sur le Stade Olympique de Montréal*. http://www.stadeolympiquemontreal.ca/les-revenus-du-tabac.aspx. Accessed 28 February 2016.

Physicians for a Smoke-Free Canada. (2015). Taxes on cigarettes in Canadian jurisdictions. http://www.smoke-free.ca/factsheets/pdf/taxrates.pdf. Accessed 11 December 2015.

Pomeranz, J. L. (2012). Advanced policy options to regulate sugar-sweetened beverages to support public health. *Journal of Public Health Policy, 33*(1), 75–88. doi:10.1057/jphp.2011.46.

Pomeranz, J. L. (2014). Sugary beverage tax policy: Lessons learned from tobacco. *American Journal of Public Health, 104*(3), e13–e15. doi:10.2105/AJPH.2013.301800.

Preece, R. (2013). The effective contribution of excise taxation on non-alcoholic beverages to government revenues and social objectives: a review of the literature. *World Customs Journal, 7*(1), 21–38.

Prince Edward Island Department of Finance. (2013). Transition to the Harmonized Sales Tax (HST). http://www.taxandland.pe.ca/index.php3?number=76948&lang=E. Accessed 22 January 2016.

Public Health Agency of Canada (PHAC). (2004). The role and organization of public health. http://www.phac-aspc.gc.ca/publicat/sars-sras/naylor/3-eng.php#s3b. Accessed 23 September 2015.

Quebec Government. (1976). Loi modifiant la Loi de l'Impôt sur le tabac. Chapitre 21. Sanctionnée le 30 Juin 1976.: http://www.bibliotheque.assnat.qc.ca/DepotNumerique_v2/AffichageFichier.aspx?idf=99424. Accessed 25 February 2016.

Quebec Government. (2015a). Loi instituant le Fonds pour la promotion des saines habitudes de vie. http://www2.publicationsduquebec.gouv.qc.ca/dynamicSearch/telecharge.php?type=2&file=/F_4_0021/F4_0021.html. Accessed 25 February 2016.

Quebec Government. (2015b). Loi instituant le fonds pour le développement du sport et de l'activité physique. http://www2.publicationsduquebec.gouv.qc.ca/dynamicSearch/telecharge.php?type=2&file=/F_4_003/F4_003.html. Accessed 25 February 2016.

Quebec Government. (2015c). Loi sur le tabac. http://www2.publicationsduquebec.gouv.qc.ca/dynamicSearch/telecharge.php?type=2&file=/T_0_01/T0_01.html. Accessed 25 February 2016.

Régie des Installations Olympiques (RIO). (2006). Données et statistiques. http://parcolympique.qc.ca/le-parc-olympique/notre-organisme/donnees-et-statistiques/. Accessed 28 February 2016.

Revenu Quebec. (2010). *La TVQ, la TPS/TVH et l'alimentation*. www.revenu.gouv.qc.ca. Accessed 1 March 2012.

Revenu Quebec. (2012a). Renseignements généraux sur la TVQ et la TPS/TVH. IN-203. http://www.revenuquebec.ca/fr/sepf/publications/in/in-203.aspx. Accessed 1 March 2012.

Revenu Quebec. (2012b). Taxes. Produits alimentaires de base. http://www.revenuquebec.ca/fr/citoyen/taxes/tpstvq/alimentation/alimentationbase.aspx. Accessed 1 March 2012.

Revenu Quebec. (2014). Tobacco tax rate table (applicable as of June 5, 2014). Tobacco Tax Act. http://www.revenuquebec.ca/documents/en/formulaires/ta/TA-1-V(2014-06).pdf. Accessed 22 January 2016.

Revenu Quebec. (2015a). Specific tax on alcoholic beverages—billing. http://www.revenuquebec.ca/en/citoyen/taxes/autres-taxes/alcool.aspx. Accessed 22 January 2016.

Revenu Quebec. (2015b). Tobacco tax. Obligations respecting the tobacco tax by activity. http://www.revenuquebec.ca/en/entreprises/taxes/tabac/default.aspx. Accessed 22 January 2016.

Revenu Quebec, & Canada Revenue Agency. (2013). *The QST and the GST/HST: how they apply to foods and beverages*. http://www.revenuquebec.ca/documents/en/publications/in/in-216-v(2013-05).pdf. Accessed 22 January 2016.

Ries, N. M., & von Tigerstrom, B. (2011). Legal interventions to address obesity: Assessing the State of the Law in Canada. *The University of British Columbia Law Review 361*, *43*(2).

Saskatchewan Ministry of Finance. (2015). *The provincial sales tax act. Information for grocery and drug stores* (No. PST-2 Issued November 1991; Revised: June 2015). http://finance.gov.sk.ca/revenue/pst/bulletins/PST-2%20Grocery%20and%20Drug%20Stores.pdf. Accessed 22 January 2016.

Servicio de Impuestos Internos (Chili). (2016). Impuestos indirectos. Información vigente, actualizada el 4-01-2016. http://www.sii.cl/aprenda_sobre_impuestos/impuestos/impuestos_indirectos.htm. Accessed 22 January 2016.

Sharma, A., Hauck, K., Hollingsworth, B., & Siciliani, L. (2014a). The effects of taxing sugar-sweetened beverages across different income groups: The impact of SSB taxes on different income groups. *Health Economics, 23*(9), 1159–1184. doi:10.1002/hec.3070.

Sharma, A., Vandenberg, B., & Hollingsworth, B. (2014b). Minimum pricing of alcohol versus volumetric taxation: Which policy will reduce heavy consumption without adversely affecting light and moderate consumers? *PLoS ONE, 9*(1), e80936. doi:10.1371/journal.pone.0080936.

Sinckler, C. P. (2015). Presentation of the financial statement and budgetary proposals 2015 delivered by the hon. Christopher P. Sinckler to the House of Assembly, Monday June 15, 2015, 4 pm. http://www.barbadosparliament.com/uploads/document/d1efb84aac6a7abe4c6c0efcf8ceedd2.pdf. Accessed 17 December 2015.

Skerrit, R. (2015). *Budget address 2015–2016—"keeping it real" by Honourable Roosevelt Skerrit*. http://finance.gov.dm/index.php/budget/budget-addresses/file/22-budget-address-2015-2016-keeping-it-real-by-honourable-roosevelt-skerrit-july-24-2015. Accessed 17 December 2015.

Société des alcools du Quebec (SAQ). (2016). Terms and conditions of outlet sales. https://www.saq.com/content/SAQ/en/a-propos/infos-achat/achat-succursale.html. Accessed 22 January 2016.

Stockwell, T., Leng, J., & Sturge, J. (2006). *Alcohol pricing and public health in Canada: issues and opportunities. A discussion paper prepared for the National Alcohol Strategy Working Group*. Centre for Addictions Research of BC. http://www.uvic.ca/research/centres/carbc/assets/docs/report-alcohol-pricing-public-health-cnd.pdf. Accessed 28 February 2016.

The Canadian Press. (2015, April 16). How alcohol is sold in provinces across Canada. http://www.citynews.ca/2015/04/16/how-alcohol-is-sold-in-provinces-across-canada/. Accessed 22 January 2016.

Thomas, G., & Canadian Centre on Substance Abuse. (2012). *Price policies to reduce alcohol-related harm in Canada. (Alcohol Price Policy Series: Report 3)*. Ottawa (Ontario): Canadian Centre on Substance Abuse.

Thow, A. M., Quested, C., Juventin, L., Kun, R., Khan, A. N., & Swinburn, B. (2011). Taxing soft drinks in the Pacific: Implementation lessons for improving health. *Health Promotion International, 26*(1), 55–64. doi:10.1093/heapro/daq057.

Toronto Star. (2012, October 23). Ontario's doctors call for a war on junk food. *Toronto Star*. Toronto. http://www.thestar.com/opinion/editorials/2012/10/23/ontarios_doctors_call_for_a_war_on_junk_food.html. Accessed 28 February 2016.

VicHealth. (2015). VicHealth: our history. VicHealth 25th Anniversary. https://www.vichealth.vic.gov.au/about/our-history. Accessed 20 December 2015.

von Tigerstrom, B. (2012). Taxing sugar-sweetened beverages for public health: Legal and policy issues in Canada. *Alberta Law Review, 50*(1), 37–64.

World Cancer Research Fund (WCRF) (Ed.). (2015). WCRF international nourishing framework. Use economic tools to address food affordability and purchase incentives. Table last updated 24 November 2015. http://www.wcrf.org/sites/default/files/3_Economic%20Tools_Final.pdf. Accessed 17 December 2015.

World Health Organization (WHO). (2012). *Tobacco taxation and innovative health-care financing*. New Delhi, India: World Health Organization. http://www.searo.who.int/tobacco/documents/2012-pub1.pdf. Accessed 22 January 2016.

World Health Organization (WHO). (2015a). *Using price policies to promote healthier diets*. Copenhagen (Denmark): World Health Organization (WHO).

World Health Organization (WHO). (2015b). *WHO report on the global tobacco epidemic, 2015. Raising taxes on tobacco. Fresh and alive. MPOWER. Annexes*. http://www.who.int/tobacco/global_report/2015/appendix2.pdf?ua=1. Accessed 20 December 2015.

Yale Rudd Center for Food Policy and Obesity. (2012). Sugar-sweetened beverage taxes and sugar intake: Policy statements, endorsements, and recommendations. http://www.childhoodobesity-foundation.ca/stakeholders-taxation-of-sugar-sweetened-beverages. Accessed 1 March 2013.

Chapter 13
Social and Political Acceptability of Sugar-Sweetened Beverage Taxation

Abstract An overview of statements and arguments confirm that the relevancy of health-related food taxes remains controversial at international level and in Canada. Soda tax proposals are often subject to intense advocacy efforts. Public health considerations, economic concerns and partisan politics influence the debate on sugar-sweetened beverage (SSB) taxation and may influence the outcome of the policy process. The propensity of a soda tax to be adopted also depends on budgetary, fiscal and administrative considerations. Among stakeholders and within the population, food taxes often appear to be less popular than other nutritional policies. However, the support for such taxes tends to increase if the proceeds are to be earmarked for health promotion initiatives or social programs. In Canada, over the last years, SSB taxation has been more intensively debated in Quebec than in other provinces, which may favour the adoption of a tax should a political window of opportunity open.

The acceptability of an intervention refers to the "degree of acceptance of the intervention by the various stakeholders including: parents and carers; teachers; healthcare professionals; the general community; policy makers; the private sector; government and other third party funders" (Swinburn et al. 2005, p. 31). It is an important dimension, since low political support for a policy and associated controversies may foil its adoption (see Fig. 13.1), complicate its implementation and possibly alter desired effects on the public's health (Morestin et al. 2011).

In this chapter, we will put the emphasis on social and political acceptability concerns surrounding sugar-sweetened beverage (SSB) taxation, starting with international evidence. Secondly, we will look over the positions taken by a broad diversity of stakeholders on soda taxation across Canada. Finally, we will conclude on the potential interest to initiate a constructive and multistakeholder dialogue to inform decisions.

Y. Le Bodo et al., *Taxing Soda for Public Health*,
DOI 10.1007/978-3-319-33648-0_13

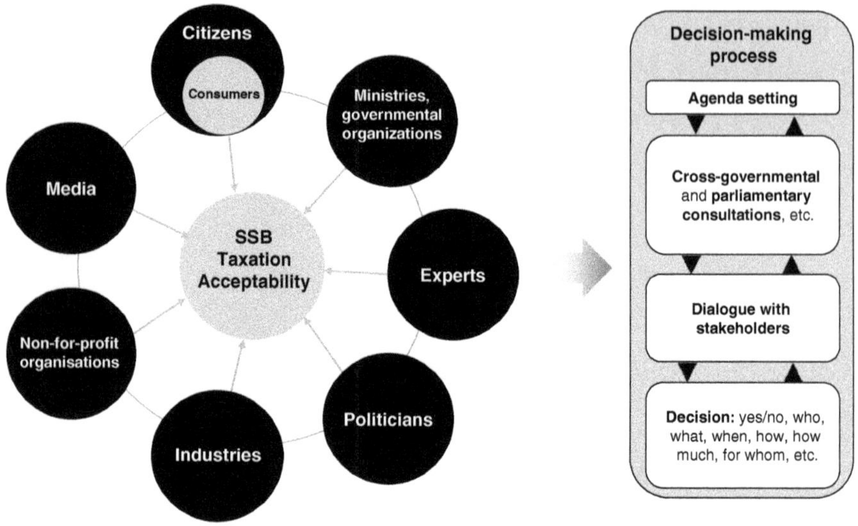

Fig. 13.1 Potential concerns expressed by various stakeholders may strongly influence the decision-making process and jeopardize SSB taxation acceptability. *Source* Inspired from Fafard (2008)

13.1 International Perspective

At international level, evidence indicates that SSB taxation often remains controversial in the scientific community, in the civil society, and among politicians.

In the scientific literature, editorials and commentaries often advocate for considering SSB taxation as a valuable prevention strategy. Arguments encompass: the alarming prevalence of overweight, obesity and their comorbidities; the contribution of SSB intakes to added sugars overconsumption; the need to implement comprehensive and multipronged prevention strategies to which food-related taxes could contribute; the very low nutritional value of SSBs and the existence of healthy substitutes; optimistic simulations of SSB tax effects on prices, demand and health; the possibility to invest SSB tax revenues in health promotion; and encouraging results from countries where SSB taxes have recently been enacted. The need to carefully monitor impacts on equity effects, potential substitution effects, as well as industries' and consumers' reactions is also highlighted, but is generally not considered as sufficient to justify inaction. The need for pilot testing and careful monitoring is sometimes emphasized (Mytton et al. 2012; Blakely et al. 2014; Block and Willett 2013; Block 2013; Howard and Davies 2014; Mozaffarian et al. 2014).

Conversely, scientific editorials, commentaries and brief reviews questioning taxes' benefits also exist. It should be noticed that these articles sometimes present evidence not limited to SSB taxation but rather address food taxes in general. Arguments include probable undesirable effects of such taxes on lower-income groups; probable substitution effects jeopardizing positive impact on energy balance; probable lower effectiveness on heaviest SSB consumers, likely to be less responsive to price changes; probable minimal contribution to improve public health and negative impact on consumers' welfare; imperfect (or absence of any) control for industries' and consumers' reactions and preferences in tax simulations; possible downturn on the local economy, especially for small businesses; existence of other strategies potentially much more effective in addressing SSB overconsumption and other risk factors (e.g. greater informational and educational efforts, behavioural strategies and collaboration with the industry) (Just and Wansink 2012; Devisch 2013a, b; Cornelsen et al. 2013; Lusk 2014; Winkler 2013a, b; Preece 2013; Cornelsen et al. 2015).

Finally, balanced scientific articles sometimes consider the pros and cons and do not strongly take sides (Hawkes 2012; Conkle and Carter 2013; Papadopoulou et al. 2014). Recently, a group of Canadian and American experts in cardiovascular health reviewed the evidence on food pricing policies aimed at preventing NCDs. Although highlighting the numerous uncertainties as regards their effectiveness and potential undesirable effects, the authors stressed that such approaches had been recommended by leading international organizations and should continue to be a focus point for researchers, health professionals and policy makers (Duhaney et al. 2015).

Among stakeholders, in Europe, fiscal policies aimed at obesity prevention have proved particularly unpopular in comparison with other options such as increased access to sports facilities, improved nutritional labelling, or advertising regulations (Lobstein and Millstone 2006; Holdsworth et al. 2007). In the public opinion, nutrition taxes do not rank highly acceptable either. The EATWELL survey conducted over 3000 citizens in five countries (UK, Italy, Belgium, Denmark and Poland) positioned fiscal measures among the "most opposed policies". However, 55 % of respondents supported the proposal to "impose taxes on unhealthy food and to use the proceeds to promote healthier eating" (UREAD 2012). The results of a study based on 12 focus groups conducted in the London area in 2011–12 (n = 94 participants) confirmed that public support for food taxes was generally low. This was essentially driven by doubts as regards tax effectiveness to change behaviours, by mistrust in the governments' actual intentions, and by the way the proceeds of the tax would be used. The perception that food taxes would mainly be used as a way to raise revenue rather than to promote the population's health was particularly emphasized (Somerville et al. 2015).

The beverage industry generally steps into that breach, arguing that SSB taxes "have not proved to achieve any public health objectives and destroy jobs and economic value" (Fooddrinktax.eu 2013). The counteradvocacy efforts of the beverage industry have been successful in many places where soda taxes have not been enacted, but also in countries where such a tax was in place. In Denmark, for instance, the tax on soft drinks and juices was actually abolished in 2014, probably dragged down with the rollback of the world-first tax on saturated fat, deemed ineffective and administratively burdensome (Scott-Thomas 2013; Bødker et al. 2015). Such industry arguments were not always successful and even sometimes proved counterproductive. In France, where the beverage industry strongly opposed the public health motivation of the soda tax proposed by the prime minister in August 2011, the main soda manufacturer even announced his intention to suspend a 17 M€ regional investment as a sign of protest. Facing unanimous critics from the political sphere, the group retracted a few days later invoking a communication mistake. In the meantime, a poll ordered by the beverage industry indicated that the public opinion was mostly unfavourable to this tax and did not trust its obesity prevention objective (Le Bodo et al. 2014). Notwithstanding, a few months after the introduction of the French soda tax on 1 January 2012, the measure seemed to be rather favourably perceived by the population. According to a survey conducted over a subsample of nearly 2000 individuals participating in the Nutrinet-Santé cohort study, half of the respondents were "generally supportive" of the tax and 57.7 % perceived it as "helpful in improving population health". Interestingly, these results are independent of SSB consumption level,[1] but support for the tax is significantly lower in respondents with lower education. Respondents were more likely to support the tax if the revenue generated would be used for "healthcare system improvement" (72.7 %) and "if the prices of other foods and beverages (which are good for health) would go down" (71.5 %) (Julia et al. 2015).

In the USA, where SSB taxation has been much debated, similar acceptability patterns have been observed. In 2010, results for a random-digit dialled telephone survey (n = 592) analysed reactions to a 20 % tax announcement on SSBs. It showed that only a minority of respondents (36 %) were supportive of the tax, although 39 % indicated that they would reduce their consumption if SSBs would be taxed (Rivard et al. 2012). Results from a large US representative survey on SSB taxation acceptability conducted in 2011 also indicate that a majority of respondents disagree with the proposal. Most frequently picked anti-tax arguments were that such a tax "does not affect consumption of other unhealthy foods" (60 %), "is a quick way for politicians to fill budget holes" (58 %), "is an unacceptable intrusion of government into people's lives" (53.8 %), "is opposed by most Americans" (53 %) and "is harmful to the poor" (51 %). On the other hand, almost half agreed that "sugar-sweetened beverages were the single largest contributor to obesity" (49 %) and that "it would raise revenue for obesity prevention" (41 %) (Barry et al. 2013, p. 158). Another web-based public opinion survey on SSB reduction policies conducted in 2012 from a nationally representative sample of the US population shows that calorie labelling and removing drinks from schools are among the policies receiving the strongest support, whereas taxes or portion size restrictions were the most unpopular (Gollust et al. 2014). The local economic burden caused by revenue and job losses related to SSB taxation is part of the commonly conveyed anti-SSB tax messages (Niederdeppe et al. 2013; Jou et al. 2014). It echoes the strong opposition from the beverage industry who spent substantial funds defeating

[1]Although equally supportive of SSB taxation, moderate and high SSB consumers (vs. low SSB consumers) are less likely to believe that the measure will improve the population's health (Julia et al. 2015).

state taxation bills or local ballot measures such as in New York (2008, 2010), Washington (2010), California and Colorado (2012) (Powell et al. 2014; Nixon et al. 2015; Nestle 2015). To convey its arguments, the beverage industry has particularly relied on industry trade associations (e.g. the American Beverage Association) and local non-profit groups funded on purpose, also known as "anti-soda tax coalitions". Messages were spread via multiple channels (TV, radio, billboards, spokespersons in the community) (Nixon et al. 2015, Nestle 2015). Between 2009 and 2012, all SSB taxation bills submitted in 18 US states were rejected (Yale Rudd Center 2014). In the city of Berkeley, where the first US health-related soda tax was enacted in 2014, several contextual factors apparently helped to convince a majority of voters. This includes a strong political commitment, an early-started community-based advocacy campaign anticipating industry's efforts to dissuade the public, financial support from Bloomberg philanthropies and other organizations, and precedents in the adoption of pioneering public health measures (e.g. in tobacco control) (Dinkelspiel 2014; Nestle 2015).

Elsewhere in the world, the opposition from economic actors was also common and sometimes was at the origins of a soda tax roll back or replacement. In Fiji, for example, the domestic excise tax introduced in 2006 was replaced in 2007 with a 3 % fiscal import duty on raw materials "for the purpose of fairer administration and collection of the duty across industry". The industry certainly played a key role in this regard, since manufacturers advocated that the tax had negatively impacted their competitiveness (Thow et al. 2011, p. 57, 60). In India, the enactment of an extra 5 % excise duty on aerated drinks in 2014 was followed by important counteradvocacy efforts led by the Indian Beverage Association, which highlighted its significant contribution to agricultural growth and employment as well as the value of carbonated soft drinks in a country where "options of safe, convenient and hygienic beverages are rather limited" (Indian Beverage Association 2014; The Economic Times 2014). These movements likely contributed to replacing this duty by a less soda-stigmatizing surtax in the 2015 budget. Maybe not for a long time: in December 2015, India's Chief Economic Advisor proposed a 40 % tax on aerated beverages, which has been welcomed by public health advocates (India Resource Center 2015; Ambwani 2015). Finally, in Mexico, in 2014, despite a strong opposition from the beverage industry, the soda tax was adopted in an unusual context where political groups had to join forces to support a large fiscal reform. The president of Mexico himself supported the measure. The strong advocacy efforts of the *Allianza por la Salud Alimentaria*, financially supported by the Bloomberg philanthropies, also played an important role. Among the numerous non-governmental organizations part of this Alliance, the pro-taxation campaign conducted by the consumer group *El poder del Consumidor* was deemed particularly decisive to increase the social acceptability of the tax. Strategic partnerships with academics and politicians as well as the argument to use the tax proceeds to promote access to clean water also appeared to be key for success (Astudillo 2014; Rosenberg 2015; Nestle 2015).

In other countries (e.g. in the UK, in Australia, in South Africa), debates are still in progress and, for reasons similar to those quoted above, acceptability proves to be an important challenge (Campbell 2015; Back 2015; Myers et al. 2015).

Across countries, it is noteworthy that tax increases (in general) consistently receive greater public support when revenues from the tax are earmarked for education, social and health causes. However, studying the tobacco case, Doetinchem warns that "(…) to retain such support and realize the benefits in terms of accountability and public trust", the earmarking of tax revenues "(…) must be strict, i.e. no topping-up from general taxation and no siphoning off to other purposes" (2010, p. 6). This raises important but not insurmountable feasibility challenges (see Sect. 12.5).

In Canada, SSB taxation has also proved to be controversial. The pros and cons positions advocated by experts, civil society actors, political representatives and in the public opinion are exposed in the next sections.

13.2 Canadian Experts

Over the last few years, several Canadian initiatives including a discussion between experts and/or stakeholders have issued favourable recommendations on SSB taxation. To name a few, in 2010, a panel of academics considered the potential impact of economic policies targeting obesity, diet and physical activity. On the basis of scoping reviews and consultations, the group concluded that "overall, the evidence is not sufficiently strong to provide clear policy direction" but observe that the needed research to provide more evidence in this domain is "likely to be complex and potentially unfeasible", which is not a reason "to take no action" (Faulkner et al. 2011, p. 1). Despite several concerns, the panel mostly agreed to support the implementation of a caloric sweetened beverage tax as part of a comprehensive multilevel approach. Examining how to implement corollary fruit and vegetable subsidies targeted at children and low-income households was also part of the recommendations (Faulkner et al. 2011). Similar conclusions were issued during another consensus conference, which gathered researchers, policy experts and practitioners in 2011. Based on the available evidence including potential impacts and political feasibility, SSB taxation was also presented as an "opportunity to promote healthy eating" within "a multipronged comprehensive approach" (Buhler et al. 2013, p. 100). It was particularly recommended that "provincial governments undertake a synchronized adoption of an excise (per volume) tax of $0.05/100 ml applied to all sugar-sweetened beverages sold in any form and in any setting, and indicate the amount of such taxes on the label of the product" (Buhler et al. 2013, p. 100). The group also recommended that half of the revenue from the tax be allocated to healthy eating promotion. Finally, in 2014, an applied research project led by the Evaluation Platform on Obesity Prevention (EPOP) and the Quebec National Public Health Institute (*Institut national de santé publique du Québec*) worked on the identification of priorities to improve healthy eating and physical activity behaviours in the province of Quebec. The project consisted in the following: (1) a macroscopic portrait of existing initiatives; (2) a comparison of these initiatives to international recommendations; (3) a deliberative forum involving key experts and stakeholders; (4) a knowledge synthesis combining scientific, contextual and informal evidence. In conclusion, "enacting a tax on sugar-sweetened beverages and earmarking revenues for health promotion" was one of the 50 actions suggested to improve healthy lifestyles in the province of Quebec (Le Bodo et al. 2016, free translation).

Recently, two expert committees directly mandated by Quebec's public authorities have also recommended considering soda taxes as an option that could be part of large prevention strategies. On the one hand, in March 2015, the final report of the Quebec tax review Board[2] (CEFQ) included a recommendation to "evaluate the possibility to use taxation in order to reduce the consumption of some food products "harmful" to health, including sugar-sweetened beverages". The report states that "only some empirical evidence are available to confirm the existence of an association between this type of taxes and the fight against obesity and related diseases" but adds that such a tax could be used as a "signal demonstrating the attention devoted to the issue". Finally, the report also mentioned that "generated revenues should be reinvested in the fight against obesity to have a true impact" and suggested earmarking revenues for school-based informational and educational initiatives or broader informational measures such as nutrition labelling improvements in restaurants (CEFQ 2015, p. 152, free translations). On the other hand, the report of the Quebec Commissioner for sustainable development, part of Auditor General annual report[3] issued in June 2015, includes a section devoted to the promotion of healthy eating behaviours. This report criticizes the absence of a governmental position on the relevancy (or not) to implement measures promoting healthier eating behaviours, such as the taxation of "junk food" (including carbonated beverages) (Vérificateur Général du Québec 2015).

At the opposite, not-for-profit "pro-business" research organizations such as the Fraser Institute (Korstrom 2013) and the Montreal Economic Institute (Gratzer and Guénette 2012; Kelly-Gagnon 2013) have explicitly disapproved SSB taxation policies. Their reasoning highlight, among other things, that SSB consumption has tended to decrease in Canada over the last years, that such a policy focused on a single food category is unlikely to have a meaningful impact on obesity prevalence (in particular because of substitution effects towards other calorie-dense foods and beverages), that SSB taxation would most probably be regressive and would affect all consumers whatever their lifestyle and health status are, that fat taxes (e.g. in

---

[2]The Quebec Tax Review Board has been officially appointed by decree in June 2014. The mandate given by the government was to consider a potential reform of the Quebec tax system in order to make it more competitive, efficient and equitable (CEFQ 2015).

[3]Quebec Auditor General annual report draws attention of the Members of Parliament on key issues related to the use of public funds and goods. Since 2008, a volume of the report refers to the activities of the Quebec Commissioner for sustainable development. It covers some educational, social and health matters.

Denmark) have proved economically damaging and administratively burdensome, and that many other options are preferable to encourage better eating and physical activity (e.g. via physical education in schools, urban planning, medical counselling). Several of these arguments are also commonly used by the SSB industry to oppose soda tax proposals (see next section).

13.3 Canadian Civil Society

In parallel, in the civil society, various organizations or their representatives publicly expressed opinions supporting SSB taxation over the last years (see Table 13.1). Most frequent pro-tax arguments include the rising overweight, obesity and type-2 diabetes prevalence across the country (particularly in youth) and alarming related costs for the society; the need to combine multiple actions at

Table 13.1 Examples of Canadian civil society organizations having publicly issued pro- or anti-taxation statements at federal or provincial/territorial levels

Pro-taxation statements	Federal	Canadian Medical Association (PostmediaNews 2011a) Chronic Diseases Prevention Alliance of Canada (CDPAC 2011) Heart and Stroke Foundation of Canada (CNW 2013b; HSF 2014) Canadian Diabetes Association (CDA 2015) Dieticians of Canada (2016)
	Provincial	Quebec Weight Coalition (CQPP 2014) Alberta Policy Coalition for Chronic Disease Prevention (APCCP 2015). Quebec Public Health Association (*Association pour la santé publique du Québec*) (ASPQ 2012) Ontario Medical Association (Toronto Star 2012) Childhood Obesity Foundation (Stueck 2012) British Columbia Healthy Living Alliance (BCHLA 2011) British Columbia Medical Association Council on Health Promotion (PostMediaNews 2011b) Quebec Association of Medical Doctors specialized in Community Health (*Association des médecins spécialistes en santé communautaire du Québec*) (Lalonde 2014).
Anti-taxation statements	Federal	Canadian Beverage Association (CBA 2013; CNW 2012) Food and Consumer Products of Canada (Guelf Mercury 2012)
	Provincial	Quebec Soft Drinks Bottlers Association (*Association des embouteilleurs de boissons gazeuses du Québec*) (CNW 2013a) Quebec Food Retailers Association (*Association des détaillants en alimentation du Québec*) (Protégez-vous 2012) Quebec Chambers of Commerce Federation (*Fédération des chambres de commerce du Québec*) (Protégez-vous 2012)

multiple levels to tackle obesity and NCDs; health risks related to SSB consumption; low SSB nutritive value; strong youth-oriented SSB marketing; high SSB intakes (particularly in adolescents); increasing consumption of new types of sugary beverages over the last years (e.g. energy drinks, flavoured waters); the potential effectiveness of taxation on behaviours as part of a multipronged approach including other strategies (as proved successful in tobacco control); the potential use of tax revenues to fund healthy living initiatives (especially towards lower-income households); the fact that SSB taxation already took effect and provided encouraging results in other countries (e.g. Mexico, France); and support for soda taxes expressed by renowned scientific organizations.

At the opposite, SSB taxation has also been strongly contested, especially by economic actors (see Table 13.1). Frequent anti-tax arguments include that an educational approach is preferable to taxation; that SSBs are already taxed at a high rate; that simplistic solutions focused on one single food or beverage cannot address complex issues such as obesity; that SSB consumption decreased over the last years in Canada while obesity rates continued to rise; that taxing SSBs would mainly consist in raising additional fiscal revenues; that such a tax would not benefit public health, would be regressive and could damage the local economy; and that the beverage industry has already made significant nutritional improvements, has adopted guidelines in terms of marketing to children and reduced SSB availability in schools. To a much lower extent, consumers and citizen groups such as the Canadian Taxpayers Federation (Le Journal de Québec 2011) and the Quebec Not-for-Profit Association *Option Consommateurs* (Protégez-vous 2010; Le Nouvelliste 2010) have punctually expressed hesitations and stated that SSB taxation would be too interventionist and likely regressive.

This controversy was already reflected by an opinion survey conducted among key policy influencers in Alberta and Manitoba in 2009 (provincial government, municipal government, school boards, print media companies, and large workplaces; $n = 236$; response rate of 15 %): 57.4 % of the respondents were strongly or somewhat supportive of a tax on unhealthy food and beverage. Greater support was clearly obtained for individual-focused policy approaches (e.g. education programs on healthy eating) and for some other environmental policies (e.g. policies for school nutrition programs; banning the use of trans-fat in all food products). The authors mentioned such results were expectable, considering that most respondents were from governments and that, in these provinces, taxation is traditionally not popular among policy makers (Raine et al. 2014).

13.4 Canadian Elected Representatives

In such a controversial climate, it is not surprising that the Federal Health Minister (PostMediaNews 2012; La Presse Canadienne 2011) as well as provincial political leaders from Quebec (Bissonnet 2015; Metro Montréal 2012), Ontario (Ferguson 2013), British Columbia (Korstrom 2013), New Brunswick (Leyral 2012),

Newfoundland and Labrador (The Telegram 2013) and Nunavut (Greer 2010) so far expressed doubts and reluctance to adopt new taxes on SSBs. A notable exception is the pro-tax position adopted by the Quebec Health Minister in 2013, who highlighted the health concerns related to SSB consumption and the potential effectiveness of a tax to reduce SSB consumption (Teisceira-Lessard 2013). But political turns did not let much time for the proposal to eventually expand beyond a simple declaration of interest. At municipal level, only a few political leaders took a stance on soda taxes. It probably relates to the fact that, in Canada, soda taxes are unlikely to be part of municipal jurisdiction (see Sect. 12.2). However, in December 2014, several Montreal city councillors submitted a motion asking the provincial government to adopt a new special tax on SSBs as part of obesity prevention efforts (CNW 2014; Rotrand et al. 2014). The proposal has been supported by several medical and civil society organizations but has been contested by private organizations, including the soft drink bottlers and retailers, who strongly called the mayor's majority to reject it (Lalonde 2014). Finally, the soda tax proposal was removed from the motion, which still draws attention of the provincial government to the importance of inciting manufacturers to decrease significantly the amount of calories consumed from their beverages (Ville de Montréal 2014).

13.5 Canadian Public Opinion

As elsewhere, a few polls sponsored by civil society groups show ambivalent positions in the public opinion, depending on how the issue is framed. For instance, according to an Ipsos Descarie survey for the Quebec Weight Coalition (CQPP) in May 2010, 44 % of Canadian respondents (56 % of Quebecers) were in favour of a special tax on sugary beverages but 70 % (77 % of Quebecers) supported an SSB tax if revenues were "invested in preventing obesity and in promoting healthy lifestyle choices" (Buhler et al. 2013, p. 98). Conversely, according to an Ipsos survey for Refreshments Canada in January 2011, 87 % of Canadians (89 % of Quebecers) felt that "the government should be educating the public about changing their behaviour, not taxing them" (CNW 2011).

An Ipsos-Reid survey on childhood obesity prevention commissioned by the Public Health Agency of Canada (PHAC) in February 2011 brings congruent evidence. The telephone survey (representative sample of 1222 Canadian adults) has shown that the support for taxation of junk food and sugary drinks was not as high as for other options. Only 53 % supported (somewhat or strongly) the idea of "adding a special tax on sugary drinks (e.g. soft drinks), to discourage people from buying them". However, the support climbed to 67 % when it was proposed to use tax revenues "to fund other programs that fight childhood obesity". The report notes that low- and middle-income (vs. high-income) individuals were more likely to be favourable. The qualitative research (based on eight focus groups nationwide in Toronto, Halifax, Winnipeg and Montreal) has shown that food taxes were not very popular either with two emerging positions: a minority considering that such taxes may have to be part of the solution, and a majority questioning the appropriateness, effectiveness and equity of this measure (Ipsos-Reid 2011).

13.6 Importance to Consider Jointly Public Health and Budgetary Rationales

The diversity of positions advocated on SSB taxation in Canada and elsewhere suggests that public health considerations, economic concerns, tax design and partisan politics influence the political debate on SSB taxation. Elsewhere in the world, economic arguments ultimately proved more influential than public health concerns for soda taxes to be adopted (Mosier 2013, Le Bodo et al. 2014).

Box 13.1: Sociopolitical circumstances having influenced the adoption of a soda tax in France

France is one of the countries where a new tax on non-alcoholic beverages recently came into force (see Table 1.1). To better understand the conditions that may have influenced the formulation and the adoption of such a controversial policy, a case study was built from a scoping review, newspaper articles, and documents from public, academic and other sources. It reveals that members of the French Parliament had regularly submitted health-related food taxation bills (including SSBs) over the previous years. Various scenarios had also been elaborated at ministerial level, which suggests fiscal and technical predispositions for a tax to be enacted. However, until 2010, no proposals had received sufficient support (especially from the government) to cross the legislative filter. Unexpectedly, during the summer 2011, in a context of a severe economic crisis urging public authorities to reduce public deficits, the government announced the inclusion of an excise soda tax (€3.58 cents/L) as part of a package of fiscal measures included within the 2012 social security finance bill. The proposal made explicit the intention to reduce SSB consumption (excluding water, juices with no added sugars, and NCSBs) and to contribute to obesity prevention and healthcare funding. A cascade of events followed this announcement. In October 2011, the position adopted by influential members of the Parliament was particularly critical: they pushed for the tax bill to be adopted, but did not adhere to the public health perspective proposed by the Government. Instead, their main objective was to raise fiscal revenue for a much different issue: relieving wage costs in the farming sector. To these ends, this group of parliamentarians proposed to double the tax rate and to extend the tax scope to NCSBs. After political tergiversations in both chambers of the Parliament, two different taxes were finally voted in order to preserve—to all appearances—both the public heath objectives of the Government and the budgetary objective advocated at the Parliament: (1) the first tax (€7.16 cents/l, eq. CAD 10.5 cents/l) targeted

SSBs with half of its revenue earmarked for the National Health Insurance System; (2) the second tax (€7.16 cents/l) targeted NCSBs with all of its revenues earmarked for the general budget. The French Constitutional Council judged that, along the legislative process, budgetary considerations had taken precedence over the initial public health objectives of the tax. From this point of view, paradoxically, the inclusion of NCSBs in the tax scope may have precluded a constitutional breach: since SSBs and NCSBs were part of a same objectively and rationally defined group of beverages, a budget-oriented tax on both types of beverages became more defensible than could have been a public-health-oriented tax. The two taxes came into force on 1 January 2012 (Le Bodo et al. 2014).

For example, the French case (see Box 13.1) illustrates that the soda tax proposal was first officially announced by the Prime Minister and then strongly supported by the Minister of Budget all along the legislative process. There are similarities with the case of the Mexican soda tax, adopted in 2013 within a package of fiscal reforms under the leadership of the Finance Ministry (Rosenberg 2015). Case studies of soda taxes implemented in the Pacific (Fiji, Samoa, Nauru and French Polynesia) also show how the interaction between the Ministries of Health, Finance and Revenue was critical at every stage of the policy-making process for the enactment of theses taxes. Actually, in most cases, the governmental budgetary and fiscal priorities appeared to be as much (or even more) decisive as (than) public health considerations to put the tax on the political agenda, to design it in a feasible way and to make it adopted (Thow et al. 2011). Indeed, we observe that ministries of finance and revenue generally manage the implementation and the administration of soda taxes. The context of announcement of soda taxes makes it clear: it is more often part of annual budgetary speeches pronounced by the Ministers of Finance or Revenue or included in fiscal/budgetary laws and documentation rather than introduced as part of public health plans (see references in Table 1.1). Therefore, the propensity of a soda tax to be adopted clearly depends on budgetary, fiscal and administrative considerations. On the one hand, soda taxation advocacy efforts would certainly benefit from addressing such considerations rather than disregarding it and only focus on public health arguments (see also Chap. 12). On the other hand, minimizing the public health rationale of a tax may put it at greater risk of repeal in case of undesirable effects, as shown in the case of the Danish fat tax (Bødker et al. 2015).

Ultimately, the implementation of a soda tax is an intervention of "political" nature in its etymological sense.[4] In the above-mentioned cases, political decision-makers were certainly convinced that this policy would produce more good than harm, and they made decisions within a window of political opportunity.

[4]The word "politic" comes from the Greek "politikos", meaning the matters "pertaining to the state and its administration or to the public life" (Etymonline 2015).

The "human factor" in the final steps of such a complex and lengthy decision process cannot be underestimated.

A constructive dialogue with all stakeholders may be an essential way to address concerns and prevent implementation failures (Fafard 2008). For example, a consensus-based approach could be a first step in that direction. In Australia, a panel of 12 Australian experts in childhood obesity including nutrition professionals, academics and clinicians was formed to consider the rational for food taxes. Discussions addressed issues of food categorization and definitions, degree of consensus on the evidence base, complexity of tax designs, potential opposition and adaptations by the industry, and equity concerns. Although the overconsumption of several food and drink categories was highlighted, sugar-sweetened beverages were almost the only category considered "amenable" to taxation (as part of a comprehensive strategy). The panel also suggested complementary subsidies on healthier foods (Pitt et al. 2014).

Alternatively, a stakeholders' consultation may prove relevant. For example, in 2012, the Institute of Public Health in Ireland (IPHI) was requested by the Department of Health to conduct a health impact assessment of a 10 % SSB taxation policy. It included a one-day consultation with 50 stakeholders including representatives of industry groups, health organizations, the civil society (consumers, youth, disadvantaged groups) and academic experts. Main objectives were to share knowledge and draw recommendations. It appeared to add value to the whole process, which also included a literature review, economic simulations and a survey of opinion (IPHI 2012).

Finally, in Brisbane (Australia), a citizens' jury was invited over two days to learn, debate and deliberate on the relevancy of a food and drink taxation to reduce childhood obesity. This democratic approach provided some unanimous positions (particularly in favour of SSB taxation and earmarking of tax revenues for health promotion) and was deemed useful in engaging the public in the decision-making process (Moretto et al. 2014).

In Canada, as suggested by analyses of the public discourse, the prospect for SSB taxation under a conservative government was unlikely, up until the political landscape changed in October 2015 with the election that put the liberal party in power. At provincial level, enforcing an excise tax is technically difficult but a special tax on SSBs sold at retail according to the tobacco and alcohol special tax models may be a feasible alternative (see Sect. 12.3). Over the last years, data suggest that SSB taxation has been increasingly and more intensively debated in Quebec than in other provinces. This includes advocacy efforts of a vocal coalition group on weight-related issues, a strong opposition by other actors including the beverage industry, the pros and cons statements from academic experts, contributions from a research group at Laval University (*Université Laval*) working in collaboration with the *Quebec National Public Health Institute* (*Institut national de santé publique du Québec*) opinion polls showing support for taxation conditional to proper earmarking of revenues, an increasing interest at the public health administrative level, and varying statements by successive health ministers. All these elements suggest a maturation of SSB taxation proposals, which may favour the adoption of a tax should a political window of opportunity open (Le Bodo et al. 2015).

Finally, we have previously stressed the fact that SSB taxation should not be considered as a "magic bullet" but rather as a contributing instrument within a comprehensive prevention strategy (see Sect. 3.1). This also suggests that instead of considering SSB taxation political acceptability in isolation, such a tax should rather be debated with the intention (or not) to make it part of a governmental intersectoral strategy aimed at promoting healthy eating, preventing obesity and related non-communicable diseases.

Key messages

- The relevancy of health-related food taxes remains debated at international level and in Canada. It often appears to be less popular than other nutritional policies.
- Soda tax proposals are often subject to intense advocacy efforts. Public support for soda taxes generally increases if tax revenues are to be earmarked for health promotion initiatives and social programs.
- In the political arena, the propensity of a "soda tax" to be adopted seems to depend highly on political leadership as well as on budgetary, fiscal and administrative considerations.
- In Canada, SSB taxation has been more intensively debated in Quebec than in other provinces, which may favour the adoption of a tax should a political window of opportunity open.
- A constructive dialogue with stakeholders seems necessary to support decision-making and, eventually, to prevent implementation failures.

References

Alberta Policy Coalition for Chronic Disease Prevention (APCCP). (2015). *Taxing sugar sweetened-beverages: The case for public health*. Edmonton: Alberta Policy Coalition for Chronic Disease Prevention.

Ambwani, M. V. (2015, December 11). Coca Cola India says may shut some factories if "sin tax" is implemented. *The Hindu Business Line*. New Delhi. http://www.thehindubusinessline.com/companies/sin-tax-on-aerated-beverages-will-make-industry-unviable-cocacola-india/article7976813.ece. Accessed January 5, 2016.

Association pour la santé publique du Québec (ASPQ). (2012, November 20). L'Association pour la santé publique du Québec salue le Budget du Québec 2013–2014. Montréal (Québec). http://www.aspq.org/fr/salle-de-presse/communiques-et-publications/48/l-association-pour-la-sante-publique-du-quebec-salue-le-budget-du-quebec-2013-2014. Accessed February 28, 2016.

Astudillo, O. (2014). Country in focus: Mexico's growing obesity problem. *The Lancet Diabetes & Endocrinology, 2*(1), 15–16. doi:10.1016/S2213-8587(13)70160-8.

Back, A. (2015, October 13). Sugar tax inevitable in Australia's fight against obesity, says public health researcher Boyd Swinburn. *The Canberra Times*. http://www.canberratimes.com.au/act-news/sugar-tax-inevitable-in-australias-fight-against-obesity-says-public-health-researcher-boyd-swinburn-20151013-gk7ryw.html. Accessed January 22, 2016.

Barry, C. L., Niederdeppe, J., & Gollust, S. E. (2013). Taxes on sugar-sweetened beverages: Results from a 2011 national public opinion survey. *American Journal of Preventive Medicine, 44*(2), 158–163. doi:10.1016/j.amepre.2012.09.065.

Bissonnet, J. (2015, February 17). *Le Coût de l'Obésité. Nouvelles, Société. Radio Canada*. http://ici.radio-canada.ca/nouvelles/societe/2015/02/17/003-embonpoint-obesite-coutent-un-milliard-et-demi-de-dollars-par-annee-en-sante-au-quebec.shtml. Accessed February 28, 2016.

Blakely, T., Wilson, N., & Kaye-Blake, B. (2014). Taxes on sugar-sweetened beverages to curb future obesity and diabetes epidemics. *PLoS Medicine, 11*(1), e1001583. doi:10.1371/journal.pmed.1001583.

Block, J. P. (2013). A substantial tax on sugar sweetened drinks could help reduce obesity. *BMJ, 347* (oct31 4), f5947–f5947. http://doi.org/10.1136/bmj.f5947.

Block, J. P., & Willett, W. C. (2013). Taxing sugar-sweetened beverages: Not a "holy grail" but a cup at least half; comment on "food taxes: a new holy grail?". *International Journal of Health Policy and Management, 1*(2), 183–185. doi:10.15171/ijhpm.2013.33.

Bødker, M., Pisinger, C., Toft, U., & Jørgensen, T. (2015). The rise and fall of the world's first fat tax. *Health Policy (Amsterdam, Netherlands), 119*(6), 737–742. doi:10.1016/j.healthpol.2015.03.003

British Columbia Healthy Living Alliance (BCHLA). (2011). *Healthy living in BC—The next generation* (Policy Paper) (p. 12). Vancouver.

Buhler, S., Raine, K. D., Arango, M., Pellerin, S., & Neary, N. E. (2013). Building a strategy for obesity prevention one piece at a time: The case of sugar-sweetened beverage taxation. *Canadian Journal of Diabetes, 37*(2), 97–102. doi:10.1016/j.jcjd.2013.03.025.

Campbell, D. (2015, October 22). Ministers urged to introduce sugar tax in childhood obesity report. *TheGuardian.com*. http://www.theguardian.com/society/2015/oct/22/sugar-report-delayed-jeremy-hunt-tax-radical-action-obesity. Accessed January 22, 2015.

Canadian Beverage Association (CAB). (2013, September). *In response to recommendations for the creation of a sugar sweetened beverage tax*. http://www.canadianbeverage.ca/resources/media/in-response-to-recommendations-for-the-creation-of-a-sugar-sweetened-beverage-tax/. Acce ssed February 16, 2016.

Canadian Diabetes Association (CDA). (2015, September 3). *Why a sugar-sweetened beverage tax will help prevent type 2 diabetes*. https://www.diabetes.ca/newsroom/search-news/why-a-sugar-sweetened-beverage-tax-will-help. Accessed February 15, 2016.

Chronic Disease Prevention Alliance of Canada (CDPAC). (2011). *CDPAC position statement extra sugar, extra calories, extra weight more chronic disease. The case for a sugar-sweetened beverage tax.*

CNW Telbec. (2011, February 3). *Education not taxation to combat the complex issue of obesity.* Montréal (Québec). http://www.newswire.ca/fr/news-releases/education-not-taxation-to-combat-the-complex-issue-of-obesity-507596161.html. Accessed February 24, 2016.

CNW Telbec. (2012, October 23). OMA misguided on call for taxation. Canadian Beverage Association says education is the answer. Montréal (Québec).

CNW Telbec. (2013a, February 19). Le communiqué de l'Association des embouteilleurs de boissons gazeuses du Québec en réponse à la Coalition Poids - L'éducation plutôt que les taxes pour lutter contre l'enjeu complexe de l'obésité. Montréal (Québec). http://www.newswire.ca/fr/news-releases/le-communique-de-lassociation-des-embouteilleurs-de-boissons-gazeuses-du-quebec-en-reponse-a-la-coalition-poids—leducation-plutot-que-les-taxes-pour-lutter-contre-lenjeu-complexe-de-lobesite-51 1972281.html. Accessed February 24, 2016.

CNW Telbec. (2013b, October 17). *Prescription for a healthy nation. Montréal (Québec)*. http://www.newswire.ca/news-releases/prescription-for-a-healthy-nation-513111941.html. Accessed February 24, 2016.

CNW Telbec. (2014, November 28). Taxe sur les boissons sucrées - « La Ville de Montréal doit inspirer de nouvelles initiatives en santé publique » - Marvin Rotrand, conseiller de Snowdon. Montréal (Québec). http://www.newswire.ca/fr/news-releases/taxe-sur-les-boissons-sucrees—la-ville-de-montreal-doit-inspirer-de-nouvelles-initiatives-en-sante-publique—marvin-rotrand-conseiller-de-snowdon-516574141.html. Accessed September 22, 2015.

Commission d'examen sur la fiscalité québécoise (CEFQ). (2015). *Se tourner vers l'avenir du Québec. Une réforme touchant tous les modes d'imposition. Rapport final de la Commission d'examen sur la fiscalité québécoise. Volume 2.* http://site.ebrary.com/lib/ualberta/Doc?id=11060128. Accessed February 25, 2016.

Conkle, J., & Carter, M. (2013). Taxing sugar-sweetened beverages: The fight against obesity. *The Nurse Practitioner, 38*(5), 1–4. doi:10.1097/01.NPR.0000428818.57256.0c.

Cornelsen, L., Green, R., Dangour, A. D., & Smith, R. D. (2013). Is a tax on sugary drinks too bitter to swallow? *BMJ, 347*(3), f7039–f7039. doi:10.1136/bmj.f7039

Cornelsen, L., Green, R., Dangour, A., & Smith, R. (2015). Why fat taxes won't make us thin. *Journal of Public Health, 37*(1), 18–23. doi:10.1093/pubmed/fdu032.

Devisch, I. (2013a). Food taxes: A new holy grail? *International Journal of Health Policy and Management, 1*(2), 95–97. doi:10.15171/ijhpm.2013.15.

Devisch, I. (2013b). Food taxes: How likely are likely effects? *International Journal of Health Policy and Management, 1*(3), 241. doi:10.15171/ijhpm.2013.49.

Dieticians of Canada. (2016). *Taxation and sugar-sweetened beverages. Position of dietitians of Canada.* http://www.dietitians.ca/Downloads/Public/DC-Position-SSBs-and-taxation.aspx. Accessed February 12, 2016.

Dinkelspiel, F. (2014, November 5). Why Berkeley passed a soda tax while other cities failed. *Berkeleyside.* http://www.berkeleyside.com/2014/11/05/why-berkeley-passed-a-soda-tax-where-others-failed/. Accessed January 22, 2016.

Doetinchem, O. (2010). *Hypothecation of tax revenue for health* (Background Paper No. 51). Geneva, Switzerland: World Health Organization. http://www.who.int/healthsystems/topics/financing/healthreport/51Hypothecation.pdf. Accessed December 20, 2015.

Duhaney, T., Campbell, N., Niebylski, M. L., Kaczorowski, J., Tsuyuki, R. T., Willis, K., et al. (2015). Death by diet: The role of food pricing interventions as a public policy response and health advocacy opportunity. *The Canadian Journal of Cardiology, 31*(2), 112–116. doi:10.1016/j.cjca.2014.09.005.

Etymonline. (2015). Online etymology dictionnary. http://www.etymonline.com/index.php?term=politic. Accessed December 1, 2015.

Fafard, P. (2008). *Données probantes et politiques publiques favorables à la santé : pistes fournies par les sciences de la santé et la science politique.* National Collaborating Centre for Healthy Public Policy. http://www.cprn.org/doc.cfm?doc=1904&l=fr. Accessed February 1, 2012.

Faulkner, G. E., Grootendorst, P., Nguyen, V., Andreyeva, T., Arbour-Nicitopoulos, K., Auld, M. C., et al. (2011). Economic instruments for obesity prevention: Results of a scoping review and modified delphi survey. *International Journal of Behavioral Nutrition and Physical Activity, 8* (1), 109. doi:10.1186/1479-5868-8-109.

Ferguson, R. (2013, March 4). Ban marketing of junk food to kids under 12, report recommends. *Toronto Star.* Toronto. http://www.thestar.com/news/queenspark/2013/03/04/ban_marketing_of_junk_food_to_kids_under_12_report_recommends.html. Accessed February 28, 2016.

Fooddrinktax.eu. (2013). *Why food & drink taxes won't work. The facts about food and drink taxes.* http://www.fooddrinktax.eu/. Accessed March 1, 2015.

Gollust, S. E., Barry, C. L., & Niederdeppe, J. (2014). Americans' opinions about policies to reduce consumption of sugar-sweetened beverages. *Preventive Medicine, 63*, 52–57. doi:10.1016/j.ypmed.2014.03.002.

Gratzer, D., & Guénette, J. (2012). *Peut-on contrer l'obésité en taxant les boissons sucrées?* (Les Notes Économiques). Montréal (Québec). http://www.iedm.org/fr/41506-peut-on-contrer-lobesite-en-taxant-les-boissons-sucrees. Accessed December 1, 2012.

Greer, D. (2010, March 17). Pop tax just fool's gold. *Kivalliq News.* http://www.nnsl.com/frames/newspapers/2010-03/mar29_10edit.html. Accessed February 28, 2016.

Guelf Mercury. (2012, October 24). Doctors target obesity; call for tobacco-like warnings on high-calorie, low-nutritional value foods. Guelf (Ontario).

Hawkes, C. (2012). Food policies for healthy populations and healthy economies. *BMJ (Clinical research ed.), 344*, e2801.

Heart and Stroke Foundation of Canada (HSF). (2014). *Sugar, heart disease and stroke. Position statement.* http://www.heartandstroke.com/atf/cf/%7B99452D8B-E7F1-4BD6-A57D-B136CE6C95BF%7D/Sugar-Eng.pdf. Accessed February 16, 2016.

Holdsworth, M., Kameli, Y., & Delpeuch, F. (2007). Stakeholder views on policy options for responding to the growing challenge from obesity in France: findings from the PorGrow project. *Obesity Reviews: An Official Journal of the International Association for the Study of Obesity, 8*(Suppl 2), 53–61. doi:10.1111/j.1467-789X.2007.00359.x.

Howard, S. J., & Davies, S. C. (2014). Chief medical officer urges action to tackle overweight and obesity. *BMJ (Clinical research ed.), 348*, g2375.

Indian Beverage Association (IBA). (2014, July 10). Indian Beverage Association reacts to the excise duty hike on aerated drinks in the union budget 2014–2015. *Coca-Cola India.* https://www.coca-colaindia.com/indian-beverage-association-reacts-excise-duty-hike-aerated-drinks-union-budget-2014-2015/. Accessed January 5, 2015.

India Resource Center. (2015, December 8). Move to tax sugar beverages in India welcomed by public health advocates. http://www.indiaresource.org/news/2015/1032.html. Accessed January 5, 2016.

Institute of Public Health in Ireland (IPHI). (2012). *Proposed sugar-sweetened drinks tax: Health impact assessment (HIA)* (Technical report). http://health.gov.ie/wp-content/uploads/2014/03/PSSDT_HIA_TechReport.pdf. Accessed November 25, 2016.

Ipsos Reid. (2011). Canadians' perceptions of, and support for, potential measures to prevent and reduce childhood obesity. Final Report. Prepared for: Public Health Agency of Canada.

Jou, J., Niederdeppe, J., Barry, C. L., & Gollust, S. E. (2014). Strategic messaging to promote taxation of sugar-sweetened beverages: Lessons from recent political campaigns. *American Journal of Public Health, 104*(5), 847–853. doi:10.2105/AJPH.2013.301679.

Julia, C., Méjean, C., Vicari, F., Péneau, S., & Hercberg, S. (2015). Public perception and characteristics related to acceptance of the sugar-sweetened beverage taxation launched in France in 2012. *Public Health Nutrition, 18*(14), 2679–2688. doi:10.1017/S1368980014003231.

Just, D. R., & Wansink, B. (2012). Clinical decisions. Regulation of sugar-sweetened beverages. *The New England Journal of Medicine, 367*(15), 1464–1466. doi:10.1056/NEJMclde1210278.

Kelly-Gagnon, M. (2013, May 17). Taxes will make obesity even costlier. *Toronto Sun.* http://www.iedm.org/fr/44408-taxes-will-make-obesity-even-costlier?print=yes. Accessed February 28, 2016.

Korstrom, G. (2013, July 23). Push for sugary drinks tax ignites heated debate. *Business in Vancouver*, 9.

Lalonde, M. (2014, December 11). Soft drink industry tries sweet talk to nix tax on sugary drinks. *The Montréal Gazette.* Montréal (Québec). http://montrealgazette.com/news/local-news/soft-drink-industry-tries-sweet-talk-to-nix-tax-on-sugary-drinks. Accessed September 22, 2015.

La Presse Canadienne. (2011, October 6). Boissons énergisantes: Ottawa règlemente, mais peu devront baisser leur caféine. Montréal (Québec).

Le Bodo, Y., Blouin, C., Dumas, N., Laguë, J., & De Wals, P. (2016). *L'Expérience québécoise en promotion des saines habitudes de vie et prévention de l'obésité : COMMENT FAIRE MIEUX?* (Plateforme d'évaluation en prévention de l'obésité (PEPO) and Institut national de santé publique du Québec (INSPQ).). Québec (Québec): Presses de l'Université Laval. http://iucpq.qc.ca/fr/comment-faire-mieux. Accessed February 28, 2016.

Le Bodo, Y., Dionne, M., & De Wals, P. (2015, November 2). *Is it time for a sugar-sweetened beverages tax in Canada and may Québec come first?* Poster presentation presented at the Obesity Week 2015, Los Angeles (California).

Le Bodo, Y., Étilé, F., & De Wals, P. (2014, April 8). *Conditions influençant l'adoption d'une politique de taxation des boissons sucrées en prévention de l'obésité : réflexions et perspectives à partir du cas français*. Presented at the 2ème Journée scientifique de la Faculté des Sciences Infirmières de l'Université Laval, Québec (Québec).

Le Journal de Québec. (2011, February 4). Taxer les boissons pour contrer l'obésité. *Le Journal de Québec*. Québec (Québec).

Le Nouvelliste. (2010, February 16). Une taxe sur les boissons gazeuses. Qui va payer? *Le Nouvelliste (Québec)*.

Leyral, P. (2012, November 27). Des diététiciens veulent taxer les boissons sucrées. *L'Acadie Nouvelle*. http://www.acadienouvelle.com/actualites/2012/11/26/des-dieteticiens-veulent-taxer-les-boissons-sucrees/?pgnc=1. Accessed February 25, 2016.

Lobstein, T., & Millstone, E. (2006). *Policy options for responding to obesity: Evaluating the options : Summary report of the EC-funded project to map the views of stakeholders involved in tackling obesity—the PorGrow project* (University of Sussex, SPRU : Science and Technology Policy Research.). Brighton: SPRU.

Lusk, J. L. (2014). Are sugar-sweetened beverage taxes a cost-effective means of reducing weight? *Canadian Journal of Diabetes, 38*(1), 9–10. doi:10.1016/j.jcjd.2013.08.259.

Metro Montréal. (2012, May 18). Redevances réclamées à l'industrie des boissons gazeuses. *Métro Montréal*. Montréal (Québec).

Morestin, F., Gauvin, F.-P., Hogue, M.-C., & Benoit, F. (2011). *Method for synthesizing knowledge about public policies*. [Montréal]: Centre de collaboration nationale sur les politiques publiques et la santé, Institut national de santé publique Québec. http://collections.banq.qc.ca/ark:/52327/2066075. Accessed February 28, 2016.

Moretto, N., Kendall, E., Whitty, J., Byrnes, J., Hills, A., Gordon, L., et al. (2014). Yes, the government should tax soft drinks: Findings from a citizens' jury in Australia. *International Journal of Environmental Research and Public Health, 11*(3), 2456–2471. doi:10.3390/ijerph110302456.

Mosier, S. L. (2013). Cookies, candy, and coke: Examining state sugar-sweetened-beverage tax policy from a multiple streams approach. *International Review of Public Administration, 18*(1), 93–120. doi:10.1080/12294659.2013.10805242.

Mozaffarian, D., Rogoff, K. S., & Ludwig, D. S. (2014). The real cost of food: Can taxes and subsidies improve public health? *JAMA, 312*(9), 889–890. doi:10.1001/jama.2014.8232.

Myers, A., Fig, D., Tugendhaft, A., Mandle, J., Myers, J., & Hofman, K. (2015). Sugar and health in South Africa: Potential challenges to leveraging policy change. *Global Public Health*, 1–18. doi:10.1080/17441692.2015.1071419

Mytton, O. T., Clarke, D., & Rayner, M. (2012). Taxing unhealthy food and drinks to improve health. *BMJ (Clinical research ed.), 344*, e2931.

Nestle, M. (2015). *Soda politics: Taking on big soda (and winning)*. Oxford [UK]: Oxford University Press.

Niederdeppe, J., Gollust, S. E., Jarlenski, M. P., Nathanson, A. M., & Barry, C. L. (2013). News coverage of sugar-sweetened beverage taxes: Pro- and antitax arguments in public discourse. *American Journal of Public Health, 103*(6), e92–e98. doi:10.2105/AJPH.2012.301023.

Nixon, L., Mejia, P., Cheyne, A., & Dorfman, L. (2015). Big Soda's long shadow: News coverage of local proposals to tax sugar-sweetened beverages in Richmond, El Monte and Telluride. *Critical Public Health, 25*(3), 333–347. doi:10.1080/09581596.2014.987729.

Papadopoulou, E., Spiro, A., & Stanner, S. (2014). Food taxes—What role might they have in the battle against obesity?: Food taxes. *Nutrition Bulletin, 39*(4), 374–378. doi:10.1111/nbu.12110.

Pitt, E., Kendall, E., Hills, A. P., & Comans, T. (2014). Listening to the experts: is there a place for food taxation in the fight against obesity in early childhood? *BMC Obesity, 1*(1). doi:10.1186/s40608-014-0015-3

PostmediaNews. (2011a, February 28). Growing debate over fat tax. *PostmediaNews*. Don Mills, Ont.

PostmediaNews. (2011b, March 18). Is a "fat tax" the answer to an obesity epidemic? *PostmediaNews*. Don Mills, Ont.

PostmediaNews. (2012, January 9). Canadians back ban on junk food ads targeting kids: Survey. *PostmediaNews*. Don Mills, Ont.

Powell, L. M., Wada, R., Persky, J. J., & Chaloupka, F. J. (2014). Employment impact of sugar-sweetened beverage taxes. *American Journal of Public Health, 104*(4), 672–677. doi:10.2105/AJPH.2013.301630.

Preece, R. (2013). The effective contribution of excise taxation on non-alcoholic beverages to government revenues and social objectives: A review of the literature. *World Customs Journal, 7*(1), 21–38.

Protégez-vous. (2010, May). Alimentation: Boissons gazeuses: Coupables ? *Protégez-vous.*

Protégez-vous. (2012, December 7). Les experts en santé publique sont unanimes: le régime alimentaire occidental est désastreux. Faut-il pour autant taxer ou interdire la malbouffe? *Protégez-vous.*

Quebec Weight Coalition (CQPP). (2014). *Freiner la croissance des dépenses publiques : une priorité économique qui passe par la prévention de l'obésité. Mémoire déposé dans le cadre des consultations prébudgétaires sur le budget 2014–2015.* http://www.cqpp.qc.ca/fr/salle-de-presse/communiques-et-publications/205/freiner-la-croissance-des-depenses-publiques-une-priorite-economique-qui-passe-par-la-prevention-de-l-obesite. Accessed February 27, 2014.

Raine, K. D., Nykiforuk, C. I. J., Vu-Nguyen, K., Nieuwendyk, L. M., VanSpronsen, E., Reed, S., & Wild, T. C. (2014). Understanding key influencers' attitudes and beliefs about healthy public policy change for obesity prevention. *Obesity (Silver Spring, Md.), 22*(11), 2426–2433. doi:10.1002/oby.20860

Rivard, C., Smith, D., McCann, S. E., & Hyland, A. (2012). Taxing sugar-sweetened beverages: A survey of knowledge, attitudes and behaviours. *Public Health Nutrition, 15*(8), 1355–1361. doi:10.1017/S1368980011002898.

Rosenberg, T. (2015, November 3). How one of the most obese countries on earth took on the soda giants. *TheGuardian.com.* http://www.theguardian.com/news/2015/nov/03/obese-soda-sugar-tax-mexico. Accessed December 16, 2015.

Rotrand, M., Dorais, B., & Shanahan, S. (2015). Motion concernant les risques associés à la consommation de boissons sucrées et demandant au gouvernement du québec d'étudier la possibilité d'une taxe additionnelle sur ces produits. Motion déposée en vue de l'assemblée du conseil municipal du 15 décembre 2014 - 28 novembre 2014 (2014). https://coalitionmtl.com/fr/nouvelles/article/178/. Accessed September 22, 2015.

Scott-Thomas, C. (2013, April 25). Denmark to scrap decades-old soft drink tax. *FoodNavigator.com.* http://www.foodnavigator.com/Legislation/Denmark-to-scrap-decades-old-soft-drink-tax. Accessed 27 February 2014.

Somerville, C., Marteau, T. M., Kinmonth, A. L., & Cohn, S. (2015). Public attitudes towards pricing policies to change health-related behaviours: A UK focus group study. *European Journal of Public Health, 25*(6), 1058–1064. doi:10.1093/eurpub/ckv077.

Stueck, W. (2012, September 3). Happy teachers and fewer students in B.C. schools this year. *The Globe and Mail.* Vancouver. http://www.theglobeandmail.com/news/british-columbia/happy-teachers-and-fewer-students-in-bc-schools-this-year/article4517214/. Accessed February 25, 2016.

Swinburn, B., Gill, T., & Kumanyika, S. (2005). Obesity prevention: A proposed framework for translating evidence into action. *Obesity Reviews: An Official Journal of the International Association for the Study of Obesity, 6*(1), 23–33. doi:10.1111/j.1467-789X.2005.00184.x.

Teisceira-Lessard, P. (2013, March 9). Taxe sur les boissons gazeuses: le ministre Hébert «y croit». *LaPresse.Ca.* http://www.lapresse.ca/actualites/sante/201303/09/01-4629452-taxe-sur-les-boissons-gazeuses-le-ministre-hebert-y-croit.php. Accessed February 28, 2016.

The Economic Times. (2014, July 10). Budget 2014: Excise duty hike would lead to slowdown in the sector: Indian Beverage Association. *The Times of India.* New Delhi. http://articles.economictimes.indiatimes.com/2014-07-10/news/51300799_1_duty-hike-excise-duty-pepsico-india. Accessed January 5, 2016.

The Telegram. (2013, May 23). Does one's penchant for a bag of chips, a can of pop, or a package of sour candies merit the introduction of a new tax to curb the consumption of such sugary goods? *The Telegram.*

Thow, A. M., Quested, C., Juventin, L., Kun, R., Khan, A. N., & Swinburn, B. (2011). Taxing soft drinks in the Pacific: Implementation lessons for improving health. *Health Promotion International, 26*(1), 55–64. doi:10.1093/heapro/daq057.

Toronto Star. (2012, October 23). Ontario's doctors call for a war on junk food. *Toronto Star*. Toronto. http://www.thestar.com/opinion/editorials/2012/10/23/ontarios_doctors_call_for_a_war_on_junk_food.html. Accessed February 28, 2016.

UREAD. (2012). *Effectiveness of policy interventions to promote healthy eating and recommendations for future action: Evidence from the Eatwell project. Deliverable 5.1.* http://www.eatwellproject.eu/en/. Accessed February 28, 2016.

Vérificateur Général du Québec. (2015). *Rapport du commissaire au développement durable – Rapport du Vérificateur général du Québec à l'Assemblée nationale pour l'année 2015-2016, printemps 2015. Chapitre 2 : Promotion d'une saine alimentation comme mesure de prévention en santé.* http://www.vgq.qc.ca/fr/fr_publications/fr_rapport-annuel/fr_index.aspx?Annee=2015. Accessed September 8, 2015.

Ville de Montréal. (2014). Procès-verbal de l'assemblée ordinaire du conseil municipal du 15 décembre 2014. http://ville.montreal.qc.ca/documents/Adi_Public/CM/CM_PV_ORDI_2014-12-15_14h00_FR.pdf. Accessed September 22, 2015.

Winkler, J. T. (2013a). Tax on soft drinks to reduce obesity. Lowering the price of "good" foods would be better than taxing "bad" foods in reducing obesity. *BMJ (Clinical research ed.), 347*, f7038.

Winkler, J. T. (2013b). Brutal pragmatism on food. *BMJ, 346*(1), f3728–f3728. doi:10.1136/bmj.f3728

Yale Rudd Center for Food Policy and Obesity. (2014). Sugar-sweetened beverages: Taxes. http://www.yaleruddcenter.org/what_we_do.aspx?id=271. Accessed February 27, 2014.

Chapter 14
Evaluability of Sugar-Sweetened Beverage Taxation

Abstract Taxes on soda/sugar-sweetened beverages (SSBs) implemented across the world are still little evaluated. Yet, evaluating the effects and impact of soda taxes is critical to provide evidence with more confidence and external validity than what currently provide simulations and experimental trials. Planning a detailed and robust evaluation prior to implementation is essential. In the case of a price-induced behavioural logic, evaluation should focus on SSB price, SSB intakes, energy balance and eventually health outcomes. Key methodological aspects have to be considered in order to be able to disentangle tax-specific effects from other factors. In the case of a fiscal logic, the evaluation would minimally consist in the collection and monitoring of tax revenues as well as in the monitoring of how tax proceeds are earmarked for specific uses, such as health promotion initiatives. Finally, in the case of a signal-induced denormalization logic, evaluation may focus on the public perception of the tax as well as on reactions of manufacturers.

14.1 Importance of Sugar-Sweetened Beverage (SSB) Taxation Evaluation

Although governments increasingly consider the implementation of nutrition-related taxes, the importance to plan the evaluation of these policies is rarely discussed (Mytton et al. 2014; WHO 2015). Yet, it is critical for several reasons:

1. The effects of SSB taxes implemented across the world are still little documented. Mytton et al. (2014) highlight that the evaluation of ongoing large-scale natural experiments is needed to provide evidence with more external validity than what currently provides simulations and experimental trials.
2. Even if an ex-ante evaluation is conducted, in the complex area of chronic disease prevention policies, decisions often have to be made despite incomplete evidence: it makes ex-post-evaluation even more essential for summative and formative purposes.

Y. Le Bodo et al., *Taxing Soda for Public Health*,
DOI 10.1007/978-3-319-33648-0_14

3. Because of the uncertainties surrounding SSB taxation effects (see part II of this book), it seems even more essential to plan a detailed and robust evaluation in advance.

Hence, allocating resources for the evaluation of an eventual SSB taxation policy in Canada would be essential. For example, some have suggested that government use a part of the revenues from nutrition taxes (e.g. 1 %) for evaluation purposes (Smed and Robertson 2012). The evaluation plan of an SSB taxation policy should be carefully designed in advance. On purpose, an evaluability assessment is desirable to confirm if the policy can be evaluated in a reliable and credible way (Leclerc 2012). A full assessment includes many components (e.g. policy context and conception, observation of the policy if ongoing, demonstration of reasons why evaluation is necessary), which is beyond the scope of this chapter. But logic models (see Chap. 4), at least, are helpful to identify key variables, to assess whether expected outcomes are realistic, achievable and measurable, and to identify the data to be measured and collected. The following sections address some of these concerns.

As a preamble, we have previously argued that SSB taxation should rather be considered as a contributing instrument within a comprehensive intersectoral strategy aimed to decrease SSB overconsumption and prevent NCDs rather than an isolated measure (see Sect. 3.2). Therefore, it is essential to plan the evaluation of a soda tax as part of a larger evaluation strategy accounting for the multiple healthy eating promotion interventions concurrently pursued. Such a strategy falls out of the scope of this book but, in this regard, we encourage the reader to consult the WHO framework to monitor and evaluate the implementation of the global strategy on diet, physical activity and health (WHO 2008). It illustrates how an action can be monitored and evaluated as a component part of a larger evaluation strategy. Keeping this perspective in mind, the following sections describe how the contribution of a soda tax can be specifically assessed.

14.2 Evaluation of a Price-Induced Behavioural Logic

In the case of a price-induced behavioural logic, evaluation should focus on SSB price, SSB intakes, energy balance and eventually health outcomes (see Chaps. 4–7). However, unlike clinical trials using experimental and quasi-experimental designs, a "real-world" population-based policy such as SSB taxation cannot be controlled and evaluated the same way (Mytton et al. 2014).

Mytton et al. (2014) highlight that taxation effects on behaviours and health may be disrupted by a myriad of contextual factors and confounders whose precise influence is difficult to identify and account for (see Sects. 5.1, 6.2 and 7.1). Additionally, SSB taxation would probably occur concomitantly with other large-scale interventions (mass communication campaigns, nutrition labelling, etc.), which complicates the demonstration of causal inferences and tax-specific effects

(Colchero et al. 2016). As presented in part II of this book, empirical evidence confirms that a significant tax resulting in a 10–20 % price increase can significantly reduce SSB consumption, but effects on energy intakes, obesity and health may be negligible and, therefore, hardly detectable. In such circumstances, although the surveillance of obesity and chronic diseases should remain a public health priority, as others (Mytton et al. 2014; WHO 2015), we consider that the evaluation of SSB taxes should primarily focus on intermediary effects (e.g. on price and consumption). Provided such effects are well documented, their contribution to health outcomes may be assessed. Multiple sources of evidence should be used and triangulation efforts should be made to increase confidence in conclusions (Chatterji et al. 2014; Mytton et al. 2014).

To do so, databases should provide disaggregated and objective information on SSB marketing practices in general, and on SSB pricing in particular:

- The degree to which the tax has been effectively passed on to consumer prices should be estimable (see Chap. 5). In the case of an excise tax (generally applied on the manufacturer or the first importer), a thorough monitoring of the manufacturing price (when out of the factory) and the retail price (paid by the consumer in the store) gives an indication of how the private sector reacted to the tax and at which level(s) the tax may have been under- or overshifted (Berardi et al. 2012; Ecorys 2014).
- Obviously, the categorization used in the data set should match as far as possible the fiscal nomenclature used to distinguish taxed and untaxed products (INSP 2014; Ecorys 2014). Most importantly, the data set should cover the period of implementation of the tax, should provide a high temporal resolution, and should allow robustness checks to discriminate tax effects from other contextual influences in the long term (i.e. data available for a broad range of products (taxed and untaxed) across the territory, several years before and after the enactment of the tax, and at sufficiently close time intervals) (Mytton et al. 2014; Grogger 2015).

Databases should also ideally provide disaggregated and objective information on SSB purchases (date, quantity, price, category, nutrition facts, packaging format, point of sale characteristics, promotions, etc.) (see Chaps. 6 and 7) (Sharma et al. 2014).

- Data on a broad range of foods and beverages would be desirable to fully account for potential substitution effects with untaxed products (e.g. diet drinks, sparkling and still plain water, 100 % juices, flavoured water with non-caloric sweeteners, milk) (Nghiem et al. 2013). As mentioned before, the categorization of the data set should make possible the distinction between taxed and untaxed products (INSP 2014; Ecorys 2014). It may also be interesting to detect whether consumers switch from premium to non-premium products, since the latter are generally cheaper but present a similar nutritional profile. For the same reasons, datasets distinguishing branded products and private labels are valuable (Ecorys 2014; Berardi et al. 2012). Finally, as previously mentioned, the dataset should provide a high temporal resolution to discriminate as far as possible price effects on SSB demand from other contextual factors (Mytton et al. 2014; Colchero et al. 2016).

- The dataset should be based on a large and representative sample of consumers. Ideally, detailed characteristics should be individually documented (age, income, baseline SSB consumption, eating habits, health status, etc.) since these may work as effect modifiers at various stages of the intervention logic (Mytton et al. 2014). In particular, such data are necessary to evaluate potential inequities resulting from the tax (e.g. degree to which vulnerable households may suffer a larger tax burden and lower consumption changes) (see Sect. 4.4) and to analyse tax effectiveness across population subgroups (e.g. in youth, in large SSB consumers, in overweight or obese people) (see Chap. 8). Alternatively, aggregated data (i.e. total consumer demand on a given territory) may be used, provided the total number of inhabitants is accounted for (Ecorys 2014).
- Commercial panels of consumers primarily used for business purposes generally provide the aforementioned information, but these are unfortunately very expensive until they fall outdated. Another drawback is that they may only include scanner data from supermarket purchases and do not account for other purchases (e.g. take away, corner stores), which may represent the biggest part of SSB purchases. Alternatively, aggregated statistics on average prices and total consumer demand may provide workable information on a broad scope of items, although analyses are more restricted and consumers' profile is generally missing in these datasets. In complement, population-based nutrition and health surveys may be used to document consumption changes, but they present several disadvantages (large time intervals, self-reported data, beverage categories weakly differentiated) (Nghiem et al. 2013; Mytton et al. 2014; Sharma et al. 2014; Ecorys 2014). For example, the Canadian Community Health Survey (CCHS)—Nutrition is a national health survey collecting information from a representative sample of Canadians. The survey encompasses self-reported data on eating habits, including SSB consumption. Data collection instruments consist in a 24-h dietary recall operated by a computer-assisted personal interview (CAPI) and, for approximately 30 % of respondents, in a second dietary recall using a computer-assisted telephone interview (CATI) (Statistics Canada 2014). Unfortunately, CCHS-Nutrition was last done in 2004 and is only currently being repeated (data collection in 2015, data release in 2017): this large time interval would definitely have to be narrowed in the future if such a survey were to be used for SSB taxation evaluation purposes (e.g. for analyses of interrupted time series).

As far as possible, when using price and demand data, sophisticated models should enable to correct for seasonality and other important confounding factors (Bahl et al. 2003; Freithem et al. 2009). The recent evaluation of the Mexican soda tax impact on SSB purchases provides a pioneering example in this regard (Colchero et al. 2016). In Canada, interrupted times series have been used to evaluate the impact of alcohol taxation at provincial level and may also be adapted to the evaluation of an SSB taxation policy (Stockwell et al. 2012). Although it is almost impossible to have a control site, territories not concerned by the tax on the same period may be used as imperfect comparison sites (e.g. a neighbouring country or province). For example, in Canada, Stockwell et al. (2012) have used such methods to perform an impact evaluation of a minimum alcohol price increase in Saskatchewan in 2010. In this case, the Saskatchewan Liquor and Gaming Authority (SLGA) sales and price data were computed 2 years before and after the minimum price increase of different types of alcoholic beverages, on a monthly basis. The model included adjustment for household income, cost of living, socio-demographic variables, consumption of other alcoholic beverages, trends in the data, and seasonality. It has demonstrated that a 10 % increase in minimum prices significantly reduced the consumption of various beverages in the range of 5–25 %. In the meantime, no changes were observed in the neighbouring province of Alberta, which reinforces the assumption that the effects observed are linked with the minimum pricing policy (Stockwell et al. 2012).

To increase confidence in the results, undesirable or competing effects should be estimated in parallel (see Chap. 9). This may include a monitoring of major influences on food and beverage prices (e.g. macroeconomic changes in sugar price policies), potential market distortion effects on SSB consumption (e.g. increased cross-border shopping in adjacent regions) and other large-scale public health interventions (e.g. removal of vending machines within and nearby schools).

14.3 Evaluation of a Fiscal Logic

In the case of a fiscal logic, the evaluation would minimally consist in the collection and monitoring of tax revenues as well as in the monitoring of how tax revenues are earmarked for specific uses, such as health promotion initiatives and social programs (see Chap. 10). Eventually, the process and impact of those initiatives could also be evaluated since it may document indirect tax effects on the behaviours and health of the population. The monitoring of tax revenues may be easier when the tax policy instruments are already routinely used by the government (e.g. sales or excise taxes), but has proved to be much more complex when this was not the case, as shown for the Danish "fat tax" in 2010 (see Sect. 9.1) (Smed 2012; Bødker et al. 2015). It should be emphasized that, so far, the evaluation of the earmarking of tobacco taxes across the world is considered to be weak (see Sect. 10.4). As a result, the WHO recently recommended allocating funds for the evaluation of tobacco taxes' earmarking in order to improve existing mechanisms and introduce innovative ones where earmarking is not already used (WHO 2012). Similar challenges should be anticipated in the case of soda taxes.

Finally, from a fiscal and economic viewpoint, the benefits of a new tax should be weighed against potential undesirable effects, e.g. in terms of competitiveness (see Chap. 9). Such evaluation requires accessing and analysing data on administrative costs, manufacturing prices, retail prices, number of companies, number of persons employed, investments, gross income, profits and trade flows (Ecorys 2014a). Aggregated statistics are sometimes publicly administered and available upon special request, but this may not be sufficient. It may require contacting various organizations on purpose (chambers of commerce, food and beverage industry associations, etc.) to obtain complementary information and datasets (Ecorys 2014).

14.4 Evaluation of a Signal-Induced Denormalization Logic

In the case of a signal-induced denormalization logic, evaluation may focus on the public awareness of the tax, as well as on the characteristics that may favour tax acceptance and behaviour changes in the population (e.g. socio-demographic and dietary profiles of respondents) (see Chap. 11). Interesting evaluation data from the French Nutrinet-Santé cohort study have been published in this regard (Julia et al. 2015). It would also be relevant to monitor the perception of the tax (and its impact) by all stakeholders, in particular those highly involved in the implementation process (e.g. manufacturers, retailers). Finally, monitoring reformulation and marketing practices that may be related to the tax would also be of interest (Ecorys 2014) (see Chap. 11). More globally, it appears that social norms on SSB consumption (and diet in general) are poorly explored. It has been recommended to conduct exploratory research and targeted surveys on diet perceptions and representations, on the meanings given to own and others' eating behaviours, and on how these fit into social connections (Baril et al. 2014; Robinson et al. 2014).

14.5 The Challenge of Economic Evaluations

Finally, provided distal effects of SSB taxation on health outcomes and well-being can be estimated satisfactorily, cost-effectiveness, cost-utility and cost–benefit evaluations may be undertaken. Depending on the data available, such evaluations could weigh tax-related costs (i.e. administrative burden, consumers' welfare losses, potential negative impact on the economy) against tax-related benefits (i.e. revenues generated, population health status improvements, cost reductions from fewer obesity-related diseases, increased productivity at work) (Smed and Robertson 2012; Étilé 2012; Gortmaker et al. 2011; Sassi 2010; Moodie et al. 2013). Such evaluations are generally conducted ex-ante, although ex-post economic evaluations would definitely be relevant for accountability purposes.

Key messages

- Evaluating the effects and impact of soda taxes is critical to provide evidence with more external validity than what currently provide simulations and experimental trials.
- Uncertainties surrounding soda tax effects make even more essential to plan a robust evaluation. Appropriate methods should be used and corresponding databases should be available.
- In the case of a price-induced behavioural logic, evaluation should focus on SSB price, SSB intakes, energy balance and eventually health outcomes. As far as possible, tax-specific effects should be disentangled from other factors.
- In the case of a fiscal logic, the evaluation would minimally consist in the collection and monitoring of tax revenues as well as in the monitoring of how tax proceeds are earmarked for specific uses.
- In the case of a signal-induced denormalization logic, evaluation may focus on the public awareness of the tax as well as on reformulation and marketing practices of manufacturers.

References

Bahl, R., Bird, R., & Walker, M. B. (2003). The uneasy case against discriminatory excise taxation: Soft drink taxes in Ireland. *Public Finance Review, 31*(5), 510–533. doi:10.1177/1091142103253753.

Baril, G., Trudel, C., Le May, D., Mongeau, L., & Noreau, C. (2014). *Les saines habitudes de vie, c'est bien normal!* (p. 46). [Montréal, Qué.]: Rapport sous la responsabilité de la Coordination conjointe (Québec en Forme-Directions régionales de santé publique-Ministère de la Santé et des Services sociaux du Québec). http://veilleaction.org/documents/Les-saines-habitudes-de-vie-cest-bien-normal.pdf

Berardi, N., Sevestre, P., Tepaut, M., & Vigneron, A. (2012). *The impact of a "soda tax" on prices: Evidence from French micro data* (Banque de France Working Paper No. 415) (p. 29). Banque de France Eurosystème. Direction Générale des Études et des Relations internationales. http://ssrn.com/abstract=2192470. Accessed 24 February 2016.

Bødker, M., Pisinger, C., Toft, U., & Jørgensen, T. (2015). The rise and fall of the world's first fat tax. *Health Policy (Amsterdam, Netherlands), 119*(6), 737–742. doi:10.1016/j.healthpol.2015.03.003.

Chatterji, M., Green, L. W., & Kumanyika, S. (2014). L.E.A.D.: A framework for evidence gathering and use for the prevention of obesity and other complex public health problems. *Health Education and Behavior, 41*(1), 85–99. doi:10.1177/1090198113490726.

Colchero, M. A., Popkin, B. M., Rivera, J. A., & Ng, S. W. (2016). Beverage purchases from stores in Mexico under the excise tax on sugar sweetened beverages: Observational study. *BMJ*, h6704. doi:10.1136/bmj.h6704.

Ecorys. (2014). Food taxes and their impact on competitiveness in the agri-food sector. Final report (No. Ref. Ares(2014)2365745—16/07/2014). Rotterdam: Client: DG Enterprise and Industry.

Étilé, F. (2012). La taxation nutritionnelle comme outil de santé publique: justifications et effets attendus. *Cahiers de Nutrition et de Diététique, 47*(1), 25–34. doi:10.1016/j.cnd.2011.12.002.
Fretheim, A., Oxman, A. D., Lavis, J. N., & Lewin, S. (2009). SUPPORT tools for evidence-informed policymaking in health 18: Planning monitoring and evaluation of policies. *Health Research Policy and Systems, 7*(Suppl 1), S18. doi:10.1186/1478-4505-7-S1-S18.
Gortmaker, S. L., Swinburn, B. A., Levy, D., Carter, R., Mabry, P. L., & Finegood, D. T., et al. (2011). Changing the future of obesity: Science, policy, and action. *Lancet (London, England), 378*(9793), 838–847. doi:10.1016/S0140-6736(11)60815-5.
Grogger, J. (2015). Soda taxes and the prices of sodas and other drinks: Evidence from Mexico (No. w21197). Cambridge, MA: National Bureau of Economic Research. http://www.nber.org/papers/w21197.pdf. Accessed 28 February 2016.
Instituto Nacional de Salud Pública (INSP). (2014). Resultados preliminares sobre los efectos del impuesto de un peso a bebidas azucaradas en México. www.insp.mx/epppo/blog/preliminares-bebidas-azucaradas.html. Accessed 18 February 2015.
Julia, C., Méjean, C., Vicari, F., Péneau, S., & Hercberg, S. (2015). Public perception and characteristics related to acceptance of the sugar-sweetened beverage taxation launched in France in 2012. *Public Health Nutrition, 18*(14), 2679–2688. doi:10.1017/S1368980014003231.
Leclerc, B.-S. (2012). Utilité de l'évaluation de l'évaluabilité des politiques gouvernementales de lutte contre le tabagisme: l'expérience québécoise des centres d'abandon du tabagisme. *The Canadian Journal of Program Evaluation, 24*(3), 61–73.
Moodie, M., Sheppard, L., Sacks, G., Keating, C., & Flego, A. (2013). Cost-effectiveness of fiscal policies to prevent obesity. *Current Obesity Reports, 2*(3), 211–224. doi:10.1007/s13679-013-0062-y.
Mytton, O. T., Eyles, H., & Ogilvie, D. (2014). Evaluating the health impacts of food and beverage taxes. *Current Obesity Reports, 3*(4), 432–439. doi:10.1007/s13679-014-0123-x.
Nghiem, N., Wilson, N., Genç, M., & Blakely, T. (2013). Understanding price elasticities to inform public health research and intervention studies: Key issues. *American Journal of Public Health, 103*(11), 1954–1961. doi:10.2105/AJPH.2013.301337.
Robinson, E., Thomas, J., Aveyard, P., & Higgs, S. (2014). What everyone else is eating: A systematic review and meta-analysis of the effect of informational eating norms on eating behavior. *Journal of the Academy of Nutrition and Dietetics, 114*(3), 414–429. doi:10.1016/j.jand.2013.11.009.
Sassi, F. (2010). *Obesity and the economics of prevention: Fit not fat* (p. 265). Organisation for Economic Co-operation and Development (OECD).
Sharma, A., Hauck, K., Hollingsworth, B., & Siciliani, L. (2014). The effects of taxing sugar-sweetened beverages across different income groups: The impact of SSB taxes on different income groups. *Health Economics, 23*(9), 1159–1184. doi:10.1002/hec.3070.
Smed, S. (2012). Financial penalties on foods: The fat tax in Denmark: The fat tax in Denmark. *Nutrition Bulletin, 37*(2), 142–147. doi:10.1111/j.1467-3010.2012.01962.x.
Smed, S., & Robertson, A. (2012). Are taxes on fatty foods having their desired effects on health? *BMJ, 345*(oct16 1), e6885–e6885. doi:10.1136/bmj.e6885.
Statistics Canada. (2014). Canadian Community Health Survey—Nutrition (CCHS). http://www23.statcan.gc.ca/imdb/p2SV.pl?Function=getSurvey&SDDS=5049. Accessed 29 October 2015.
Stockwell, T., Zhao, J., Giesbrecht, N., Macdonald, S., Thomas, G., & Wettlaufer, A. (2012). The raising of minimum alcohol prices in Saskatchewan, Canada: Impacts on consumption and implications for public health. *American Journal of Public Health, 102*(12), e103–e110. doi:10.2105/AJPH.2012.301094.
World Health Organization (WHO). (2008). *A framework to monitor and evaluate implementation WHO global strategy on diet, physical activity and health.* Geneva, Switzerland: World Health Organization. http://www.who.int/dietphysicalactivity/M&E-ENG-09.pdf. Accessed 29 February 2016.

World Health Organization (WHO). (2012). Tobacco taxation and innovative health-care financing. New Delhi, India: World Health Organization. http://www.searo.who.int/tobacco/documents/2012-pub1.pdf. Accessed 22 January 2016.
World Health Organization (WHO). (2015). Using price policies to promote healthier diets. Copenhagen (Denmark): World Health Organization (WHO).

Part IV
Conclusion

Chapter 15
Conclusion

Abstract The evidence gathered in this book indicates that, provided several considerations are accounted for, a tax on soda/sugar-sweetened beverages (SSBs) could be part of a portfolio of nutrition-enhancing policies in Canada. Considering that public health objectives should be as specific, measurable, achievable, realistic and time-bound as possible, the focus of a tax could be the following: in a short term, to raise awareness of the population on the importance to avoid SSB overconsumption; in a medium term, to raise fiscal revenues truly earmarked for health-oriented causes; and in a medium to longer term, to contribute reducing SSB overconsumption while increasing healthier beverage choices in the population. Persistent concerns and uncertainties in terms of SSB taxation rationale, potential impact and implementation should be further discussed prior to implementation. It may not justify inaction. If a tax is to be adopted, a detailed and robust evaluation plan should be designed ex-ante. Preliminary effects should be carefully monitored to fully assess the relevancy of this policy in Canada as well as for the benefit of other jurisdictions.

Although life expectancy in Canada is among the highest worldwide, the burden of non-communicable diseases (NCDs) is threatening. Cardiovascular diseases, cancer, diabetes and chronic respiratory diseases are predominant causes of illnesses and deaths. About one in five Canadian adults live with at least one of these diseases, including one in 10 suffering from diabetes (PHAC 2015). Obesity is known to increase the risk of a number of these conditions (AICR/WCRF 2007; WHO 2011). According to statistical data collected in 2012–2013, about one in four Canadian adults and above one in 10 Canadian children are obese (PHAC 2015). These figures have significantly increased over the last decades. The Canadian economic burden related to obesity and its comorbidities has been estimated to be between $4.6 and $7.1 billion in 2008 (CIHI/PHAC 2011). Although strong evidence indicates that a healthful diet is a key protective factor (AICR/WCRF 2007; Mozzafarian 2014), eating habits of many Canadians are not in line with the food guide's recommendations (Garriguet 2007; PHAC 2015). To tackle the issue, healthy eating promotion is crucial. So far, nutritional policies aimed at informing

Y. Le Bodo et al., *Taxing Soda for Public Health*,
DOI 10.1007/978-3-319-33648-0_15

consumers about the benefits of a healthful diet have led to modest impacts on consumers' behaviours and health. Therefore, it has been increasingly suggested (including in Canada) to implement policies and regulations targeting the market environment, such as food taxes (Réquillart and Soler 2014; Tourigny et al. 2014; Duhaney et al. 2015).

As recommendations and evidence supporting "unhealthy food" taxation have grown over the last 10 years, sodas[1] were increasingly targeted because of their high sugar content, their poor nutritive value, the health risks related to their overconsumption, the high intakes often observed in youth and the aggressive marketing practices of manufacturers (Brownell and Frieden 2009; Brownell et al. 2009; Lustig et al. 2012; Block and Willett 2013; Mytton 2015). The introduction of this book shows that, since 2006, taxes on soda or sugar-sweetened beverage (SSB) have been reinforced or newly introduced for public health reasons in over 20 jurisdictions across the world (see Chap. 1). However, SSB taxation remains controversial (Chaloupka et al. 2011; Fletcher et al. 2011; Sarlio-Lähteenkorva and Winckler 2015) and many governments remain hesitant or reluctant to tax SSB: it is the case at federal and provincial levels in Canada. Therefore, this book attempted to advance knowledge on the following question: *Is it relevant to tax soda for public health in Canada?*

To tackle this endeavour, we have conducted a realist review of literature and used knowledge synthesis techniques (Pawson et al. 2005; Morestin et al. 2011) in order to consider the rationale (WHY), the potential impact (WHAT) and the applicability (HOW) of an SSB taxation policy. This knowledge-breaking perspective builds on conceptual frameworks aimed to guide decision-making in public health and in the complex field of obesity prevention (Swinburn et al. 2005; Morestin et al. 2011; Chatterji et al. 2014). Such questions called for an interdisciplinary approach including various types of data from nutrition, public health, health economics, health policy, law and communication. International evidence—especially from taxes implemented in the "real world"—was considered essential. Therefore, while this book focuses on how the evidence fits into the Canadian context, many of the ideas put forward may be informative for other jurisdictions. Results have been subdivided into three main sections focused on the following: SSB taxation rationale (Part I, Chaps. 2–4), SSB taxation impacts (Part II, Chaps. 5–11) and SSB taxation applicability (Part III, Chaps. 12–14).

This last chapter builds on all key messages formulated along the book (see also Appendix 1) and proposes a multidimensional graphic representation articulating 14 components that appear to us worth considering to answer the research question (see Fig. 15.1). In conclusion, we briefly highlight the lessons learnt on each of these components.

---

[1]In the context of nutrition policies, the term "soda", although originally designating carbonated beverage, frequently and commonly refers to sugar-sweetened beverage (SSB). In this book, both terms are used in an interchangeable manner.

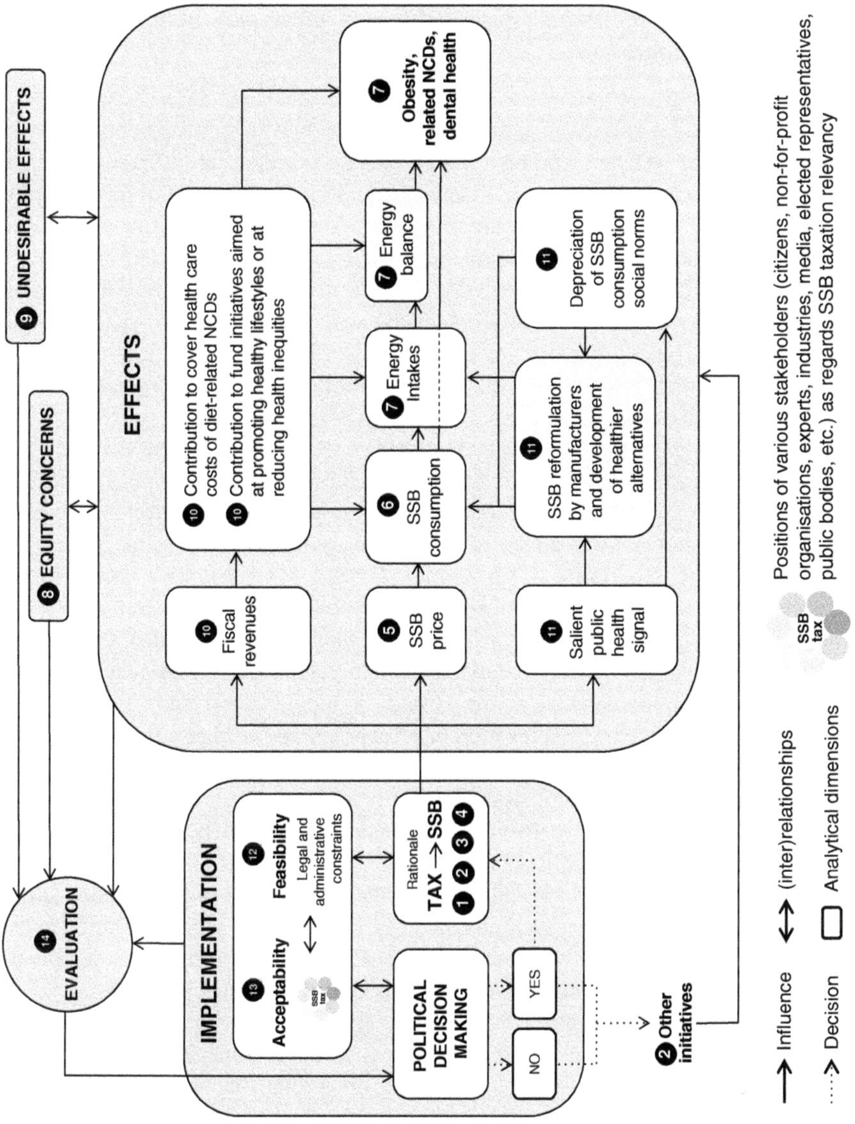

Fig. 15.1 Multidimensional perspective aimed at supporting evidence gathering and decision-making to the question: *Is it relevant to tax sugar-sweetened beverages?* (See also Appendix 1 for all key messages corresponding to the numeration). Dimension titles: ❶ Sugar-sweetened beverages (SSBs) and health; ❷ Taxation value as a public health instrument; ❸ Ethical concerns on SSB taxation; ❹ SSB taxation logics; ❺ Taxation effects on SSB prices; ❻ Taxation effects on SSB demand; ❼ SSB taxation effects on energy balance and health; ❽ SSB taxation distributional effects; ❾ SSB taxation undesirable effects; ❿ Fiscal revenues from SSB taxation; ⓫ Signal effects of SSB taxation; ⓬ Feasibility of SSB taxation; ⓭ Acceptability of SSB taxation; and ⓮ Evaluability of SSB taxation

Part I of the book has highlighted **why soda taxes may be justified from a public health perspective**:

- First, the evidence reviewed in Chap. 2 indicates that SSB consumption is certainly not the only risk factor to focus on to achieve significant changes in the prevention of NCDs: a combination of various strategies at different levels and in multiple settings is necessary. However, an important burden of proof justifies specific prevention efforts towards SSB overconsumption in Canada: (1) SSBs represent a significant proportion of sugar intake and daily calories in some groups of the Canadian population, youth in particular; (2) SSB overconsumption has been consistently associated with various health risks, including weight gain, type-2 diabetes and dental caries; (3) SSBs have a particularly low nutritive value; and (4) SSB overconsumption is stimulated by intense marketing practices ❶.
- What about the solutions? To tackle SSB overconsumption, the evidence presented in Chap. 3 indicates that taxation is certainly not the only policy option to focus on. It is generally recommended to implement a portfolio of various congruent interventions aimed at raising awareness, educating the public and changing the environment (in terms of SSB offering, price, accessibility, advertising, etc.). Within this portfolio of actions, which would gain to be developed within an intersectoral governmental health promotion strategy, we argue that taxation deserves particular attention. Indeed, it may reach a critical mass of the population at a low cost of implementation and generate fiscal revenues. Altogether, SSB taxation has the potential for high cost-effectiveness. Additionally, taxation has proved useful as part of other public health efforts, e.g. within tobacco control strategies ❷.
- Notwithstanding the aforementioned justifications, SSB taxation is a controversial policy and ethical concerns (e.g. in terms of freedom of choice, autonomy and equity) need to be addressed collectively prior to a tax taking effect. Several arguments have been advanced in this book to indicate that SSB taxation would be ethically defensible (see Chap. 4). Ultimately, a democratic dialogue involving all stakeholders may help to inform decisions ❸.
- To contribute disentangling the debate, Chap. 4 argues that it is essential to make clear which public health objective(s) would be pursued by SSB taxation in order to properly assess its relevancy on a Canadian ground. Three SSB taxation logics (not mutually exclusive) are generally distinguished: (1) a price-induced behavioural logic (i.e. taxing SSBs to increase prices and reduce consumption), (2) a fiscal logic (i.e. taxing SSBs to raise revenues to be earmarked for public health causes) and (3) a signal-induced denormalization logic (i.e. taxing SSBs to send an "alarm signal" to consumers and a "coercive signal" to manufacturers) ❹.

Part II of this book reviewed evidence on the **potential impacts of these three taxation logics**:

- **As regards a price-induced behavioural logic**: Chap. 5 illustrates that factors other than a tax may influence price changes: disentangling the influence of tax-specific effects from these other factors is challenging. However, international evidence clearly suggests that a tax applied on SSBs on a large territory (eq. to a province, state or country) tends to be shifted to prices, possibly below or beyond the increase theoretically expected from the tax ❺. Evidence also indicates that significant price hikes (10–20 %) would most probably be followed by a reduction in SSB purchases, although the magnitude and sustainability of this effect as well as its variability across beverage types remain unclear ❻. As regards SSB tax impact on energy balance and health, objectives should be stated very carefully: there is consistent evidence indicating that the reduction in SSB intakes resulting from a tax could be compensated (at least to some extent) by increased consumption of other caloric foods and drinks (the so-called "substitution effects"). In any cases, we should remember that considered in isolation, the effects of most obesity prevention policies on health outcomes are difficult to detect due to many competing factors and interventions. Therefore, it seems inappropriate to expect tax-induced skyrocketing improvements in the population's health in a short to medium term. Instead, a careful monitoring of intermediary tax effects on prices, demand and substitutions is essential ❼. Other potential pitfalls should be considered. First, evidence remains unclear about the extent to which the core target (youth, high SSB consumers) would be reached by the policy. More research and evaluations will be crucial in this area. Secondly, SSB taxation raises equity concerns: the tax burden would likely be modest and acceptable for most of the population but may impact more negatively low-income households consuming large quantities of SSBs and not prone to reduce their consumption. This could be somewhat compensated if tax revenues can be truly earmarked for health promotion initiatives and social programmes ❽. Finally, potential undesirable effects on business and industry are sometimes mentioned but remain poorly documented. Simulations and anecdotal evidence suggest that such effects would likely be limited, in particular if the tax scope is restricted to SSBs ❾. Finally, as for many policies, all the impacts of a tax (and their magnitude) cannot be fully anticipated before large-scale natural experimentation. This may not justify inaction. If a tax is to be adopted in Canada, a detailed and robust evaluation plan should be designed ex-ante, and preliminary effects should be carefully monitored before extending implementation in a longer term ⓮.
- **As regards the fiscal SSB taxation logic** (see Chap. 10): from a public health perspective, evidence indicates that taxing SSBs as a way to raise fiscal revenues could be acceptable, realistic and rapidly achievable, provided important clarifications and commitments are made: (1) targeting SSBs specifically should be clearly motivated from a nutritional perspective; (2) this fiscal logic could reasonably be accompanied with a behavioural objective to reduce SSB consumption in the population; (3) if SSB demand decreases significantly overtime, a new tax will at worst generate fewer revenues, but it will always be more than if the tax was absent; and (4) earmarking tax revenues for health care or health

promotion activities could bring additional public health benefits, counterbalance inequity concerns and favour the social acceptability of the tax. However, it appears desirable to ensure that tax revenues will be truly earmarked for such initiatives ❿.

- **As regards the signal-induced denormalization logic** (see Chap. 11): the evidence gathered confirms that social norms are a significant driver of SSB consumption habits, especially in adolescents. Parallel data indicate that SSB taxation, within a comprehensive strategy, may work as a signal altering SSB consumption norms, but concrete evidence is limited and counteracting effects cannot be ruled out. The degree of acceptability of the tax in the public opinion should be explored and could be used as a proxy to measure "signal effects" in the population. SSB taxation may also be used as a "coercive signal" pushing manufacturers to develop healthier products ⓫. Here again, if a tax is to be adopted, preliminary impacts should be carefully monitored ⓮.

Finally, Part III of the book raises several **applicability challenges:**

- Administratively speaking (see Chap. 12), taxing SSBs via an excise duty levied on the manufacturer is preferentially recommended in the literature: this mechanism is likely feasible at the federal level only in Canada. Since most proposals have been made at the provincial/territorial level, it is worth exploring other options. For example, a special tax directly levied on the consumer at the point of purchase could be an alternative in these jurisdictions. Specific taxes applied on alcohol and tobacco in several provinces may be used as a model. Also, defining the taxation scope would likely raise some issues since the Canadian fiscal nomenclature currently does not distinguish carbonated beverages according to calorie content and composition. Finally, Chap. 12 highlights that a robust mechanism facilitating the administration and the earmarking of tax revenues for health promotion initiatives and social programmes would likely be desirable and feasible in Canada ⓬.
- Chapter 13 questions the social and political acceptability of an SSB taxation policy. As in many countries, SSB taxation is controversial among Canadian stakeholders and in the general population. Earmarking tax revenues for health promotion initiatives may partly solve this issue. Additionally, the human factor in the decision-making process at federal or provincial level should not be underestimated. In other countries, several conditions seem to have contributed to opening a window of political opportunity for the adoption of a soda tax, such as a government in favour of social policies, a difficult budgetary situation and a convinced champion at the highest political level. In all cases, initiating a dialogue with all stakeholders seems valuable to support decision-making and, eventually, to prevent implementation failures. In this regard, deliberative processes involving experts, decision-makers and civil society actors have proved to be useful: it can enrich scientific and contextual evidence with experiential and colloquial knowledge and it can help disentangle controversies (Lomas et al.

2005; Morestin et al. 2011). We hope this book could be an informative material for such initiatives ⓭.

- Finally, Chap. 14 underlines that evaluating the effects and impact of soda taxes is critical to provide evidence with more external validity than what currently provide simulations and experimental trials. Allocating resources on purpose is a first step in that direction. Depending on the SSB taxation logic(s) pursued, a proper evaluation should be planned in advance, appropriate methods should be used and corresponding databases should be identified and available. This is indispensable to assess the degree of implementation, the effectiveness and the overall impact of an SSB taxation policy ⓮.

It is noteworthy that some of our conclusions are relatively congruent with recent considerations advanced by Hawkes et al. (2015) as regards "Smart food policies for obesity prevention". Among many other orientations, these authors provide specific guidance on food taxes and recommend: (1) a prioritization of taxes on "unhealthy foods that are popular and habit-forming" in youth, (2) a prioritization of taxes on food likely to encourage positive market reactions (e.g. through reformulation and shift in product portfolios), (3) taxes large enough to "stimulate consumers to reassess their purchase decisions", (4) the possibility to improve tax effects via awareness raising actions at the point of purchase and (5) having close, healthier and cheaper substitutes to avoid substitution towards less healthy untaxed food.

As indicated earlier, on the one hand, our realist review of evidence presents several limitations: (1) it heavily relies on the ability and skills of the research team to make relevant connexions across readings and consultation of various pieces of evidence. Recurrent presentations of the work in progress to experts and stakeholders as well as discussions within the research team were used as a mechanism to control for the consistency and robustness of the process; and (2) the mass and diversity of information consulted as well as the long-term developmental process of writing made it difficult to provide comprehensive data extraction tables. However, the book provides detailed and specific bibliographic references for presented evidence. Ultimately, it reflects the "archaeology" of decisions that we have made and the "interpretative trail" that we have followed (Pawson et al. 2005, p. 31). Finally, we pretend in no way that this review is definitive per se. Researchers' role is not to decide on the adoption and implementation of policies, but to inform decision-makers of the pros and cons of various options under consideration. From this perspective, we consider this research as a starting point for further discussions, consultations, contextualization and deliberation.

On the other hand, as mentioned in the preface of the book, this work presents several advantages. First, summarizing scientific, contextual and informal evidence on interventions deemed controversial and where the volume of evidence is rapidly growing seems essential: this is precisely the case of soda taxes. Secondly, SSB taxation has often been addressed using a diversity of perspective such as in public health (soda and health), ethics (tax justification), economics (impact evaluations and behavioural simulations), law (tax adoption and implementation) and communication (political and social acceptability). Although each domain brings a critical piece of evidence, this book attempts to comprehensively assemble and articulate evidence from an interdisciplinary and multidimensional standpoint (Morin 2003). Thirdly, the debate on soda taxation may vary from one country to another because of different public health priorities, dietary norms and behaviours, environmental characteristics, on-going interventions, political circumstances, etc. Therefore, it is essential to consider both international evidence and contextual implications. As a result, this book addresses SSB taxation from a Canadian perspective, although many of the ideas put forward can be of interest in other places.

In conclusion, the evidence gathered in this book suggests that, provided several considerations are accounted for, a soda tax could be meaningful within a portfolio of nutrition-enhancing policies in Canada. It also suggests that the three aforementioned SSB taxation logics may be concurrently pursued. Considering that public health objectives should be as specific, measurable, achievable, realistic and time-bound as possible, the focus of a tax could be the following: in a short term, to raise awareness of the population on the importance to avoid SSB overconsumption; in a medium term, to raise fiscal revenues truly earmarked for health-oriented causes; and in a medium to longer term, to contribute reducing SSB overconsumption while increasing healthier beverage choices in the population. Persistent concerns and uncertainties in terms of SSB taxation rationale, potential impact and implementation should be further discussed prior to implementation. It may not justify inaction. If a tax is to be adopted, a detailed and robust evaluation plan should be designed ex-ante. Preliminary effects should be carefully monitored to fully assess the relevancy of this policy in Canada as well as for the benefit of other jurisdictions.

References

American Institute for Cancer Research (AICR), & World Cancer Research Fund (WCRF) (Eds.). (2007). *Food, nutrition, physical activity and the prevention of cancer: A global perspective: A project of World Cancer Research Fund International*. Washington, D.C: American Institute for Cancer Research.

Block, J. P., & Willett, W. C. (2013). Taxing sugar-sweetened beverages: Not a "holy grail" but a cup at least half comment on "food taxes: A new holy grail?". *International Journal of Health Policy and Management, 1*(2), 183–185. doi:10.15171/ijhpm.2013.33.

Brownell, K. D., Farley, T., Willett, W. C., Popkin, B. M., Chaloupka, F. J., Thompson, J. W., & Ludwig, D. S. (2009). The public health and economic benefits of taxing sugar-sweetened beverages. *New England Journal of Medicine, 361*(16), 1599–1605. doi:10.1056/NEJMhpr0905723.

Brownell, K. D., & Frieden, T. R. (2009). Ounces of prevention—The public policy case for taxes on sugared beverages. *New England Journal of Medicine, 360*(18), 1805–1808. doi:10.1056/NEJMp0902392.

Canadian Institute for Health Information (CIHI), & Public Health Agency of Canada (PHAC). (2011). Obesity in Canada a joint report from the Public Health Agency of Canada and the Canadian Institute for Health Information. [Ottawa]: Public Health Agency of Canada: Canadian Institute for Health Information.

Chaloupka, F. J., Powell, L. M., & Chriqui, J. F. (2011). Sugar-sweetened beverages and obesity prevention: Policy recommendations. *Journal of Policy Analysis and Management: [The Journal of the Association for Public Policy Analysis and Management], 30*(3), 662–664.

Chatterji, M., Green, L. W., & Kumanyika, S. (2014). L.E.A.D.: A framework for evidence gathering and use for the prevention of obesity and other complex public health problems. *Health Education and Behavior, 41*(1), 85–99. doi:10.1177/1090198113490726.

Duhaney, T., Campbell, N., Niebylski, M. L., Kaczorowski, J., Tsuyuki, R. T., Willis, K., et al. (2015). Death by diet: The role of food pricing interventions as a public policy response and health advocacy opportunity. *The Canadian Journal of Cardiology, 31*(2), 112–116. doi:10.1016/j.cjca.2014.09.005.

Fletcher, J. M., Frisvold, D. E., & Tefft, N. (2011). Are soft drink taxes an effective mechanism for reducing obesity? *Journal of Policy Analysis and Management: [the Journal of the Association for Public Policy Analysis and Management], 30*(3), 655–662.

Garriguet, D. (2007). Canadians' eating habits. *Health Reports, 18*(2), 17–32.

Hawkes, C., Smith, T. G., Jewell, J., Wardle, J., Hammond, R. A., Friel, S., et al. (2015). Smart food policies for obesity prevention. *The Lancet, 385*(9985), 2410–2421. doi:10.1016/S0140-6736(14)61745-1.

Lomas, J., Culyer, T., McCutcheon, C., McAuley, L., & Law, S. (2005). *Conceptualizing and combining evidence for health system guidance: Final report.* Canadian Health Services Research Foundation. http://www.cfhi-fcass.ca/migrated/pdf/insightAction/evidence_e.pdf. Accessed 28 February 2016.

Lustig, R. H., Schmidt, L. A., & Brindis, C. D. (2012). Public health: The toxic truth about sugar. *Nature, 482*(7383), 27–29. doi:10.1038/482027a.

Morestin, F., Gauvin, F.-P., Hogue, M.-C., & Benoit, F. (2011). *Method for synthesizing knowledge about public policies.* [Montréal]: Centre de collaboration nationale sur les politiques publiques et la santé, Institut national de santé publique Québec. http://collections.banq.qc.ca/ark:/52327/2066075. Accessed 28 February 2016.

Morin, E. (2003). Sur l'interdisciplinarité. *L'Autre Forum, Mai 2003*, 5–10.

Mozaffarian, D. (2014). The promise of lifestyle for cardiovascular health. *Journal of the American College of Cardiology, 64*(13), 1307–1309. doi:10.1016/j.jacc.2014.06.1191.

Mytton, O. (2015). Time for a sugary drinks tax in the UK? *Journal of Public Health (Oxford, England), 37*(1), 24–25. doi:10.1093/pubmed/fdu033.

Pawson, R., Greenhalgh, T., Harvey, G., & Walshe, K. (2005). Realist review—a new method of systematic review designed for complex policy interventions. *Journal of Health Services Research and Policy, 10*(Suppl 1), 21–34. doi:10.1258/1355819054308530.

Public Health Agency of Canada (PHAC). (2015). Chronic disease and injury indicator framework. Quick stats overview. http://infobase.phac-aspc.gc.ca/cdiif/. Accessed 16 February 2016.

Réquillart, V., & Soler, L.-G. (2014). Is the reduction of chronic diseases related to food consumption in the hands of the food industry? *European Review of Agricultural Economics, 41*(3), 375–403. doi:10.1093/erae/jbu010.

Sarlio-Lähteenkorva, S., & Winkler, J. T. (2015). Could a sugar tax help combat obesity? *BMJ*, h4047. doi:10.1136/bmj.h4047.

Swinburn, B., Gill, T., & Kumanyika, S. (2005). Obesity prevention: A proposed framework for translating evidence into action. *Obesity Reviews: An Official Journal of the International Association for the Study of Obesity, 6*(1), 23–33. doi:10.1111/j.1467-789X.2005.00184.x.

Tourigny, A., Namoro, A., Arsenault, P.-É., Héronneau, C., Guilbert-Couture, A., & Anstett, É., et al. (2014). *La prévention en obésité au Québec : un portrait croisé et analytiques des interventions et politiques gouvernementales visant les jeunes.* Québec (Québec): École nationale d'administration publique (ENAP).

World Health Organization (WHO). (2011). *Global status report on noncommunicable diseases 2010.* Geneva, Switzerland: World Health Organization.

Appendix 1
Key Messages About the Relevancy to Tax Sugar-Sweetened Beverages as Part of a Portfolio of Nutrition-Enhancing Policies in Canada[a]

#	Why—rationale
1	**Sugar-sweetened beverages (SSBs) and health** • Most recent systematic reviews, meta-analyses and experimental studies confirm that SSB overconsumption increases the risk of weight gain, type 2 diabetes and dental caries • SSB consumption is stimulated by an attractive and diversified offer, a widespread distribution, low prices and a strong promotion (especially towards youth) • SSB consumption represents a significant proportion of sugar intake and daily energy intake in some groups of the Canadian population, youth in particular • From a public health perspective, preventing SSB overconsumption in Canada requires specific efforts at a large scale
2	**Taxation as a public health instrument** • Rather than questioning whether SSB taxation alone is effective in preventing obesity and related chronic diseases, it is more realistic to question whether it can bring a valuable contribution to the issue, as part of a large and comprehensive prevention strategy • Many educational initiatives and environmental changes in public places already contribute to prevent SSB overconsumption across Canada. International evidence and recommendations suggest that policies targeting the market environment could bring additional benefits • Considering that sugar-sweetened beverages are particularly affordable, taxation has been increasingly suggested as a way to increase price and discourage demand at a large scale • The success of taxes applied to other public health domains (e.g. tobacco control), as well as their potential cost-effectiveness, make SSB taxation a policy instrument deserving particular attention
3	**Ethical concerns on SSB taxation** • The ethical justification of SSB taxation remains controversial among stakeholders and in the population: this should be carefully considered and debated before implementation on a Canadian ground • Several arguments suggest SSB taxation may be ethically defensible: it addresses an acute public health issue; it targets a specific category of food and beverages which is not part of basic groceries; taxation is potentially among the most cost-effective options and may not raise insurmountable concerns in terms of equity, undesirable effects and feasibility
4	**SSB taxation logics** • Three SSB taxation intervention logics generally emerge in the literature and could be concurrently pursued • The price-induced behavioural logic consists in taxing SSBs as a way to increase their price, decrease consumption and impact the health of the population • The fiscal logic consists in taxing SSBs as a way to raise fiscal revenues that may be earmarked for health care, for health promotion activities or for the reduction of health inequities • The signal-induced denormalization logic consists in taxing SSBs as a way to send a salient signal to the population and the manufacturers in order to encourage, in a longer term, social norms changes and product reformulation
#	**What—impact**
5	**Taxation Effects on SSB Prices** • Overall, there is convincing evidence that the implementation of a significant SSB tax levied on the manufacturers is generally followed by price increases in the short term, sometimes beyond or below the increase theoretically generated by the tax

(continued)

Y. Le Bodo et al., *Taxing Soda for Public Health*,
DOI 10.1007/978-3-319-33648-0

(continued)

#	What—impact
	• In many cases, it is difficult to disentangle precisely tax effects from other contextual factors. However, evidence from France, Mexico and Berkeley (California) strongly suggests the existence of tax-specific effects on SSB prices
6	**Taxation effects on SSB demand** • Experimental studies in controlled conditions indicate that SSB consumers are sensitive to price change and that a tax has the potential to favour SSB substitution towards healthier beverages • In usual market conditions, data analyses from several countries estimate that a 1 % SSB price increase is generally accompanied by a 0.8–1.3 % decrease in SSB demand. Exploring Canadian data is necessary • In places where a 10–20 % price hike followed the enactment of soda taxes, SSB demand generally decreased. In most cases, it remains difficult to isolate precisely tax-specific effects from other factors • An evaluation of the Mexican soda tax, however, provides sound evidence that an average 9 % price increase was associated with SSB purchases that were 6% lower than if no tax had been imposed • These encouraging trends will have to be confirmed over the long term
7	**SSB taxation effects on energy balance and health** • A taxed-induced decrease in SSB consumption may be partially compensated by a higher consumption of other caloric foods and drinks. Therefore, SSB taxation objectives on energy balance and health should be stated very carefully • Studies simulating a 20 % price increase on SSBs and accounting for substitution effects generally predict a positive—although modest—impact on energy balance, but they are based on fragile assumptions. Therefore, substitution effects should be explored from soda taxes implemented across the world • Few "real-world" evaluations have been conducted, with the exception of Mexico, where evidence suggests that the tax has significantly reduced SSB purchases while favouring water consumption • If a tax is to be adopted in Canada, a detailed and robust evaluation plan should be designed ex ante, and preliminary results should be carefully monitored in terms of price change, consumer demand and substitution effects before extending implementation in the longer term.
8	**SSB taxation distributional effects** • Evidence suggests that the financial contribution to tax revenues would be relatively modest for most consumers and not much different across income groups • Low-income households consuming large SSB amounts and not prone to reduce their consumption may be more negatively impacted than other consumers • As regards SSB taxation health benefits, these are hardly predictable (see point 7) and it remains unclear whether these could be greater in lower-income households and the heaviest SSB consumers. Simulation studies are encouraging, but more research and evaluations will be crucial in this area • Inequity concerns surrounding SSB taxation could be somewhat mitigated if the proceeds of a soda tax are truly earmarked for health promotion causes
9	**SSB taxation undesirable effects** • Decision-makers have to weigh all SSB taxation potential benefits against potential risks. Among other things, this includes taking into account undesirable effects in terms of administration costs, employment, investment, competitiveness and cross-border shopping • Overall, such effects have been weakly documented, but observational studies, simulations and anecdotal evidence suggest that effects of a soda tax on these variables may be relatively limited • Provided that other uncertainties are addressed (see other points) and an evaluation of preliminary effects is well designed in advance, these risks may not justify inaction
10	**Fiscal revenues from SSB taxation** • From a public health perspective, taxing SSBs as a way to raise fiscal revenues could be acceptable, realistic and rapidly achievable • This fiscal logic could reasonably be accompanied with a behavioural objective to reduce SSB consumption in the population • If SSB demand decreases significantly over time, a new tax will at worst generate fewer revenues, but it will always be more than if the tax was absent • Earmarking tax revenues for health care or for health promotion initiatives could bring additional public health benefits, somewhat mitigate inequity concerns, and increase the acceptability of the tax. It is desirable to ensure that tax revenues will be truly earmarked for such initiatives
11	**Signal effects of SSB taxation** • Social norms are a significant driver of SSB consumption behaviours, especially in adolescents • Parallel data (e.g. from tobacco control) indicate that SSB taxation may work as a signal altering SSB consumption norms, but evidence is limited and counterproductive effects cannot be ruled out

(continued)

(continued)

#	What—impact
	• Rather than an isolated "signal", SSB taxation would be better positioned within a comprehensive "denormalization" strategy aimed at influencing knowledge, opinions, attitudes and intentions • SSB taxation may also work as a "coercive signal" pushing manufacturers to reformulate existing products or to develop healthier options. Such effects should be further documented. A tax linearly indexed on the sugar content of the beverages or based on a threshold of sugar content above which beverages are taxed may be particularly inciting
#	**How—applicability**
12	**Feasibility of SSB taxation** • Excise duties generally are the more documented and recommended option to tax soda. Introducing such a tax in Canada is likely feasible, but at the federal level only • At provincial level, the introduction of a special tax on SSBs sold at retail according to the tobacco and alcohol special tax models is an avenue deserving attention. In particular, the special tax imposed on wine and beer in Quebec grocery and convenience stores suggests transferability to SSB. Nonetheless, the burden it may represent for retailers should be carefully considered • Defining the taxation scope would likely raise some issues in Canada since, in the nomenclature, all carbonated beverages are considered regardless of their calorie content and composition • Finally, a robust mechanism facilitating the administration and the earmarking of tax revenues for health promotion initiatives and social programs would likely be desirable and feasible in Canada
13	**Acceptability of SSB taxation** • The relevancy of health-related food taxes remains debated at international level and in Canada. It often appears to be less popular than other nutritional policies • Soda tax proposals are often subject to intense advocacy efforts. Public support for soda taxes generally increases if tax revenues are to be earmarked for health promotion initiatives and social programs • In the political arena, the propensity of a "soda tax" to be adopted seems to depend highly on political leadership as well as on budgetary, fiscal and administrative considerations • In Canada, SSB taxation has been more intensively debated in Quebec than in other provinces, which may favour the adoption of a tax should a political window of opportunity open • A constructive dialogue with stakeholders seems necessary to support decision-making and, eventually, to prevent implementation failures
14	**Evaluability of SSB taxation** • Evaluating the effects and impact of soda taxes is critical to provide evidence with more external validity than what currently provide simulations and experimental trials • Uncertainties surrounding soda tax effects make even more essential to plan a robust evaluation. Appropriate methods should be used and corresponding databases should be identified and available • In the case of a price-induced behavioural logic, evaluation should focus on SSB price, SSB intakes, energy balance and eventually health outcomes. As far as possible, tax-specific effects should be disentangled from other factors • In the case of a fiscal logic, the evaluation would minimally consist in the collection and monitoring of tax revenues as well as in the monitoring of how tax proceeds are earmarked for specific uses • In the case of a signal-induced denormalization logic, evaluation may focus on the public awareness of the tax as well as on reformulation and marketing practices of manufacturers

[a]The numeration corresponds to the graphic representation of the multidimensional perspective presented in the conclusion chapter (see Fig. 15.1).

Index

Y. Le Bodo et al., *Taxing Soda for Public Health*,
DOI 10.1007/978-3-319-33648-0

GPSR Compliance
The European Union's (EU) General Product Safety Regulation (GPSR) is a set of rules that requires consumer products to be safe and our obligations to ensure this.

If you have any concerns about our products, you can contact us on

ProductSafety@springernature.com

In case Publisher is established outside the EU, the EU authorized representative is:

Springer Nature Customer Service Center GmbH
Europaplatz 3
69115 Heidelberg, Germany

www.ingramcontent.com/pod-product-compliance
Ingram Content Group UK Ltd.
Pitfield, Milton Keynes, MK11 3LW, UK
UKHW021033050925
462611UK00011B/858
* 9 7 8 3 3 1 9 8 1 5 6 5 7 *